Polycyclic Aromatic Hydrocarbon Carcinogenesis: Structure-Activity Relationships

Volume I

Editors

Shen K. Yang, Ph.D.
Professor of Pharmacology
F. Edward Hébert School of Medicine
Uniformed Services University of the Health Sciences
Bethesda, Maryland

B. D. Silverman, Ph.D.
Research Staff Member
Physical Sciences Department
IBM T. J. Watson Research Center
Yorktown Heights, New York

CRC Press, Inc.
Boca Raton, Florida

LIBRARY OF CONGRESS
Library of Congress Cataloging-in-Publication Data

Polycyclic aromatic hydrocarbon carcinogenesis : structure-activity
relationships / editors, Shen K. Yang, B. D. Silverman.
p. cm.
Includes bibliographies and index.
ISBN 0-8493-6730-1 (v. 1). ISBN
0-8493-6731-X (v. 2.)
1. Polycyclic aromatic hydrocarbons--Structure-activity
relationships. 2. Carcinogenesis. I. Yang, Shen K., 1941-
II. Silverman, B. D.
[DNLM: 1. Neoplasms--chemically induced. 2. Polycyclic
Hydrocarbons--adverse effects. 3. Structure-Activity Relationship.
QZ 202 P7818]
RC268.7.P64P64 1988
616.99'4071--dc19

87-27722
CIP

Direct all inquiries to CRC Press, Inc., 2000 Corporate Blvd., N.W., Boca Raton, Florida, 33431.

International Standard Book Number 0-8493-6730-1 (Volume I)
International Standard Book Number 0-8493-6731-x (Volume II)

Library of Congress Card Number 87-27722
Printed in the United States

PREFACE

The induction of carcinogenesis by polycyclic aromatic hydrocarbons (PAH) is currently understood to be a complex multistep process that is dependent upon many factors. One intriguing aspect has been the wide range of chemical potency that depends upon PAH molecular geometry. This variation in potency is found to depend, not only on different salient structural features of the parent PAH, but also on the different detailed stereochemical geometries of potentially reactive metabolites and their precursors. Investigations over the past decade have led to a clearer appreciation of factors that are important during the "initiation stage" of PAH induced carcinogenesis. Such factors include the metabolic fate of parent PAHs, the extent of formation and reactivity of electrophilic species and subsequent enzymatic repair of PAH-DNA adducts formed from such species. All of these processes depend sensitively upon molecular geometry. The present volume is, therefore, devoted to investigations aimed at elucidating the role played by PAH structure in the initiation of tumorigenesis. Hopefully, an understanding of the structural factors resulting in different carcinogenic potencies of PAHs and their metabolites can assist practical considerations of cancer prevention.

THE EDITORS

Shen K. Yang is a Professor of Pharmacology at the F. Edward Hébert School of Medicine, Uniformed Services University of the Health Sciences, Bethesda, Maryland. Dr. Yang received his B.S. degree in chemistry from National Taiwan University in 1964, his M.A. degree in physical chemistry from Wesleyan University (Middletown, Connecticut) in 1969, his M.Ph. degree (1970) and Ph.D. degree (1971) in biophysical chemistry from Yale University. Following postdoctoral fellowship at Yale University (1971—73), he was a research fellow at California Institute of Technology (1973—75), a senior staff fellow at the National Cancer Institute (1975—77), and has been a faculty member at the Uniformed Services University since 1975. His current research interest is on the metabolism, mutagenicity, and carcinogenicity of polycyclic aromatic hydrocarbons and their structure-activity relationships. Dr. Yang is a member of American Association for Cancer Research, American Society of Biological Chemists, American Society of Pharmacology and Experimental Therapeutics, and International Society for the Study of Xenobiotics.

B. D. Silverman is a research staff member at the IBM Thomas J. Watson Research Center, Yorktown Heights, New York.

Dr. Silverman received his Ph.D. in theoretical solid state physics at Rutgers University in 1959. He was a member of the theory group at the Raytheon Research Division, Waltham, Massachusetts until 1966. He then joined the NASA Electronics Research Center, Cambridge, Massachusetts, where he became manager of the Advanced Research Department. In 1969 he joined the IBM Research Laboratory in San Jose, California, where he managed a group responsible for developing materials for optical computer memories. He transferred to Yorktown Heights in 1973 managing a group involved with the physics of conducting organic materials.

He has worked in such divergent areas as ferroelectrics, ultrasonics, quantum electronics, optical computer memories, organic metals, chemical carcinogenesis, and polymers. During 1982 he spent a sabbatical year as manager of the IBM Toxicology Group in San Jose, California. He is a fellow of the American Physical Society.

CONTRIBUTORS

Volume I

Shantu G. Amin Ph.D.
Head, Organic Synthesis Section
Division of Chemical Carcinogenesis
American Health Foundation
Valhalla, New York

Allan H. Conney, Ph.D.
Professor and Chairman
Department of Chemical Biology and Pharmacognosy
Rutgers University
Piscataway, New Jersey

Maurice M. Coombs, Ph.D.
Imperial Cancer Research Foundation Laboratory
Department of Chemistry
Chemistry Laboratory
University of Surrey
Guildford, Surrey, England

Avram Gold, Ph.D.
Professor
Department of Environmental Sciences and Engineering
University of North Carolina
Chapel Hill, North Carolina

Stephen S. Hecht, Ph.D.
Director of Research
American Health Foundation
Valhalla, New York

Donald M. Jerina, Ph.D.
Chief
Section on Oxidation Mechanisms
Laboratory of Bioorganic Chemistry
National Institutes of Health
Bethesda, Maryland

Edmond J. LaVoie, Ph.D.
Associate Division Chief
Division of Environmental Carcinogenesis
American Health Foundation
Valhalla, New York

Roland E. Lehr, Ph.D.
Professor
Department of Chemistry
University of Oklahoma
Norman, Oklahoma

Wayne Levin, M.S.
Full Member
Department of Protein Biochemistry
Hoffman-La Roche, Inc.
Nutley, New Jersey

Anthony Y. H. Lu, Ph.D.
Senior Director
Department of Animal Drug Metabolism
Merck Sharp & Dohme Research Laboratories
Rahway, New Jersey

Assieh A. Melikian, Ph.D.
Scientist
Division of Environmental Carcinogenesis
American Health Foundation
Valhalla, New York

Stephen Nesnow, Ph.D.
Chief
Carcinogenesis and Metabolism Branch
Environmental Protection Agency
Research Triangle Park, North Carolina

Joseph E. Rice, Ph.D.
Head, Section of Metabolic Chemistry
Division of Environmental Carcinogenesis
American Health Foundation
Valhalla, New York

Ramiah Sangaiah, Ph.D.
Research Associate
Department of Environmental Sciences and Engineering
University of North Carolina
Chapel Hill, North Carolina

Peter G. Wislocki, Ph.D.
Senior Research Fellow
Department of Animal Drug Metabolism
Merck Sharp & Dohme Research
Laboratories
Rahway, New Jersey

Alexander W. Wood, Ph.D.
Research Leader
Department of Oncology and Virology
Hoffman-La Roche, Inc.
Nutley, New Jersey

TABLE OF CONTENTS

Volume I

TABLE OF CONTENTS

Volume II

Chapter 1

CARCINOGENICITY AND MUTAGENICITY OF PROXIMATE AND ULTIMATE CARCINOGENS OF POLYCYCLIC AROMATIC HYDROCARBONS

Peter G. Wislocki and Anthony Y. H. Lu

TABLE OF CONTENTS

I. INTRODUCTION

Kennaway and Hieger[1] and Cook et al.[2] isolated polycyclic aromatic hydrocarbons as the first pure chemicals which could cause cancer in animals. This induction of tumors by a pure chemical was a landmark in the field of chemical carcinogenesis. It afforded cancer researchers a tool by which they could examine the carcinogenic process. Since those first steps, much progress has been made in our understanding of how chemicals cause cancer. It is of interest that although the polycyclic aromatic hydrocarbons were the first pure chemical carcinogens that were identified, our understanding of the steps involved in the carcinogenicity of polycyclic aromatic hydrocarbons came later than our understanding of the steps involved in the carcinogenicity of other chemical carcinogens such as aromatic amines and amides, and nitrosamines. Drs. James and Elizabeth Miller[3] were the first to formulate the theory that the key step involved in chemical carcinogenesis was the interaction of the electrophilic chemical carcinogen or electrophilic metabolite of a chemical carcinogen with cellular nucleophiles to initiate the carcinogenic process. The idea that certain chemical carcinogens required metabolic activation in order to act as carcinogens led to the concept of proximate and ultimate carcinogens. An ultimate carcinogen is defined as that metabolite of the parent carcinogen which reacts with the critical cellular nucleophile of the cell to initiate the carcinogenic process. It is reactive of itself and requires no further metabolic activation in order to react with cellular components. The proximate carcinogen is an intermediate metabolite of the parent carcinogen which must undergo further metabolism to the ultimate carcinogen. There could be several proximate carcinogenic metabolites which are on the pathway of metabolism from the parent carcinogen to the ultimate carcinogen.

Based on these definitions of proximate and ultimate carcinogen, they should have several characteristics. The proximate carcinogenic metabolite should be more carcinogenic than its parent compound since it is closer to the ultimate carcinogen and is exposed to fewer inactivation pathways than the parent compound. In a series of proximate carcinogens, those which are closer to the ultimate carcinogenic metabolite should in theory be more carcinogenic than those which are closer to the parent compound. The ultimate carcinogenic metabolite should be more carcinogenic than both the parent compound and its proximate carcinogenic metabolites. Although this indeed should be the case, care should be exercised in the use of these criteria in the characterization of metabolites as proximate and especially ultimate carcinogens. Since ultimate carcinogens are chemically reactive, they may not be stable enough to be administered to animals by the same route as used for the parent or proximate carcinogens. Therefore, selection of the method for carcinogenicity testing of the proposed ultimate carcinogens is of critical importance.

Besides comparing the carcinogenicity of the proposed ultimate carcinogen with that of the parent compound and proximate carcinogenic metabolites, several other methods can be used to determine what the ultimate carcinogenic metabolite of a carcinogen is. Classically it was determined whether the proposed ultimate carcinogen or a closely related model compound would react with typical cellular nucleophiles to form covalently bound adducts. The presence of such covalently bound adducts, especially those bound to DNA, was sought in the tissues of animals treated with the parent compound in vivo. It was also determined whether the parent compound and its proximate carcinogenic metabolites could undergo metabolic activation using in vitro enzyme systems. More recently, with the advent of short-term in vitro tests, such as the Ames test, to detect the interaction of chemicals with DNA to cause mutations, these methods have been used to determine the ability of metabolites to cause mutations in the presence and absence of metabolic activation systems. By this means, the potential of metabolites to be proximate or ultimate carcinogens has been assessed. These in vitro tests have also been used successfully to determine what enzymes are involved in the metabolic activation of carcinogens. Ultimately, however, only carcinogenicity tests can define the carcinogenicity of a compound.

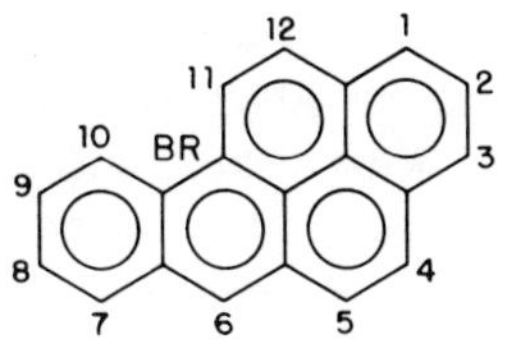

FIGURE 1. Benzo(a)pyrene with bay region (BR) indicated.

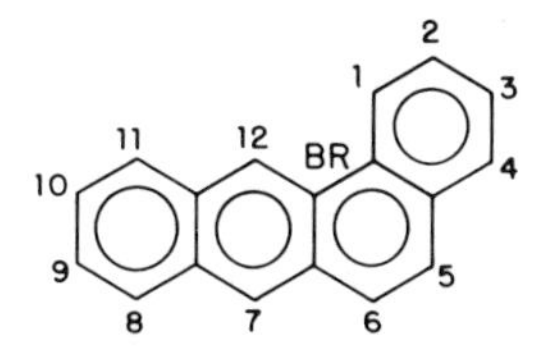

FIGURE 2. Benz(a)anthracene with bay region (BR) indicated.

In this chapter, the results of the mutagenicity and carcinogenicity testing of the proximate and ultimate carcinogenic metabolites of polycyclic aromatic hydrocarbons will be discussed. The advantages and disadvantages of these methods for determining the pathways for the metabolic activation of polycyclic aromatic hydrocarbons should become evident. Other chapters in this volume will deal with the aspects of metabolism and DNA binding of both polycyclic aromatic hydrocarbons and other types of carcinogens and therefore these subjects will not be discussed in detail in this chapter. Another chapter will also discuss in detail the methylchrysenes.

II. HISTORICAL PERSPECTIVE

In the early 1970s an understanding of the metabolic activation of several of the classes of chemical carcinogens had been obtained. This was not the case with polycyclic aromatic hydrocarbons. It was still not clear whether these compounds which induced cancer at the site of administration required metabolic activation for their carcinogenic activity. Boyland[4] had proposed earlier that epoxides of polycyclic aromatic hydrocarbons were responsible for their carcinogenic activity. Such epoxides would be the ultimate carcinogens of polycyclic aromatic hydrocarbons. For many years the K-region (that area at the ''bend'' of the molecule: the 4,5 carbons of benzo(a)pyrene, Figure 1; and the 5,6 carbons of benz(a)anthracene, Figure 2) was proposed to be the key area for metabolic activation of polycyclic aromatic hydrocarbons.[5] Indeed, in a review by Sims et al.[6] in 1973, just prior to the explosive growth of this area of chemical carcinogenesis, the major thrust of the review was concerned with the feasibility of the K-region oxide as the ultimate carcinogen of polycyclic aromatic hydrocarbons. Yet in this same review, in retrospect it was evident that other regions of the polycyclic aromatic hydrocarbons might be important in the biological activity of this class of compounds. It was reported that the 9,10-epoxide of 7,8,9,10-tetrahydrobenzo(a)pyrene was extremely toxic in the Ames test and showed strong alkylating ability. Its parent compound, the 7,8-dihydrobenzo(a)pyrene, was strongly carcinogenic. This region of the molecule later became known as the bay region. The first direct evidence of the importance of this region of the benzo(a)pyrene molecule also came from the laboratory of Sims and Grover. In 1974, Sims et al.[7] reported the metabolic formation of benzo(a)pyrene 7,8-diol-9,10-epoxide and its binding to the DNA of cultured cells. Earlier, Borgen et al.[8] showed that benzo(a)pyrene 7,8-dihydrodiol can be metabolically activated to a species which showed great ability to bind to DNA. These findings initiated more than a decade of studies on the metabolic activation of polycyclic aromatic hydrocarbons to their ultimate carcinogens. These studies were first concentrated on benzo(a)pyrene and later on other polycyclic aromatic hydrocarbons. The results of studies on the mutagenicity and carcinogenicity of the proximate and ultimate carcinogenic metabolites of many of the polycyclic aromatic hydrocarbons will be reported in this chapter.

The findings of Borgen et al. and Sims et al. led to a veritable ''land rush'' of research among several laboratories. Much of the competition between laboratories resulted in the collaboration between two or more laboratories with the capability of synthesizing the chemicals, and those which could test the chemicals for biological activity. The laboratories of Conney at Hoffmann-La Roche and Jerina at the National Institute of Arthritis, Metabolism, and Digestive Diseases had begun collaborating in 1973 in order to synthesize and test as many of the metabolites of benzo(a)pyrene as possible. By this means it was hoped that the proximate and ultimate carcinogenic metabolites could be determined. The laboratory of Gelboin and Yang at the National Cancer Institute, NCI, collaborated with Huberman and with Sugimura. The laboratory of Sims and Grover collaborated with Marquardt and with Malaveille and Bartsch at the International Agency for Research on Cancer, IARC, and with Chouroulinkov. Other laboratories such as Slaga's at Oak Ridge and Harvey's in Chicago also made significant contributions in the area of polycyclic aromatic hydrocarbons carcinogenesis with respect to the testing or synthesis of proximate or ultimate carcinogens. These collaborations proved very fruitful and the competition between laboratories enhanced both the quality and quantity of research performed.

III. METHODS

A. Mutagenicity

The main in vitro method used to determine potential proximate and ultimate carcinogenic metabolites of polycyclic aromatic hydrocarbons was the Ames test.[9] In brief, this test involves the incubation of bacteria with the chemical to be tested in the presence or absence of enzymes capable of metabolically activating the chemical to a mutagenic species. The source of the enzymes could be the microsome and cytosol containing S9 fraction isolated from the liver of rats which had been treated with inducers of microsomal enzymes (such as phenobarbital, 3-methylcholanthrene, or the polychlorinated biphenyl mixture Aroclor® 1254) or could be the purified microsomal enzymes (cytochromes P-450) themselves. The interaction of the mutagenic species with the bacteria causes the bacteria to revert from one requiring histidine for growth to one which can grow in the absence of histidine to form a colony which can be tabulated. Several Ames strains of Salmonella were used in the studies reported. As will be indicated below, they varied in their relative sensitivities to the compounds tested.

The ability of a chemical to mutate V79 cells was another indication of mutagenic potency.[10] After treatment of the Chinese hamster lung cells with the chemical, the cells were treated with 8-azaguanine or ouabain. Those cells which had mutated and had become resistant to the toxic effects of these selection agents were counted. Since V79 cells have limited capacity to metabolically activate chemicals, they were sometimes supplemented with other cell types which could metabolically activate the various polycyclic aromatic hydrocarbon metabolites, such as golden hamster embryo cells.[11] Transformation assays in vitro also were used to a small extent in these studies.

B. Carcinogenicity

Several carcinogenicity models were used to assess the oncogenic potential of the various derivatives of the polycyclic aromatic hydrocarbons. The two-stage initiation and promotion model was used extensively.[12] In this model, one applies the chemical once to the shaved backs of mice to initiate the tumorigenic process and cause the development of papillomas. This treatment was then followed with a promoter such as 12-O-tetradecanoylphorbol-13-acetate (TPA) to complete the carcinogenic process. The degree of tumorigenicity of the compounds was related to the percentage of mice with tumors and more specifically to the number of tumors/tumor-bearing mouse.

A second method of assessing the carcinogenicity of the compounds is to apply the chemical chronically by topical administration to the back of the mouse.[13] This method enables one to determine whether the compound under test is a complete carcinogen and not just an initiator. The disadvantage is that this method requires a significantly greater amount of compound than the initiation-promotion experiment. The development of carcinomas on the backs of the mice was the end point used for the assessment of the carcinogenicity of the compound.

The method which proved to be the best method for determining the carcinogenicity of the polycyclic aromatic hydrocarbon metabolites was the newborn mouse model.[14-17] This model involved the administration of small quantities of the chemical to mice at 1, 8, and 15 days of age by intraperitoneal injection in DMSO. The mice developed lung adenomas, liver tumors, and malignant lymphomas over a period of 18 to 40 weeks. The length of time required for tumor formation was dependent on the dose and the carcinogenic potential of the compound. This method used small amounts of compound, was only labor intensive at the beginning of the study, and proved to give the most meaningful results.

Other methods such as the induction of subcutaneous tumors by subcutaneous injection of the compound into the nape of the neck or i.v. administration of the compound to female rats to cause mammary tumors did not prove as successful as the methods discussed.

As indicated earlier, care must be taken in the interpretation of results since the reactivity of ultimate and in some cases proximate carcinogens could lead to unexpected results. The results with some benzo(a)pyrene metabolites can be used as an example of the care which must be taken in the interpretation of results. Based on the finding that a 7,8-diol-9,10-epoxide was the ultimate carcinogen of benzo(a)pyrene, benzo(a)pyrene 7,8-oxide, and benzo(a)pyrene 7,8-dihydrodiol were, therefore, proximate carcinogenic metabolites. However, when tested for biological activity in the various systems, they did not always behave as expected. Wislocki et al.[18] found that benzo(a)pyrene 7,8-oxide was not strongly mutagenic to V79 cells. This was probably due to the absence of activating enzymes. In the presence of purified activating enzymes, Wood et al.[19] reported that benzo(a)pyrene 7,8-oxide and benzo(a)pyrene 7,8-dihydrodiol were metabolically activated to mutagenic species to a greater extent than benzo(a)pyrene. These data indicated that further metabolism was required to form the mutagenic species. Slaga et al.[20] had found that the presumed ultimate carcinogens of several polycyclic aromatic hydrocarbons were less active than the parent compound in the initiation-promotion model. Yet these same compounds were active in the newborn mouse model. Therefore, the results from several different systems should be considered before the final conclusions of what metabolites might be the proximate or ultimate carcinogens are drawn.

IV. BENZO(a)PYRENE

The first study on the mutagenicity of benzo(a)pyrene 7,8-dihydrodiol and the 7,8-diol-9,10-epoxide was reported by Malaveille et al.[21] The 7,8-dihydrodiol of benzo(a)pyrene was more mutagenic towards *Salmonella typhimurium* strain TA100 than was benzo(a)pyrene. The diol-epoxide also displayed mutagenic activity although it was not remarkably stronger than benzo(a)pyrene 4,5-oxide. The strong mutagenic activity of a benzo(a)pyrene 7,8-diol-9,10-epoxide in both the Ames *S. typhimurium* strains and in cultured Chinese hamster V79 cells was demonstrated by Wislocki et al.[22] Since the epoxide oxygen can be either cis or trans to the 7-hydroxyl group, a pair of enantiomers exists for each diol-epoxide. In the early stage of the study of the diol-epoxides of benzo(a)pyrene, only the geometric relationship of the 7-hydroxy group and the epoxide oxygen was considered. The diol-epoxide studied by Wislocki et al.[22] had the 7-hydroxy group on the same side of the plane of the molecule as the epoxide oxygen. Based on the nomenclature of Jerina, it was designated as diol-epoxide-1. The diol-epoxide which earlier workers had studied[21] had the 7-hydroxy group on the opposite side of the plane of the ring relative to the epoxide oxygen. It was designated as diol-epoxide-2. (In Gelboin's laboratory, these epoxides were known as diol-epoxides (II) and (I), respectively.) Diol-epoxide-2 was a racemic mixture of a pair of enantiomers. It was several-fold more mutagenic than benzo(a)pyrene-4,5-oxide in the Ames strains TA98 and TA100. Likewise in V79 cells, this diol-epoxide was approximately 40 times more mutagenic than the 4,5-oxide of benzo(a)pyrene. Earlier, Huberman et al.[11] had found that both diol-epoxides-1 and -2 were more mutagenic to V79 cells than the 4,5-oxide. Diol-epoxide-2 was 20 times more active than diol-epoxide-1. The stronger activity of diol-epoxide-2 compared to diol-epoxide-1 was later confirmed by Wood et al.[23] who found that diol-epoxide-2 was 2 times more mutagenic in V79 cells than was diol-epoxide-1. Newbold and Brookes[24] also described the higher mutagenicity of the diol-epoxide-2 when compared to diol-epoxide-1 in V79 cells. Although in their studies diol-epoxide-2 was severalfold more mutagenic than diol-epoxide-1, this result was confirmed by Marquardt and Baker[25] who found a tenfold difference in mutagenicity between the two diol-epoxides. In studies of the mutagenicity of the diol-epoxides in the Ames strains, findings of high mutagenicity for both diol-epoxides were also being reported. Significantly in the Ames strains, diol-epoxide-1 was more mutagenic than diol-epoxide-2.[23,26] The difference in results obtained with the diol-epoxides in V79 cells and the Ames test led to the question of which test system best predicts the carcinogenicity of the diol-epoxides. In parallel to the mutagenicity testing of the proposed proximate and ultimate carcinogens of benzo(a)pyrene, carcinogenicity testing of the compounds was also being performed.

Wood et al.[13] in 1976 showed that when benzo(a)pyrene 7,8-oxide, a proposed proximate carcinogen of benzo(a)pyrene, was chronically applied to the backs of mice, it was a complete carcinogen. Although weaker than benzo(a)pyrene at low doses, it caused a substantial number of tumors at high doses (equal in potency to benzo(a)pyrene). The finding of insignificant tumorigenic activity of the 4,5-oxide or 9,10-oxide served to confirm the importance of activation at the 7, 8, 9, 10 positions. Chouroulinkov et al.[27] found that of the 4,5-, 7,8-, and 9,10-dihydrodiols, only the benzo(a)pyrene 7,8-dihydrodiol possessed tumor-initiating activity similar to that of benzo(a)pyrene. Likewise, at the same time, Slaga et al.[28] had determined that benzo(a)pyrene 7,8-oxide was one third as active as benzo(a)pyrene, and that benzo(a)pyrene 7,8-dihydrodiol was equally potent as benzo(a)pyrene. Levin et al.[29] reported that benzo(a)pyrene 7,8-dihydrodiol was equally potent as benzo(a)pyrene as a complete carcinogen after repeated topical administration of the compound to mouse skin. These data further confirmed the theory that benzo(a)pyrene 7,8-dihydrodiol was a proximate carcinogen of benzo(a)pyrene in the tumor-initiation system. However, Slaga in the first test of its carcinogenic potential[28] had also tested benzo(a)pyrene diol-epoxide-2 and surprisingly this compound showed only weak tumor-initiating activity. It was proposed that this may have been due to its high reactivity. The lack of greater carcinogenic activity by the diol-epoxides of benzo(a)pyrene on mouse skin was confirmed in several studies,[30-32] as was the lack of significantly greater carcinogenic activity of the 7,8-dihydrodiol of benzo(a)pyrene compared to that of benzo(a)pyrene.[32] These studies[31,32] did, however, indicate that diol-epoxide-2, although not as strong a carcinogen as benzo(a)pyrene, was more carcinogenic than diol-epoxide-1.

It was not until the newborn mouse model was used to test for the carcinogenicity of the proposed proximate and ultimate carcinogens of benzo(a)pyrene, that results consistent with

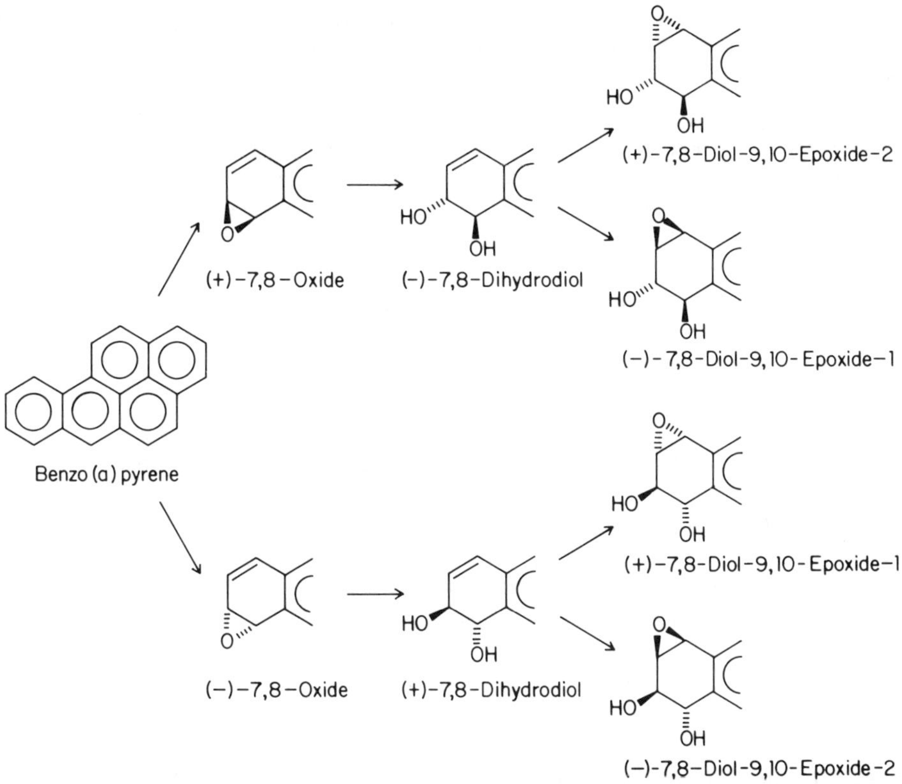

FIGURE 3. The metabolic activation of benzo(a)pyrene to optically active proximate and ultimate carcinogens.

their proposed importance were found. Kapitulnik et al.[17] reported that benzo(a)pyrene 7,8-dihydrodiol was about 12 times more carcinogenic than benzo(a)pyrene in the induction of lung adenomas and induced a high incidence of malignant lymphomas which were not observed in the benzo(a)pyrene-treated group. Benzo(a)pyrene diol-epoxide-2 could only be tested at a level of $^{1}/_{50}$ of the dose at which benzo(a)pyrene was tested. At this level diol-epoxide-2 was equally carcinogenic to benzo(a)pyrene. Therefore, the 7,8-dihydrodiol and diol-epoxide-2 appeared to be proximate and ultimate carcinogens of benzo(a)pyrene, respectively, in the newborn mouse. Diol-epoxide-1 proved to be too toxic and a sufficient number of mice were not available for tumor incidence determination. In a definitive study, Kapitulnik et al.[33] demonstrated that diol-epoxide-1 was not tumorigenic in the newborn mouse. Diol-epoxide-2 was approximately 40 times more carcinogenic than benzo(a)pyrene and greater than two times more carcinogenic than benzo(a)pyrene 7,8-dihydrodiol. Based on these studies, there was a better correlation between the mutagenicity of the diol-epoxides in V79 cells compared to their mutagenicity in the Ames bacterial strains.

It was known that benzo(a)pyrene 7,8-oxide, 7,8-dihydrodiol, and 7,8-diol-9,10-epoxides were mixtures of enantiomers (optical isomers). When metabolism studies[11,34-36] indicated that benzo(a)pyrene was metabolized stereoselectively to (−)-7,8-dihydrodiol, and furthermore that the (−)-7,8-dihydrodiol was metabolized stereoselectively to (+)-diol-epoxide-2 (Figure 3), it became essential to determine the biological activity of the enantiomers of the proximate and ultimate carcinogenic metabolites of benzo(a)pyrene. When the (+)- and (−)-enantiomers of benzo(a)pyrene *trans*-7,8-dihydrodiol were tested for mutagenicity in the Ames strain,[37] the (+)-enantiomer showed greater activity, while in the V79 cells,[38] the (−)-enantiomer showed greater activity. These data raised intriguing questions concerning the role of metabolism in the carcinogenicity of the dihydrodiols and the diol-epoxides.

Enantiomers of diol-epoxides-1 and -2 could be the metabolic products of the (+)- and (−)-7,8-dihydrodiol. It therefore remained a question as to what the carcinogenicities of the different optical isomers of the proposed proximate and ultimate carcinogenic metabolites of benzo(a)pyrene were. No carcinogenicity studies had been done previously on the enantiomers of any carcinogens, especially ultimate carcinogens. Therefore, the (+)- and (−)-enantiomers of benzo(a)pyrene 7,8-dihydrodiol were tested for carcinogenicity in both the initiation-promotion model on mouse skin and in the newborn mouse. Levin et al.[39,40] found that the (−)-7,8-dihydrodiol of benzo(a)pyrene was approximately tenfold more tumorigenic as an initiator than the (+)-isomer. It was also significantly more active then benzo(a)pyrene. These data were the first indication that the 7,8-dihydrodiol of benzo(a)pyrene could be more active than its parent compound.

This greater carcinogenicity of the (−)-isomer was also found in the newborn mouse model. In this model, the difference in carcinogenicity between the two isomers was even more dramatic. The (−)-7,8-dihydrodiol was 15 times more active in causing lung adenomas than (+)-7,8-dihydrodiol. The (−)-isomer also caused a much higher incidence of malignant lymphomas compared to the (+)-isomers. When the four diastereomers of the 7,8-diol-9,10-epoxides of benzo(a)pyrene were tested for carcinogenicity, it was found that only one of the isomers possessed significant tumor-initiating or carcinogenic activity. Buening et al.[41] showed that (+)-7,8-diol-9,10-epoxide-2 was greater than 30 times more active than any of the other three optically active diol-epoxides and approximately 100 times more carcinogenic than benzo(a)pyrene in the newborn mouse model. This was the first study which demonstrated that enantiomers of an ultimate carcinogen possessed different carcinogenic activity.

In the initiation-promotion model, Slaga et al.[42] also showed that (+)-7,8-diol-9,10-epoxide-2 was the only isomer which initiated a significant number of papillomas. However, this isomer still had lower tumor-initiating potential than benzo(a)pyrene.

When these carcinogenicity results were compared to the mutagenicity results obtained in V79 cells, and in the Ames tester strains TA100 and TA98, it was clear that the results from studies using V79 cells were a much better predictor of the carcinogenicity of these compounds than were the results with the Ames strains. Wood et al.[43] reported that in the Ames strain TA98, the (−)- and (+)-enantiomers of diol-epoxide-1 were equally mutagenic, while in the strain TA100, (−)- and (+)-7,8-diol-9,10-epoxide-1 and the (+)-isomer of benzo(a)pyrene 7,8-diol-9,10-epoxide-2 were equally mutagenic. This is in stark contrast to the results obtained in the carcinogenicity experiments. However, in V79 cells, (+)-benzo(a)pyrene 7,8-diol-9,10-epoxide-2 was the most mutagenic species. It was five- to tenfold more mutagenic than the other diol-epoxides. These latter results were in agreement with the tumorigenicity studies.

With regard to the tumorigenicity of the proximate carcinogenic metabolite, benzo(a)pyrene 7,8-oxide, studies on its optical isomers were also done. Levin et al.[44] demonstrated that (+)-benzo(a)pyrene 7,8-oxide, which is the isomer predominantly formed by the mixed function oxidase system, was 2 to 10 times more tumorigenic than the (−)-isomer in the tumor-initiation and newborn mouse models. Of interest was the finding that the racemic mixture of the (+)- and (−)-isomers possessed more tumorigenic activity than either of the enantiomers. This synergistic tumorigenic effect of the enantiomers was due, at least in part, to the differences in affinity to epoxide hydrolase and differences in the rates of hydration.

Table 1 gives the relative tumorigenic activities of the 7,8,9,10-benzo ring derivatives of benzo(a)pyrene. These data serve to indicate the differences observed between the models used to test the compounds and the optical activity of the compounds themselves.

As pointed out by Levin et al.,[44] the metabolism of benzo(a)pyrene at each point (7,8-oxide formation, hydration of this oxide, and oxidation of the 7,8-dihydrodiol) led to the formation of the proximate or ultimate carcinogenic metabolite in preference to less active

Table 1
RELATIVE TUMORIGENIC ACTIVITY OF BENZO-RING DERIVATIVES OF BENZO(a)PYRENE

Compounds	Initiation-promotion model	Newborn mouse model
Benzo(a)pyrene	66	1
(±)-7,8-Oxide	20	2
(±)-7,8-Dihydrodiol	66	9
(+)-7,8-Dihydrodiol	8	5
(−)-7,8-Dihydrodiol	100	20
(±)-7,8-Diol-9,10-epoxide-1	0	0
(±)-7,8-Diol-9,10-epoxide-2	10	28
(+)-7,8-Diol-9,10-epoxide-1	0	2
(−)-7,8-Diol-9,10-epoxide-1	0	1
(+)-7,8-Diol-9,10-epoxide-2	40	100
(−)-7,8-Diol-9,10-epoxide-2	0	0

Note: The relative tumorigenic activity was estimated from the results of a number of studies. The compound which possessed the greatest tumorigenic activity in each model was assigned a value of 100. The other compounds were assigned values from 0—100 relative to the most active compound.

or inactive metabolites. Therefore, the ultimate carcinogen of benzo(a)pyrene (+)-7,8-diol-9,10-epoxide-2 was formed in sufficient amounts to make benzo(a)pyrene a potent carcinogen.

During the course of the studies on benzo(a)pyrene, two questions became of interest:

1. Why were the 7,8-diol-9,10-epoxides so biologically active compared to other diol-epoxides, such as the 9,10-diol-7,8-oxide, which was shown to possess little biological activity?
2. Why was the (+)-7,8-diol-9,10-epoxide the only significant tumorigenic 7,8-diol-9,10-epoxide of benzo(a)pyrene?

Previous theories proposed to explain the carcinogenesis of polycyclic aromatic hydrocarbons were no longer sufficient.

In order to answer these questions, Jerina's group at the National Institutes of Health (NIH) examined the structure-activity relationships of substituted polycyclic aromatic hydrocarbons and studied the quantum mechanics involved in the formation of reactive intermediates of polycyclic aromatic hydrocarbons.

V. BAY-REGION THEORY

In 1976, Jerina et al.[45,46] proposed the ''bay-region'' theory as a concept to use in understanding the carcinogenicity of polycyclic aromatic hydrocarbons. The theory postulates that epoxides which are situated on saturated, angular benzo-rings located in the bay region of a polycyclic aromatic hydrocarbon should be highly reactive (see figures of the various polycyclic aromatic hydrocarbons). The 7, 8, 9, 10 region of benzo(a)pyrene (Figure 1) would be part of the angular bay-region benzo-ring. This ring is saturated via formation of the dihydrodiol followed by epoxidation. The reactive epoxide can yield a reactive carbonium ion. The bay-region diol-epoxides would be good candidates as ultimate carcinogenic me-

Table 2
EASE OF CARBONIUM ION FORMATION AND CARCINOGENICITY OF SELECTED POLYCYCLIC AROMATIC HYDROCARBONS

Compounds	$\Delta E_{deloc}/\beta$	Relative tumorigenicity
Benzo(c)phenanthrene	0.600	+
Chrysene	0.640	+
Phenanthrene	0.658	−
Triphenylene	0.664	−
Benzo(e)pyrene	0.714	−
Dibenz(a,c)anthracene	0.722	+
Dibenz(a,h)anthracene	0.738	+ +
Benz(a)anthracene	0.766	+
Benzo(a)pyrene	0.794	+ + +
Dibenzo(a,h)pyrene	0.845	+ + +
Dibenzo(a,i)pyrene	0.885	+ + +

tabolites if (1) they are formed metabolically and (2) if they react at the site of the cell critical to the initiation of the carcinogenic process. The bay-region theory does not predict metabolism or reaction with the critical site of the cell. This theory was based on perturbational molecular orbital calculations which predicted the ease of carbonium ion formation in various polycyclic aromatic hydrocarbons. These calculations predicted that carbonium ions would be formed more readily on saturated, angular benzo-rings in the bay region of the molecule compared to those in the nonbay regions of the molecule. The greater the ease of carbonium ion formation, the more reactive the epoxide. However, the theory does not state that the carbonium ion is the chemical species responsible for initiating the carcinogenic process. Examination of the carcinogenicity of various polycyclic aromatic hydrocarbons, which did or did not possess hindered bay regions, supported the importance of the bay region in the carcinogenicity of polycyclic aromatic hydrocarbons. Table 2 indicates the relationship between the ease of carbonium ion formation and the carcinogenicity of selected polycyclic aromatic hydrocarbons which will be discussed in this chapter. There is a good correlation between the parameters; some of the reasons why it is not better will be discussed. Methylated compounds are not included because the molecular orbital calculations used were not designed to consider the presence of substituents.

The proposal of the bay-region theory encouraged numerous studies designed to prove or disprove it. Since various polycyclic aromatic hydrocarbons differ in their tumorigenic activities and since some of the compounds in the class are even devoid of tumorigenic activity, it became important to examine the mechanistic basis for such differences in biological activities. Thus the metabolism and biological activities of other polycyclic aromatic hydrocarbons were studied. Results of those studies with regard to the mutagenicity and tumorigenicity of the bay-region dihydrodiols and diol-epoxides tested are the subject of the next sections.

VI. BENZ(a)ANTHRACENE

Compared to benzo(a)pyrene, benz(a)anthracene (Figure 2) is a very weak carcinogen. It was first postulated by the laboratory of Sims and Grover[47,48] that the activation of benz(a)anthracene to an ultimate carcinogen occurred in the same manner as did the activation of benzo(a)pyrene, i.e., the terminal benzo-ring containing the 8, 9, 10, and 11 positions was the site of activation. This would result in the formation of an 8,9-diol-10,11-epoxide.

Evidence for the involvement of this benzo-ring in the binding of benz(a)anthracene to DNA and its mutagenicity was also presented.[47,48] Unlike benzo(a)pyrene, however, benz(a)anthracene has two benzo-rings at which diol-epoxides could be formed. Besides the 8, 9, 10, and 11 positions, the 1, 2, 3, and 4 positions of benz(a)anthracene which are in the bay region of benz(a)anthracene can be metabolized to form a diol-epoxide. It was shown by Wood et al.[49] and Slaga et al.[50] that of the five dihydrodiols of benz(a)anthracene that were tested for mutagenicity in Ames strain TA100 and in Chinese hamster V79 cells, the 3,4-dihydrodiol of benz(a)anthracene had the greatest mutagenic activity. These data indicated that the dihydrodiol which could be further activated to the bay-region diol-epoxide was the most mutagenic dihydrodiol. Further confirmation of the importance of the 3,4-dihydrodiol of benz(a)anthracene came from carcinogenicity studies on the five possible dihydrodiols. Wood et al.[51] and Slaga et al.[50] showed that the 3,4-dihydrodiol was much more potent as a tumor initiator than either benz(a)anthracene or any of the other dihydrodiols. Wislocki et al.,[52] using the newborn mouse model, found that the 3,4-dihydrodiol caused 35 times more lung adenomas than benz(a)anthracene, while the other 4 dihydrodiols had little or no activity. The 3,4-diol-1,2-epoxides of benz(a)anthracene were also tested for mutagenicity[53] and carcinogenicity.[50,54,55] The mutagenicity studies indicated that both diol-epoxide-1 and diol-epoxide-2 were potential ultimate carcinogens of benz(a)anthracene, since they possessed strong mutagenic activity compared to benz(a)anthracene. Carcinogenicity studies using the racemic mixtures of the two diastereomers, 3,4-diol-1,2-epoxide-1 and 3,4-diol-1,2-epoxide-2 of benz(a)anthracene, demonstrated that diol-epoxide-2 was more carcinogenic than benz(a)anthracene or its (−)-3,4-dihydrodiol. This dihydrodiol was the more carcinogenic of the two enantiomeric 3,4-dihydrodiols. Subsequently, studies of the mutagenicity[56] and tumorigenicity[57] of the optical isomers of the diol-epoxides were reported. The (+)-diol-epoxide-2 of benz(a)anthracene, which has the same absolute configuration as that of the (+)-diol-epoxide-2 of benzo(a)pyrene, was the most mutagenic diol-epoxide in both V79 cells and in the Ames strain TA100. However, in strain TA98, (−)-diol-epoxide-2 was the most mutagenic diol-epoxide. The results from tumorigenicity studies indicated that, as was the case with the diol-epoxides of benzo(a)pyrene, (+)-diol-epoxide-2 was the most tumorigenic diol-epoxide. In the initiation-promotion model, it was fourfold more tumorigenic than the (+)-diol-epoxide-1 and 20 times more tumorigenic than benz(a)anthracene or the other 2 diol-epoxides. In the newborn mouse model, (+)-diol-epoxide-2 was 60 times more active than (+)-diol-epoxide-1 and 260 times more active than benz(a)anthracene. The other two diol-epoxides were not significantly tumorigenic.

These data (Table 3) indicated, therefore, that the bay-region diol-epoxides of benz(a)anthracene were the ultimate carcinogens of benz(a)anthracene. These results were consistent with the bay-region theory proposed by Jerina et al.[45,46] and were the first confirmation of it. They also indicated that the absolute configuration of the diol-epoxides played a major role in their biological activity.

VII. CHRYSENE

Chrysene (Figure 4) is a weak carcinogen and tumor-initiating agent similar in activity to benz(a)anthracene. Due to its simple structure, it was among the first unsubstituted polycyclic aromatic hydrocarbons studied in order to determine whether the bay-region theory could predict the proximate and ultimate carcinogens of other polycyclic aromatic hydrocarbons.

In 1977, Wood et al.[58] found that upon metabolic activation, the 1,2-dihydrodiol of chrysene was more mutagenic to *S. typhimurium* strain TA100 than were either chrysene, or its 3,4- or 5,6-dihydrodiols. These data suggested that the bay-region 1,2-diol-3,4-epoxide of chrysene was the ultimate mutagenic species and possibly the ultimate carcinogenic species

Table 3
RELATIVE TUMORIGENIC ACTIVITY OF BENZO-RING DERIVATIVES OF BENZ(a)ANTHRACENE

Compounds	Initiation-promotion model	Newborn mouse model
Benz(a)anthracene	3	1
(±)-3,4-Dihydrodiol	35	1
(+)-3,4-Dihydrodiol	10	0
(−)-3,4-Dihydrodiol	50	14
(±)-3,4-Diol-1,2-epoxide-1	10	2
(±)-3,4-Diol-1,2-epoxide-2	50	30
(+)-3,4-Diol-1,2-epoxide-1	25	1
(−)-3,4-Diol-1,2-epoxide-1	5	0
(+)-3,4-Diol-1,2-epoxide-2	100	100
(−)-3,4-Diol-1,2-epoxide-2	4	0

Note: The relative tumorigenic activity was estimated from the results of a number of studies. The compound which possessed the greatest tumorigenic activity in each model was assigned a value of 100. The other compounds were assigned values from 0—100 relative to the most active compound.

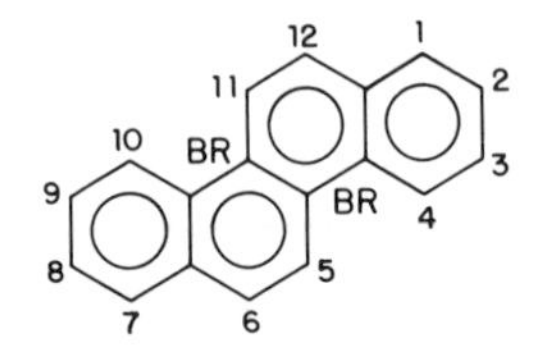

FIGURE 4. Chrysene with bay region (BR) indicated.

of chrysene. Mutagenicity studies on the diol-epoxides using bacterial and mammalian cells found that the diastereomer diol-epoxides did possess strong biological activity.[59] Diol-epoxide-2 was more mutagenic than diol-epoxide-1 in the Ames strains TA98, TA100, and in V79 cells.

Tumorigenicity studies on the three possible vicinal dihydrodiols of chrysene demonstrated that chrysene 1,2-dihydrodiol was a proximate carcinogen of chrysene.[20,60] Chrysene 1,2-dihydrodiol was approximately twice as active as a tumor initiator than was chrysene. Chrysene 3,4-dihydrodiol or 5,6-dihydrodiol were not tumorigenic. In the newborn mouse model, chrysene 1,2-dihydrodiol was 6 times more tumorigenic to the lung than was chrysene.[60] The 3,4- and 5,6-dihydrodiols were less tumorigenic than chrysene.[61] In the liver, these two dihydrodiols were not tumorigenic, while the 1,2-dihydrodiol was somewhat more active than chrysene. Tumorigenicity studies were also done using the 1,2-diol-3,4-epoxide of chrysene. Diol-epoxide-1 was as weak a tumorigen in the newborn mouse as was chrysene, and much less tumorigenic compared to diol-epoxide-2. Diol-epoxide-2 was strongly tumorigenic, causing close to a 100% tumor incidence in the lungs. It caused approximately 50 and 8 times more lung adenomas per mouse than did chrysene and its 1,2-dihydrodiol, respectively. It was also somewhat more tumorigenic to the liver than were these metabolic precursors. In the initiation-promotion model,[20] diol-epoxide-2 of chrysene was tumorigenic, but had about one third the activity of chrysene.

Table 4
RELATIVE TUMORIGENIC ACTIVITY OF BENZO-RING DERIVATIVES OF CHRYSENE

Compounds	Initiation-promotion model	Newborn mouse model
Chrysene	40	1
(±)-1,2-Dihydrodiol	80	15
(+)-1,2-Dihydrodiol	12	2
(−)-1,2-Dihydrodiol	100	50
(±)-1,2-Diol-3,4-epoxide-1	10	1
(±)-1,2-Diol-3,4-epoxide-2	45	100
(+)-1,2-Diol-3,4-epoxide-1	10	0
(−)-1,2-Diol-3,4-epoxide-1	5	1
(+)-1,2-Diol-3,4-epoxide-2	50	70
(−)-1,2-Diol-3,4-epoxide-2	12	0

Note: The relative tumorigenic activity was estimated from the results of a number of studies. The compound which possessed the greatest tumorigenic activity in each model was assigned a value of 100. The other compounds were assigned values from 0—100 relative to the most active compound.

The tumorigenicity[62] of the enantiomers of the 1,2-dihydrodiol, and the mutagenicity[63] and tumorigenicity of the enantiomers of the diastereomeric bay-region diol-epoxides of chrysene, were studied. In the newborn mouse the (+)-chrysene 1,2-dihydrodiol possessed approximately the same tumorigenicity as did chrysene, while the (−)-enantiomer caused greater than 50 times more lung adenomas per mouse than did chrysene, and approximately 6 times more liver tumors. As tumor-initiating agents, the (+)-enantiomer was less active, while the (−)-enantiomer was two times more active than chrysene. Results from tumorigenicity studies of the diol-epoxides indicated that (+)-diol-epoxide-2 was by far the most tumorigenic diol-epoxide in the newborn mouse. It was 20 to 30 times more active in causing lung tumors than either chrysene or the other diol-epoxides, and slightly more active than chrysene in causing hepatic tumors. It caused slightly more lung adenomas than its metabolic precursor, the (−)-1,2-dihydrodiol, but was less active than it in causing liver tumors. (+)-Diol-epoxide-2 was not a more potent initiator than was chrysene or its (−)-1,2-dihydrodiol. The relative tumorigenic activity of these benzo-ring derivatives are given in Table 4.

The mutagenicity data[63] on the enantiomers of the diol-epoxides indicated that in the Ames strain TA100 and in V79 cells (+)-diol-epoxide-2 was much more mutagenic, while in TA98, (−)-diol-epoxide-2 was the most mutagenic metabolite. Thus, the results from strain TA100 and the V79 cells were predictive of the tumorigenicity data. As was the case with benzo(a)pyrene,[47] the metabolism of chrysene by rat liver enzymes[62] was such that the most tumorigenic enantiomer was always formed. In comparison to benzo(a)pyrene the proximate and ultimate carcinogenic metabolites were much less tumorigenic than those of benzo(a)pyrene which had the identical stereochemistry. Indeed, in the newborn mouse model, there is a 50-fold difference in the carcinogenicity of the tumorigenic stereoisomers of benzo(a)pyrene and chrysene. This was in agreement with the quantum mechanical calculations of the bay-region theory which had predicted that the diol-epoxides of chrysene would be less reactive than the diol-epoxides of benzo(a)pyrene (Table 2).

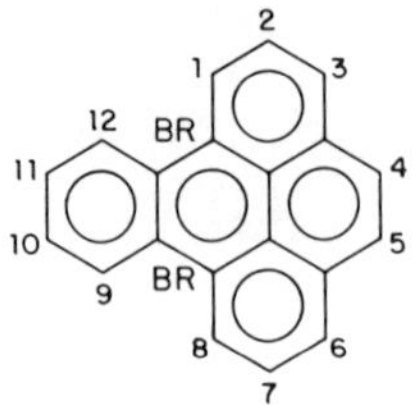

FIGURE 5. Benzo(e)pyrene with bay region (BR) indicated.

VIII. BENZO(e)PYRENE

Benzo(e)pyrene (Figure 5) is a weak tumorigenic agent. It possesses two identical bay regions centered at the 9, 10, 11, and 12 positions of the molecule. Benzo(e)pyrene was of special interest because the quantum mechanical calculations which led to the formulation of the bay-region theory predicted that the diol-epoxide of this weak carcinogen would have moderate to high reactivity. It was, therefore, of interest to determine whether the lack of carcinogenicity of benzo(e)pyrene was due to insignificant metabolism to the ultimate carcinogen, or whether the proposed ultimate carcinogen was not active. With this question in mind, the mutagenicity of the possible proximate and ultimate carcinogens of benzo(e)pyrene was studied.[64,65] It was found that the 9,10-dihydrodiol had only weak mutagenic activity. In fact, the 4,5-dihydrodiol could be activated to a more mutagenic species. This appeared to be due to the lack of further metabolism of the 9,10-dihydrodiol to the bay-region diol-epoxide. However, to confirm this point, mutagenicity studies were done on the actual diol-epoxides. They had good mutagenic activity although they were still one or two orders of magnitude less mutagenic than the corresponding diol-epoxides of benzo(a)pyrene. In order to determine whether these results could be extended to the tumorigenic activity of this compound, in vivo tumorigenicity studies were performed using both the initiation-promotion model and newborn mouse model.[20,66,67] Benzo(e)pyrene 9,10-dihydrodiol was weakly tumorigenic in both systems. However, the 9,10-diol-11,12-epoxides were also weakly tumorigenic. The lack of moderate to high tumorigenic activity as predicted by the bay-region theory could not be due to only a lack of metabolic activation. The diol-epoxides themselves must possess properties which prevented them from being as tumorigenic as predicted. The 9,10-dihydro and 9,10,11,12-tetrahydro-11,12-epoxide derivatives of benzo(e)pyrene were also tested for mutagenicity and carcinogenicity. In all systems these derivatives, which were saturated at the angular benzo-ring but did not contain hydroxyl groups, showed more biological activity. It was, therefore, concluded that the hydroxyl groups in the 9,10 position played a major role on the tumorigenicity of the diol-epoxides. In the case of benzo(e)pyrene, these dihydrodiols were diaxial (perpendicular to the plane of the molecule) as opposed to being diequatorial (parallel to the plane of the molecule) as were the hydroxyl groups of the tumorigenic diol-epoxides of benzo(a)pyrene, chrysene, and benz(a)anthracene. This conformation apparently prevents the diol-epoxides from reacting at the critical site on the target macromolecule (probably DNA) to initiate the carcinogenic process. This diaxial conformation of the 9,10-dihydrodiol coupled with the lack of metabolism at the 9,10-positions explain the low tumorigenicity of benzo(e)pyrene.

IX. BENZO(c)PHENANTHRENE

Benzo(c)phenanthrene (Figure 6) is a relatively weak carcinogen. It was of particular interest because its bay region is highly hindered sterically. This led to the assumption that the hydroxyl groups of the diol-epoxides of this compound would both be in a diequatorial

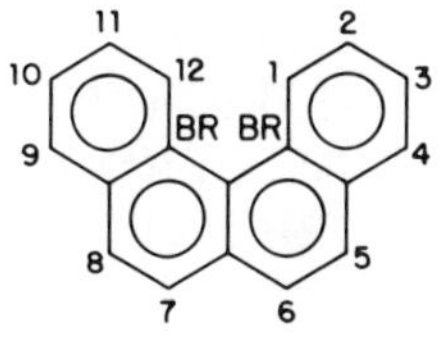

FIGURE 6. Benzo(c)phenanthrene with bay region (BR) indicated.

conformation. This is in contrast to the diol-epoxides which had been studied, in which the hydroxyl groups of diol-epoxide-1 (*cis*) were diaxial while those of the more tumorigenic diol-epoxide-2 (*trans*) were diequatorial. Therefore, it could be predicted that both diol-epoxides-1 and -2 would be tumorigenic. Mutagenicity studies on the racemic dihydrodiol of benzo(c)phenanthrene revealed that the 3,4-dihydrodiol was the most mutagenic dihydrodiol.[68] Both racemic diol-epoxides-1 and -2 were mutagenic to the Ames strains TA100 and TA98 and to V79 cells. The difference in mutagenicity in V79 cells between diol-epoxide-1 and -2 was smaller than that observed with the other polycyclic aromatic hydrocarbons. This was probably due to the fact that the hydroxyl groups of both diol-epoxides were in the diequatorial conformation. Studies to determine the tumor-initiating ability of these same diol-epoxides demonstrated that both the diol-epoxides were about equally active in initiating papillomas in the two-stage mouse skin initiation-promotion model.[69] This was consistent with the theory that the conformation of the hydroxyl groups played a key role in the tumorigenicity of the diol-epoxides. Diol-epoxide-1 of benzo(c)phenanthrene was the first diol-epoxide-1 to have a diequatorial conformation of the hydroxyl groups. It was also the first diol-epoxide to have strong tumor-initiating ability.

Further studies on the mutagenicity[70] and tumorigenicity[71] of the optical isomers of these bay-region diol-epoxides were undertaken. In contrast to the four bay-region diol-epoxides of benzo(a)pyrene, benz(a)anthracene, and chrysene, the four diol-epoxides of benzo(c)phenanthrene did not vary greatly, $<$ fivefold in their mutagenicity to the strains TA100, TA98 or to V79 cells. The (−)-diol-epoxide-2 isomer, which has the same absolute configuration as the highly tumorigenic (+)-diol-epoxides of benzo(a)pyrene, chrysene, and benz(a)anthracene, was the most mutagenic diol-epoxide to V79 cells, while (+)-diol-epoxide-2 was the most active mutagen in the Ames strains. When tested for tumorigenic activity on both mouse skin and in the newborn mouse model, surprising results were obtained. The two test systems did not give the same result. As could be predicted from the V79 data, (−)-diol-epoxide-2 was the most tumorigenic diol-epoxide in both test systems. Also, (−)-diol-epoxide-1 was the least tumorigenic in both systems. However, in the initiation-promotion model, (+)-diol-epoxide-1 was only slightly less tumorigenic than (−)-diol-epoxide-2, while (+)-diol-epoxide-2 was $^1/_4$ to $^1/_2$ as active. In the newborn mouse model, (+)-diol-epoxide-2 was also about 50% as active as (−)-diol-epoxide-2 in causing pulmonary adenomas. However, (+)-diol-epoxide-1, which had been a strong initiator on mouse skin, was only approximately $^1/_4$ as active as (−)-diol-epoxide-2 in causing the development of pulmonary adenomas. This was quite surprising, since the newborn mouse model had usually been more sensitive to the tumorigenicity of diol-epoxides compared to the initiation-promotion model. Although the DNA on mouse skin and in mouse lung should not react differently with the various diol-epoxides, it is possible that the mechanism of tumorigenesis in these two tissues may be different. The data (Table 5) from the initiation-promotion model supports the idea that the diequatorial conformation is important for tumor-initiating activity. Both models did agree that the diol-epoxide with the (R,S,S,R) absolute configuration (see (+)-7,8-diol-9,10-epoxide-2, Figure 3) was the most tumorigenic compound, as had been the case with benzo(a)pyrene, benz(a)anthracene, and chrysene. The strong tumorigenicity of the diol-epoxide is in contrast to that of the parent compound. The

Table 5
RELATIVE TUMORIGENIC ACTIVITY OF BENZO-RING DERIVATIVES OF BENZO(c)PHENANTHRENE

Compounds	Initiation-promotion model	Newborn mouse model
Benzo(c)phenanthrene	2	1
3,4-Dihydrodiol	4	3
(±)-3,4-Diol-1,2-epoxide-1	30	2
(±)-3,4-Diol-1,2-epoxide-2	30	85
(+)-3,4-Diol-1,2-epoxide-1	60	3
(−)-3,4-Diol-1,2-epoxide-1	8	2
(+)-3,4-Diol-1,2-epoxide-2	30	50
(−)-3,4-Diol-1,2-epoxide-2	100	100

Note: The relative tumorigenic activity was estimated from the results of a number of studies. The compound which possessed the greatest tumorigenic activity in each model was assigned a value of 100. The other compounds were assigned values from 0—100 relative to the most active compound.

lack of strong tumorigenicity of the parent compounds is most likely due to the small amount of compound which follows the route of metabolic activation vs. detoxification.[68]

X. DIBENZO(a,h)PYRENE AND DIBENZO(a,i)PYRENE

The diol-epoxides of both dibenzo(a,h)pyrene (Figure 7) and dibenzo(a,i)pyrene (Figure 8) would be expected to have high biological activity based on the predicted ease of carbonium ion formation. In fact, these diol-epoxides would be expected to be more active than the diol-epoxides of benzo(a)pyrene. When studied for mutagenicity[72] in both the Ames strains and in V79 cells, it was found that the diol-epoxides and the precursor dihydrodiol did possess strong mutagenic activity. Although this activity was not greater than that of the corresponding benzo(a)pyrene derivatives, these compounds were very mutagenic. As predicted from quantum mechanical theory, the 3,4-diol-1,2-epoxide-2 of dibenzo(a,i)pyrene was more mutagenic than the 1,2-diol-3,4-epoxide-2 of dibenzo(a,h)pyrene. Also consistent with this finding was the fact that the precursor dihydrodiols, the proximate mutagens, reflected the same activity relationship. Therefore, based on mutagenicity data, these dihydrodiols and the bay-region diol-epoxides derived from them were the proximate and ultimate mutagens, respectively, of dibenzo(a,i)pyrene and dibenzo(a,h)-pyrene. They were also good candidates as the proximate and ultimate carcinogens of these hexacyclic polycyclic aromatic hydrocarbons. However, when the bay-region diol-epoxides and precursor dihydrodiols were tested for tumorigenicity in both the mouse skin initiation-promotion model and in the newborn mouse model,[73] the bay-region diol-epoxides of both dibenzo(a,i)pyrene and dibenzo(a,h)pyrene failed to be more active than their precursor dihydrodiols. The 1,2-dihydrodiol of dibenzo(a,h)pyrene was also less active than dibenzo(a,h)pyrene on mouse skin. The 3,4-dihydrodiol of dibenzo(a,i)pyrene was more active than dibenzo(a,i)pyrene on mouse skin. Both of these dihydrodiols were more active than their parent compounds in the newborn mouse model. This was quite remarkable since the newborn mouse model had always shown a greater response to the proposed ultimate carcinogenic metabolites compared to the proximate carcinogenic metabolites. The diol-epoxide-2 of dibenzo(a,i)pyrene was also not more active than diol-epoxide-2 of dibenzo(a,h)pyrene, con-

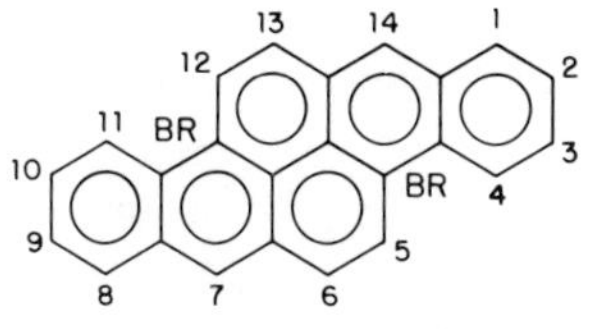

FIGURE 7. Dibenzo(a,h)pyrene with bay region (BR) indicated.

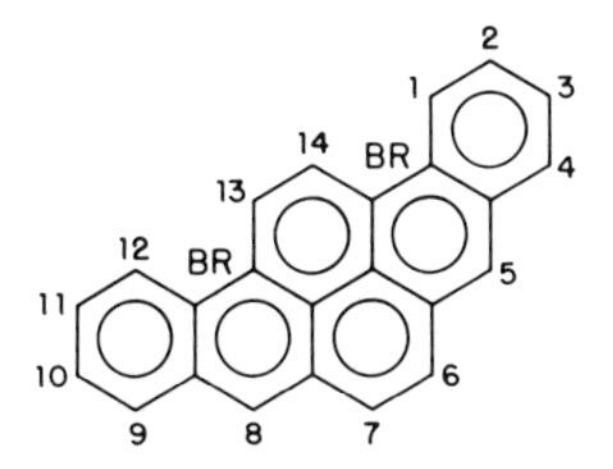

FIGURE 8. Dibenzo(a,i)pyrene with bay region (BR) indicated.

trary to prediction from quantum mechanical calculations. Furthermore, hydrogenated derivatives, in which the double bond where the bay-region epoxide would be found was reduced, showed significant biological activity indicating that another part of the molecule (possibly the other bay region) could be metabolically activated. The lower activity of the diol-epoxides compared to the dihydrodiols could be explained by their high reactivity. It would appear that other factors could also play a role in the lower activity of these compounds. However, the diol-epoxides do possess strong biological activity which is consistent with the bay-region theory. These results do emphasize the fact that the bay-region theory is only a theory and not a law. It can be used to explain and predict results. However, as with all theories, there are apparent exceptions.

XI. DIBENZ(a,h)ANTHRACENE AND DIBENZ(a,c)ANTHRACENE

Based on the bay-region theory, the bay-region diol-epoxides of dibenz(a,h)anthracene (Figure 9) should be much more reactive than those of dibenz(a,c)anthracene (Figure 10).

Dibenz(a,h)anthracene and its three dihydrodiols were tested for mutagenicity[74] and carcinogenicity.[75] In the TA100 strain of Salmonella, the 3,4-dihydrodiol could be activated to the most mutagenic metabolite compared to the other dihydrobiols. In the initiation-promotion model on mouse skin,[20,75] the 3,4-dihydrodiol of dibenz(a,h)anthracene was a slightly less potent tumor-initiating agent than its parent compound, while in the newborn mouse model it was approximately 50% as active as dibenz(a,h)anthracene in inducing pulmonary tumors. However, it was the only compound which induced hepatic tumors. The other dihydrodiols were not significantly tumorigenic. The results with hepatic tumors are consistent with the bay-region theory. It would be expected that the 3,4-dihydrodiol would be more active than its parent compound with the other tumorigenic models. Why this is not the case is unknown, but it may indicate that another mechanism of activation plays a role in the tumorigenicity of this compound.

Malaveille et al.[76] found that the 10,11-dihydrodiol of dibenz(a,c)anthracene was more mutagenic to strain TA100 than either the 1,2- or 3,4-dihydrodiols. An initiation-promotion study by Chouroulinkov et al.[77] demonstrated that both the 1,2-dihydrodiol and the 10,11-dihydrodiol have weak tumor-initiating activity which was approximately twice that of the parent compound. In contrast, Slaga et al.[20] found that both the bay region diol-epoxide and the precursor dihydrodiol were not tumorigenic. As was the case with dibenz(a,h)anthracene, these tumorigenic results are not strongly supportive of the bay-region hypothesis.

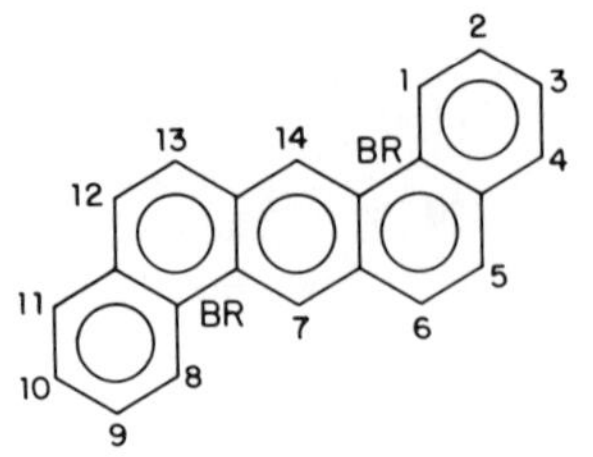

FIGURE 9. Dibenz(a,h)anthracene with bay region (BR) indicated.

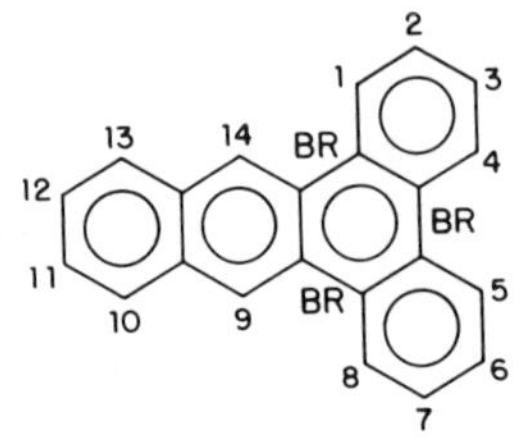

FIGURE 10. Dibenz(a,c)anthracene with bay region (BR) indicated.

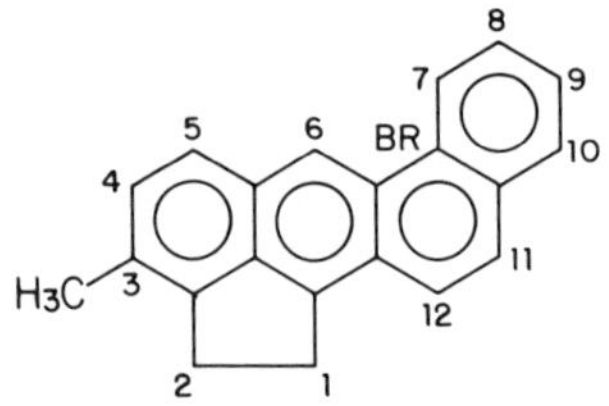

FIGURE 11. 3-Methylcholanthrene with bay region (BR) indicated.

XII. 3-METHYLCHOLANTHRENE

3-Methylcholanthrene (Figure 11) is a dialkyl substituted benz(a)anthracene derivative. Mutagenicity studies[78,79] indicated that the 9,10-dihydrodiol of both 3-methylcholanthrene and its 1-hydroxy metabolite were the most mutagenic of the dihydrodiols tested. The test systems included the Ames test and V79 cells, and also the M2 mouse fibroblast transformation assay. These data strongly suggested that the bay-region diol-epoxide could be the ultimate mutagen and carcinogen of 3-methylcholanthrene. Tumorigenicity studies on the dihydrodiol were performed to determine whether they were proximate carcinogens of 3-methylcholanthrene. The 9,10-dihydrodiol of 3-methylcholanthrene was less potent as an initiator than was 3-methylcholanthrene[80] but it was more active than the other dihydrodiol. The authors speculated that the actual proximate carcinogen might be a hydroxylated metabolite. Levin et al.[81] studied the tumorigenicity of the 1-hydroxy-3-methylcholanthrene-9,10-dihydrodiol. They found that the dihydrodiol in which the 1- and 10-hydroxyl groups were *cis* had 75% of the tumor-initiating ability of the parent compound. Although it was more tumorigenic than the 11,12-dihydrodiol of 3-methylcholanthrene, it was less or equally as tumorigenic as 2-keto- and 2-hydroxy-3-methylcholanthrene. 1-Hydroxy-3-methylcholanthrene was not strongly tumorigenic causing approximately 20% of the tumors per mouse than did its dihydrodiol derivative. Only in the newborn mouse model was the 1-hydroxy-3-methylcholanthrene-9,10-dihydrodiol more active than 3-methylcholanthrene in causing pulmonary adenomas (2 to 3 times) and hepatic tumors (10 times). It was the most active

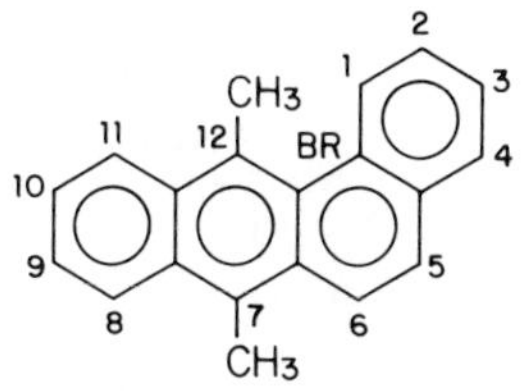

FIGURE 12. 7,12-Dimethylbenz(a)anthracene with bay region (BR) indicated.

compound tested. These data indicate that the dihydrodiol is definitely a proximate carcinogen of 3-methylcholanthrene in the newborn mouse model. It is most likely also a proximate carcinogen of 3-methylcholanthrene in the initiation-promotion model. These data indicate that the bay-region diol-epoxide is probably an ultimate carcinogen of 3-methylcholanthrene.

XIII. 7,12-DIMETHYLBENZ(a)ANTHRACENE

7,12-Dimethylbenz(a)anthracene (Figure 12) is one of the most potent initiators of tumors. It was, therefore, of interest to determine what the metabolic activation pathway of this methylated polycyclic aromatic hydrocarbon was. In collaboration with Sims and Grover, Malaveille et al.[82] and Marquardt et al.[83] studied the mutagenicity and transforming ability of dihydrodiols of 7,12-dimethylbenz(a)anthracene. In the Ames test,[82] the 3,4-dihydrodiol of 7,12-dimethylbenz(a)anthracene was severalfold more mutagenic than 7,12-dimethylbenz(a)anthracene, while the other three dihydrodiols tested (5,6-, 8,9-, and 10,11-) were less mutagenic. However, both the 3,4-dihydrodiol and the 8,9-dihydrodiol were more active than 7,12-dimethylbenz(a)anthracene as mutagens in V79 cells, and as transforming agents in mouse fibroblasts (M2).[83] Given the good record that V79 cells had for predicting the ultimate carcinogens of other polycyclic aromatic hydrocarbon, it appeared that, although not a precursor to a bay-region diol-epoxide, the 8,9-dihydrodiol of 7,12-dimethylbenz(a)anthracene might possess biological activity. Chouroulinkov et al.[80] determined the tumor-initiating ability of these same dihydrodiols of 7,12-dimethylbenz(a)anthracene. Surprisingly, none of the dihydrodiols was more potent as an initiating agent than was 7,12-dimethylbenz(a)anthracene. The next most active agent was the 5,6-dihydrodiol, followed by the 3,4-dihydrodiol. However, these results could not be reproduced by Slaga et al.[84] who found that the 3,4-dihydrodiol of 7,12-dimethylbenz(a)anthracene was two to three times more active than 7,12-dimethylbenz(a)anthracene as a tumor initiator on mouse skin. Both the 5,6- and 8,9-dihydrodiols were not tumorigenic. These results were confirmed and extended by Wislocki et al.,[85,86] who in collaboration with Yang, studied the mutagenic and tumorigenic activities of not only the dihydrodiol derivatives of 7,12-dimethylbenz(a)anthracene, but also certain of the dihydrodiols of the hydroxymethyl derivatives of 7,12-dimethylbenz(a)anthracene. The 3,4-dihydrodiol of 7,12-dimethylbenz(a)anthracene caused three to four times more tumors than did 7,12-dimethylbenz(a)anthracene in the initiation-promotion model. The 5,6-, 8,9-, and 10,11-dihydrodiols were not tumorigenic. The 3,4-dihydrodiol of the 7-hydroxymethyl derivative of 7,12-dimethylbenz(a)anthracene was more tumorigenic than its parent compound but less tumorigenic than 7,12-dimethylbenz(a)anthracene. The 7-methyl-12-hydroxymethyl benz(a)anthracene and its 3,4-dihydrodiol were equally tumorigenic, but both were much less tumorigenic than 7,12-dimethylbenz(a)anthracene. The 3,4-dihydrodiol of the dihydroxymethyl derivative of 7,12-dimethylbenz(a)anthracene was not tumorigenic. Certain of the dihydrodiols of both 7,12-dimethylbenz(a)anthracene and its 7-hydroxymethyl derivative were also tested in the newborn mouse model.[86] In this model, both 3,4-dihydrodiols of 7,12-dimethylbenz(a)anthracene and 7-hydroxymethyl-12-methyl benzanthracene were more active than 7,12-dimethyl-

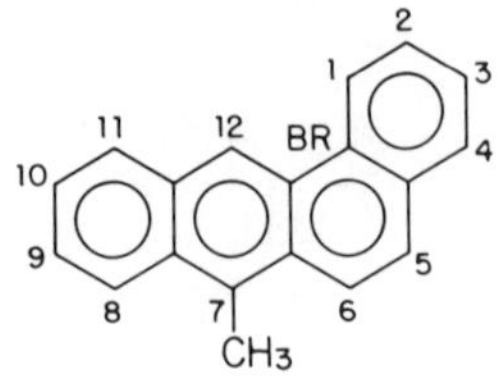

FIGURE 13. 7-Methylbenz(a)anthracene with bay region (BR) indicated.

benz(a)anthracene in the induction of lung adenomas and liver tumors. The 3,4-dihydrodiol of 7,12-dimethylbenz(a)anthracene was three times more active than that of the 7-hydroxymethyl derivative of 7,12-dimethylbenz(a)anthracene. The nonbay-region dihydrodiols of both compounds were not active. Results similar to these were found in mutagenicity studies with strain TA100.[85] In that case, these two 3,4-dihydrodiols were equally mutagenic.

From these studies, it can be concluded that the bay region is the site of activation for both 7,12-dimethylbenz(a)anthracene and its 7-hydroxymethyl derivative. The bay-region diol-epoxides of these compounds have not been tested for tumorigenicity.

XIV. 7-METHYLBENZ(a)ANTHRACENE

The laboratory of Sims and Grover, in collaboration with Malaveille, Marquardt, and Chouroulinkov studied the mutagenicity, transforming ability, and tumorigenicity of 7-methylbenz(a)anthracene (Figure 13). Malaveille et al.[87] found that the 3,4-dihydrodiol of 7-methylbenz(a)anthracene was much more mutagenic in the Ames test than was either the parent compound itself or any of the other dihydrodiols tested. Marquardt et al.[88] found similar results in both V79 cells and in the transforming ability of these compounds in M2 mouse fibroblasts. The 3,4-dihydrodiol had ten times more activity in both systems compared to 7-methylbenz(a)anthracene. Of the other dihydrodiols, only the 8,9-dihydrodiol had tumorigenic activity greater than that of the parent compound. When tested for tumor-initiating activity, Chouroulinkov et al.[89] found that the 3,4-dihydrodiol was a stronger initiator of tumors than 7-methylbenz(a)anthracene or any of the other dihydrodiols. The 8,9-dihydrodiol was less active than 7-methylbenz(a)anthracene, but was more active than the other dihydrodiols. All these data lead to the conclusion that the bay-region 3,4-diol-1,2-epoxide was the ultimate carcinogen of 7-methylbenz(a)anthracene. However, when Slaga et al.[20] tested 7-methylbenz(a)anthracene diol-epoxide-2 in the tumor-initiation model on mouse skin, it was found that this diol-epoxide was less active than 7-methylbenz(a)anthracene or its more tumorigenic 3,4-dihydrodiol. This finding may have more to do with the lack of sensitivity of the initiation-promotion model toward diol-epoxides than the intrinsic activity of this diol-epoxide.

XV. OTHER POLYCYCLIC AROMATIC HYDROCARBONS

The proposed proximate and ultimate carcinogenic metabolites of several other polycyclic aromatic hydrocarbons have been examined for mutagenicity and carcinogenicity in less detail than the previously discussed compounds. The 3,4-dihydrodiol of both 8-methylbenz(a)anthracene (Figure 14) and its 8-hydroxymethyl derivative were tested for tumor-initiating activity, along with several of the dihydrodiols.[90] While the 3,4-dihydrodiol of 8-methylbenz(a)anthracene was more than two times more active as an initiating agent than was 8-methylbenz(a)anthracene, the 3,4-dihydrodiol of 8-hydroxymethylbenz(a)anthracene was less tumorigenic than its parent compound or 8-methylbenz(a)anthracene. None of the other dihydrodiols tested possessed significant tumor-initiating activity. These data indicated that the 3,4-dihydrodiol of 8-methylbenz(a)anthracene was probably its proximate carcinogen

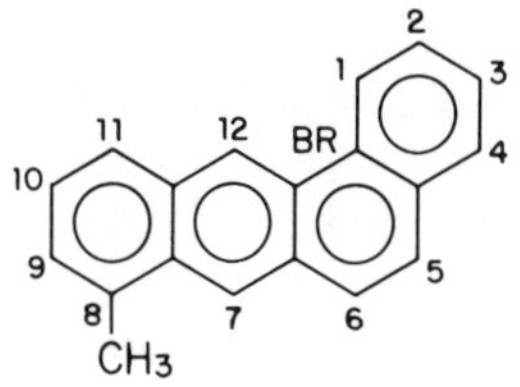

FIGURE 14. 8-Methylbenz(a)anthracene with bay region (BR) indicated.

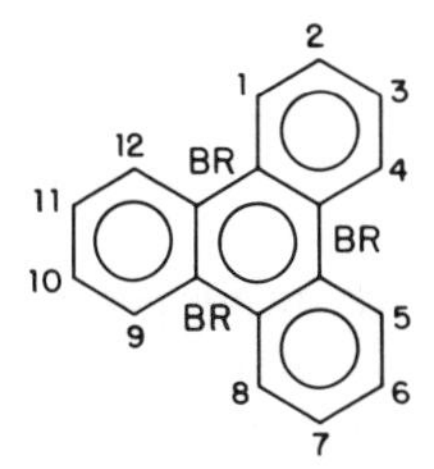

FIGURE 15. Triphenylene with bay region (BR) indicated.

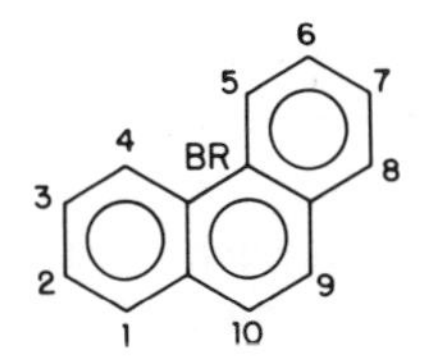

FIGURE 16. Phenanthrene with bay region (BR) indicated.

in the two-stage initiation-promotion model, implying that the bay-region diol-epoxide was the ultimate carcinogen of 8-methylbenz(a)anthracene.

The mutagenicity of the dihydrodiol metabolite of 6-methylbenz(a)anthracene has also been determined in the Ames strain TA100.[91] Again the 3,4-dihydrodiol was more mutagenic than its parent compound or the other dihydrodiol. These data also imply involvement of the bay region in the activation of 6-methylbenz(a)anthracene to a carcinogen.

The bay-region diol-epoxides and precursor dihydrodiols of the noncarcinogenic polycyclic aromatic hydrocarbon triphenylene (Figure 15) were tested for mutagenicity.[65] Although the diol-epoxide was mutagenic in the Ames test and diol-epoxide-2 was mutagenic in V79 cells, the 1,2-dihydrodiol possessed little if any mutagenic activity. It was proposed that the diaxial conformation of the dihydrodiol prevented both metabolic activation of the dihydrodiol to the diol-epoxide and also prevented the carcinogenicity of any diol-epoxide which was formed. Evidence for this theory came from studies with derivatives in which the dihydrodiol hydroxy groups were substituted with hydrogens to yield 1,2-dihydrotriphenylene or a tetrahydro-1,2-epoxide. Both of these compounds were greater than 50 times more mutagenic than their respective dihydrodiol compounds, indicating that the presence of the hydroxy groups actually inhibited the biological activity of these compounds.

Phenanthrene (Figure 16), which is also nontumorigenic, was also studied with regard to the biological activity of its bay-region diol-epoxides and precursor dihydrodiols.[59,61] Mutagenicity studies[59] in Ames strains TA98 and TA100 indicated that both bay-region 1,2-diol-3,4-epoxides-1 and -2 of phenanthrene possess mutagenic activity. These compounds were also weakly mutagenic in V79 cells with diol-epoxide-2 possessing more mutagenic activity. However, when tested for tumorigenic activity in both the initiation-promotion model[59] and newborn mouse model,[61] these diol-epoxides were not tumorigenic. These data are consistent with the finding that none of the dihydrodiols of phenanthrene, including the

1,2-dihydrodiol, possessed tumor-initiating activity. This may be due to the polar nature of the diol-epoxides preventing them from getting to the critical site (presumably by intercalation into DNA) of DNA to initiate the carcinogenic process. In support of this was the finding that the 1,2-tetrahydro-3,4-epoxide in which the dihydrodiol hydroxy groups were substituted with hydrogens was tumorigenic. However, it was only weakly tumorigenic. Thus, although mutagenic, the phenanthrene bay-region diol-epoxides are not as tumorigenic as would be predicted from the bay-region theory. Again, other factors play a role in the biological activity of the proposed ultimate carcinogenic metabolites.

XVI. STRUCTURE VS. CARCINOGENIC ACTIVITY OF POLYCYCLIC AROMATIC HYDROCARBONS MEDIATED THROUGH METABOLIC ACTIVATION AT THE BAY REGION

Based on the results discussed above, many polycyclic aromatic hydrocarbons can be metabolically activated to proximate and ultimate carcinogens at the bay region. However, the results clearly indicate that metabolism to a dihydrodiol which can be further metabolized to a bay-region diol-epoxide is not sufficient to guarantee strong tumorigenic activity. This was clearly the case when the first diol-epoxides of benzo(a)pyrene were tested for tumorigenicity in the newborn mouse model.[33] Despite the fact that both diol-epoxide-1 and -2 possessed mutagenic activity, only diol-epoxide-2 was tumorigenic. Further studies on these dihydrodiols and diol-epoxides of other polycyclic aromatic hydrocarbons confirmed this finding. The bay-region theory only predicts the chemical reactivity of the benzylic epoxide formed at the bay region. It does not take into consideration metabolic, stereochemical, conformational, or structural parameters which can influence the amount of the reactive intermediate interacting with the critical cellular nucleophiles to cause cancer. Indeed, the solvolytic reactivity of the diol-epoxides has correlated quite well with that predicted by the quantum mechanical theory used to postulate the importance of the bay region.[92,93] The tumorigenicity of the dihydrodiols is, of course, dependent upon their further metabolism to the diol-epoxides. As discussed previously with benzo(a)pyrene, whether a particular dihydrodiol would be the proximate carcinogen would depend on its stereochemistry, which determines which diol-epoxides are its metabolites. Further, the conformation of the dihydrodiol hydroxy groups as either diaxial or diequatorial play a role in its metabolism.[94,95] Both Yang et al.[94] and Slaga et al.[95] have discussed the importance of whether the dihydrodiol hydroxyl groups are either in a diaxial (approximately perpendicular to the plane of the molecule) or diequatorial (approximately within the same plane as the plane of the molecule) orientation. Consideration of the tumorigenicity of methyl-substituted polycyclic aromatic hydrocarbons and their subsequent metabolism has led to the conclusion that when a peri position (closest position on the neighboring angular benzene ring) is occupied, tumorigenicity is reduced. Dihydrodiols which are formed in the bay region when the neighboring peri position is occupied have the diaxial conformation and are not as readily metabolized to diol-epoxides. Likewise, the structure of an unsubstituted molecule itself such as benzo(e)pyrene causes the bay-region dihydrodiol hydroxy groups to be in the diaxial conformation.

At the diol-epoxide stage of activation, metabolic activation plays no further role. Steric, conformational, structural, and reactivity parameters play the major role in determining which diol-epoxides are tumorigenic and are, therefore, the ultimate carcinogens. Studies[96-98] with the diol-epoxides of benzo(a)pyrene, benz(a)anthracene, chrysene, and benzo(c)acridine have shown that only diol-epoxide-2, in which the hydroxyl groups prefer the diequatorial conformation, possess tumorigenic activity, compared to diol-epoxide-1, in which the hydroxyl groups prefer the diaxial position. This is further confirmed by the finding that the bay-region diol-epoxides of benzo(e)pyrene and triphenylene are both in-

Table 6
RELATIVE TUMORIGENIC ACTIVITY OF FOUR ENANTIOMERIC DIOL-EPOXIDES OF SELECTED POLYCYCLIC AROMATIC HYDROCARBONS

Compounds	Initiation-promotion model	Newborn mouse model
Benzo(a)pyrene		
(+)-Diol-epoxide-1	0	20
(−)-Diol-epoxide-1	0	10
(+)-Diol-epoxide-2	90	900
(−)-Diol-epoxide-2	0	0
Benz(a)anthracene		
(+)-Diol-epoxide-1	12	3
(−)-Diol-epoxide-1	3	0
(+)-Diol-epoxide-2	60	270
(−)-Diol-epoxide-2	2	0
Chrysene		
(+)-Diol-epoxide-1	2	0
(−)-Diol-epoxide-1	1	0
(+)-Diol-epoxide-2	8	10
(−)-Diol-epoxide-2	2	0
Benzo(c)phenanthrene		
(+)-Diol-epoxide-1	600	30
(−)-Diol-epoxide-1	80	20
(+)-Diol-epoxide-2	300	500
(−)-Diol-epoxide-2	1000	1000

Note: The relative tumorigenic activity was estimated from the results of a number of studies. The compound which possessed the greatest tumorigenic activity in each model was assigned a value of 1000. The other compounds were assigned values from 0—1000 relative to the most active compound.

active, and the hydroxyl groups of both prefer the diaxial conformation, while both of the bay-region diol-epoxides of benzo(c)phenanthrene are tumorigenic, and the hydroxyl groups of both prefer the diequatorial conformation. However, there are exceptions to every rule. The diol-epoxides of phenanthrene, which is not carcinogenic, are also not carcinogenic, despite the fact that the hydroxyl groups are in the diequatorial conformation.

Although the correct conformation of the hydroxyl group may be necessary for tumorigenic activity, they are not sufficient. The stereochemistry of the diol-epoxide plays a major role in its tumorigenicity. Thus, of the two optical isomers of diol-epoxide-2 of benzo(a)pyrene, only one, the (+)-diol-epoxide-2, possesses carcinogenic activity (Figure 3) (Table 6). Its absolute configuration is R,S,S,R (the stereochemistry at the optically active center is designated R or S; R,S,S,R refers to the conformation at the 7, 8, 9, and 10 positions, respectively). Likewise, of the four optically active isomers of chrysene and benz(a)anthracene, only the diol-epoxide with the R,S,S,R conformation was carcinogenic, both in the newborn mouse model and in the initiation-promotion model on mouse skin (Table 6). In the case of benzo(c)phenanthrene, all four diol-epoxides possessed some tumorigenic activity. (−)-Diol-epoxide-2 which has the R,S,S,R conformation was the most tumorigenic isomer (Table 6).

Phenanthrene, dibenzo(a,h)pyrene, and dibenzo(a,i)pyrene are all less tumorigenic than expected based on their ease of carbonium ion formation (Table 2), and their corresponding diol-epoxides, which possess diequatorial hydroxyl groups, are also less tumorigenic than expected. It is possible that the critical cellular target, which is usually considered to be DNA, may have a size and/or polarity constraint, both large and small, which may limit interaction with this critical target nucleophile. Another possible reason for the lack of tumorigenicity of the two pyrene derivatives may be their high solvolytic reactivity. They are the most reactive of the diol-epoxides tested and may degrade prior to reaching the critical target.

XVII. NONBAY-REGION PROXIMATE AND ULTIMATE CARCINOGENS

Other modes of activation of polycyclic aromatic hydrocarbons are also possible. Methylated polycyclic aromatic hydrocarbons can be activated by hydroxylation of the methyl group followed by esterification to yield a benzylic ester. If a good leaving group is present, a carbonium ion can be formed which could react with critical cellular nucleophiles. In 1973, Flesher and Sydnor[99] first proposed that the hydroxymethyl derivative of 6-methylbenzo(a)pyrene might be a proximate carcinogen of its parent compound. Its stronger carcinogenicity than its parent compound, when tested by subcutaneous injection in the mouse, strengthened this belief.[100] However, activation of the hydroxymethyl compound could still occur by the same pathway as with the parent compound, e.g., bay-region diol-epoxide. Further evidence for the involvement of benzylic esters has come from mutagenicity studies of such esters. Watabe et al.[101-103] have shown that sulfate esters of hydroxymethyl derivatives of 7,12-dimethylbenz(a)anthracene and 7-methylbenz(a)anthracene are mutagenic in Ames strain TA98. Rogan et al.[104] has also demonstrated the strong mutagenicity of sulfate or acetate esters of the hydroxymethyl derivatives of benzo(a)pyrene, 7-methylbenz(a)anthracene, 7,12-dimethylbenz(a)anthracene, and 3-methylcholanthrene. The strong tumorigenicity of the sulfate ester of 6-hydroxymethylbenzo(a)pyrene compared to that of 6-hydroxymethylbenzo(a)pyrene has also been demonstrated.[105] Further evidence for the importance of benzylic esters must come from other areas of research such as in vivo DNA binding experiments.

Another mechanism of activation of polycyclic aromatic hydrocarbons is direct one electron oxidation of the compounds to radical cations.[106] In the context of a discussion of proximate and ultimate carcinogens, it is difficult to address this subject, since no stable metabolic intermediates exist which can be tested for mutagenicity or carcinogenicity. The data which supports this mechanism of activation comes mainly from DNA-binding studies and tumorigenicity studies on methyl- or fluorinated-substituted polycyclic aromatic hydrocarbons.

XVIII. CONCLUSION

The studies reviewed indicate that there is no simple structure-activity relationship even at the level of ultimate carcinogens. Conformation, reactivity, absolute configuration, and molecular size all play a role in determining the biological activity of these compounds. Furthermore, the influence of these factors varies with the system being studied. The Ames strain appears to be least sensitive to conformation and absolute configuration, while the newborn mouse model appears to be the most sensitive. The lack of strict structure-activity relationship should not be surprising when one is dealing with such a complex phenomenon as tumor formation. For this reason, it should be remembered that theories of carcinogenesis, whether they deal with the bay region or radical cations, are theories and not laws. They are designed to explain results and encourage research. With the large number of polycyclic aromatic hydrocarbons and tissues which are susceptible to their carcinogenicity, there is clearly room for not only more than one theory, but also significant scientific contributions to the understanding of how polycyclic aromatic hydrocarbons cause cancer.

ACKNOWLEDGMENT

The authors thank Ms. Lorena Bennett for her skillful assistance in the preparation of this chapter.

REFERENCES

1. **Kennaway, E. L. and Hieger, I.,** Carcinogenic substances and their fluorescence spectrum, *Br. Med. J.*, 1, 1044, 1930.
2. **Cook, J. W., Hewitt, C. L., and Hieger, I.,** Isolation of a cancer-producing hydrocarbon from coal-tar. I. Concentration of the active substance, *J. Chem. Soc.*, London, 395, 1933.
3. **Miller, J. A.,** Carcinogenesis by chemicals: an overview, G. W. A. Clowes Memorial Lecture, *Cancer Res.*, 30, 559, 1970.
4. **Boyland, E.,** The biological significance of metabolism of polycyclic compounds, *Symp. Biochem. Soc.*, 5, 40, 1950.
5. **Pullman, A. and Pullman, B.,** Electronic structure and carcinogenic activity of aromatic molecules, *Adv. Cancer Res.*, 3, 117, 1955.
6. **Sims, P. and Grover, P. L.,** Epoxides in polycyclic aromatic hydrocarbon metabolism and carcinogenesis, *Adv. Cancer Res.*, 20, 165, 1974.
7. **Sims, P., Grover, P. L., Swaisland, A., Pal, K., and Hewer, A.,** Metabolic activation of benzo[a]pyrene proceeds by a diol-epoxide, *Nature (London)*, 252, 326, 1974.
8. **Borgen, A., Davey, H., Castagnoli, N., Crocker, T. T., Rasmussen, R. E., and Wang, I. Y.,** Metabolic conversion of benzo[a]pyrene by Syrian hamster liver microsomes and binding of metabolites to deoxyribonucleic acid, *J. Med. Chem.*, 16, 502, 1973.
9. **Ames, B. N., McCann, J., and Yamasaki, E.,** Methods for detecting carcinogens and mutagens with the Salmonella/mammalian-microsome mutagenicity test, *Mutat. Res.*, 31, 347, 1975.
10. **Chu, E. H. Y.,** Induction and analysis of gene mutations in mammalian cell cultures, in *Chemical Mutagens: Principles and Methods for Their Detection*, Hollaender, A., Ed., Plenum Press, New York, 1971, 411.
11. **Huberman, E., Sachs, L., Yang, S. K., and Gelboin, H. V.,** Identification of mutagenic metabolites of benzo[a]pyrene in mammalian cells, *Proc. Natl. Acad. Sci. U.S.A.*, 73, 607, 1976.
12. **Berenblum, I. and Shubik, P.,** The role of croton oil applications associated with a single painting of a carcinogen in tumour induction of the mouses's skin, *Br. J. Cancer*, 1, 379, 1947.
13. **Levin, W., Wood, A. W., Yagi, H., Dansette, P. M., Jerina, D. M., and Conney, A. H.,** Carcinogenicity of benzo[a]pyrene 4,5-, 7,8-, and 9,10-oxides on mouse skin, *Proc. Natl. Acad. Sci. U.S.A.*, 73, 243, 1976.
14. **Della Porta, G. and Terracini, B.,** Chemical carcinogenesis in infant animals, *Prog. Exp. Tumor Res.*, 11, 334, 1969.
15. **Grant, G. and Roe, F. J. C.,** The effect of phenanthrene on tumor induction by 3,4-benzopyrene administered to newly born mice, *Br. J. Cancer*, 17, 261, 1963.
16. **Toth, B.,** A critical review of experiments in chemical carcinogenesis using newborn animals, *Cancer Res.*, 28, 727, 1968.
17. **Kapitulnik, J., Levin, W., Conney, A. H., Yagi, H., and Jerina, D. M.,** Benzo[a]pyrene 7,8-dihydrodiol is more carcinogenic than benzo[a]pyrene in newborn mice, *Nature (London)*, 266, 378, 1977.
18. **Wislocki, P. G., Wood, A. W., Chang, R. L., Levin, W., Yagi, H., Hernandez, O., Dansette, P. M., Jerina, D. M., and Conney, A. H.,** Mutagenicity and cytotoxicity of benzo(a)pyrene arene oxides, phenols, quinones, and dihydrodiols in bacterial and mammalian cells, *Cancer Res.*, 36, 3350, 1976.
19. **Wood, A. W., Levin, W., Lu, A. Y. H., Yagi, H., Hernandez, O., Jerina, D. M., and Conney, A. H.,** Metabolism of benzo[a]pyrene derivatives to mutagenic products by highly purified hepatic microsomal enzymes, *J. Biol. Chem.*, 251, 4882, 1976.
20. **Slaga, T. J., Gleason, G. L., Mills, G., Ewald, L., Fu, P. P., Lee, H. M., and Harvey, R. G.,** Comparison of skin tumor-initiating activities of dihydrodiols and diol-epoxides of various polycyclic aromatic hydrocarbons, *Cancer Res.*, 40, 1981, 1984.
21. **Malaveille, C., Bartsch, H., Grover, P. L., and Sims, P.,** Mutagenicity of non-K-region diols and diol-epoxides of benz(a)anthracene and benzo(a)pyrene in *S. typhimurium* TA 100, *Biochem. Biophys. Res. Commun.*, 66, 693, 1975.
22. **Wislocki, P. G., Wood, A. W., Chang, R. L., Levin, W., Yagi, H., Hernandez, O., Jerina, D. M., and Conney, A. H.,** High mutagenicity and toxicity of a diol epoxide derived from benzo[a]pyrene, *Biochem. Biophys. Res. Commun.*, 68, 1000, 1976.

23. **Wood, A. W., Wislocki, P. G., Chang, R. L., Levin, W., Lu, A. Y. H., Yagi, H., Hernandez, O., Jerina, D. M., and Conney, A. H.,** Mutagenicity and cytotoxicity of benzo(a)pyrene benzo-ring epoxides, *Cancer Res.*, 36, 3358, 1976.
24. **Newbold, R. F. and Brookes, P.,** Exceptional mutagenicity of a benzo[a]pyrene diol epoxide in cultured mammalian cells, *Nature (London)*, 261, 52, 1976.
25. **Marquardt, H. and Baker, S.,** Malignant transformation and mutagenesis in mammalian cells induced by vicinal diol-epoxides derived from benzo(a)pyrene, *Cancer Lett.*, 3, 31, 1977.
26. **Malaveille, C., Kuroki, T., Sims, P., Grover, P. L., and Bartsch, H.,** Mutagenicity of isomeric diol-epoxides of benzo[a]pyrene and benz[a]anthracene in *S. typhimurium* TA98 and TA100 and in V79 Chinese hamster cells, *Mutat. Res.*, 44, 313, 1977.
27. **Chouroulinkov, I., Gentil, A., Grover, P. L., and Sims, P.,** Tumour-initiating activities on mouse skin of dihydrodiols derived from benzo[a]pyrene, *Br. J. Cancer*, 34, 523, 1976.
28. **Slaga, T. J., Viaje, A., Berry, D. L., and Bracken, W.,** Skin tumor initiating ability of benzo(a)pyrene 4,5- 7,8- and 7,8-diol-9,10-epoxides and 7,8-diol, *Cancer Lett.*, 2, 115, 1976.
29. **Levin, W., Wood, A. W., Yagi, H., Jerina, D. M., and Conney, A. H.,** *-(+)trans*-7,8-Dihydroxy-7,8-dihydrobenzo[a]pyrene: a potent skin carcinogen when applied topically to mice, *Proc. Natl. Acad. Sci. U.S.A.*, 73, 2867, 1976.
30. **Slaga, T. J., Viaje, A., Bracken, W., Berry, D. L., Fischer, S. M., Miller, D. R., and Leclerc, S. M.,** Skin-tumor-initiating ability of benzo(a)pyrene-7,8-diol-9,10-epoxide (anti) when applied topically in tetrahydrofuran, *Cancer Lett.*, 3, 23, 1977.
31. **Slaga, T. J., Bracken, W., Viaje, A., Levin, W., Yagi, H., Jerina, D. M., and Conney, A. H.,** Comparison of the tumor-initiating activities of benzo(a)pyrene arene oxides and diol-epoxides, *Cancer Res.*, 37, 4130, 1977.
32. **Levin, W., Wood, A. W., Wislocki, P. G., Kapitulnik, J., Yagi, H., Jerina, D. M., and Conney, A. H.,** Carcinogenicity of benzo-ring derivatives of benzo(a)pyrene on mouse skin, *Cancer Res.*, 37, 3356, 1977.
33. **Kapitulnik, J., Wislocki, P. G., Levin, W., Yagi, H., Jerina, D. M., and Conney, A. H.,** Tumorigenicity studies with diol-epoxides of benzo(a)pyrene which indicate that *(±)-trans*-7β,8α-dihydroxy-9α,10α-epoxy-7,8,9,10-tetrahydrobenzo(a)pyrene is an ultimate carcinogen in newborn mice, *Cancer Res.*, 38, 354, 1978.
34. **Yang, S. K., McCourt, D. W., Roller, P. P., and Gelboin, H. V.,** Enzymatic conversion of benzo[a]pyrene leading predominantly to the diol-epoxide *r*-7,*s*-8-dihydroxy-*t*-9,10-oxy-7,8,9,10-tetrahydrobenzo[a]pyrene through a single enantiomer of *r*-7,*s*-8-dihydroxy-7,8-dihydrobenzo[a]pyrene, *Proc. Natl. Acad. Sci. U.S.A.*, 73, 2594, 1976.
35. **Thakker, D. R., Yagi, H., Akagi, H., Koreeda, M., Lu, A. Y. H., Levin, W., Wood, A. W., Conney, A. H., and Jerina, D. M.,** Metabolism of benzo[a]pyrene. VI. Stereoselective metabolism of benzo[a]pyrene and benzo[a]pyrene 7,8-dihydrodiol to diol epoxides, *Chem. Biol. Interact.*, 16, 281, 1977.
36. **Thakker, D. R., Yagi, H., Lu, A. Y. H., Levin, W., Conney, A. H., and Jerina, D. M.,** Metabolism of benzo[a]pyrene: conversion of *(±)-trans*-7,8-dihydroxy-7,8-dihydrobenzo[a]pyrene to highly mutagenic 7,8-diol-9,10-epoxides, *Proc. Natl. Acad. Sci. U.S.A.*, 73, 3381, 1976.
37. **Nagao, M., Sugimura, T., Yang, S. K., and Gelboin, H. V.,** Mutagenicity of optically pure *(−)trans*-7,8-dihydroxy-dihydrobenzo[a]pyrene, *Mutat. Res.*, 58, 361, 1978.
38. **Huberman, E., Yang, S. K., McCourt, D. W., and Gelboin, H. V.,** Mutagenicity to mammalian cells in culture by (+) and *(−)trans*-7,8-dihydroxy-7,8-dihydrobenzo(a)pyrenes and the hydrolysis and reduction products of two stereoisomeric benzo(a)pyrene 7,8-diol-9,10-epoxides, *Cancer Lett.*, 4, 35, 1977.
39. **Levin, W., Wood, A. W., Chang, R. L., Slaga, T. J., Yagi, H., Jerina, D. M., and Conney, A. H.,** Marked differences in the tumor-initiating activity of optically pure (+)- and (−)-*trans*-7,8-dihydroxy-7,8-dihydrobenzo(a)pyrene on mouse skin, *Cancer Res.*, 37, 2721, 1977.
40. **Kapitulnik, J., Wislocki, P. G., Levin, W., Yagi, H., Thakker, D. R., Akagi, H., Koreeda, M., Jerina, D. M., and Conney, A. H.,** Marked differences in the carcinogenic activity of optically pure (+)- and (−)-*trans*-7,8-dihydroxy-7,8-dihydrobenzo(a)pyrene in newborn mice, *Cancer Res.*, 38, 2661, 1978.
41. **Buening, M. K., Wislocki, P. G., Levin, W., Yagi, H., Thakker, D. R., Akagi, H., Koreeda, M., Jerina, D. M., and Conney, A. H.,** Tumorigenicity of the optical enantiomers of the diastereomeric benzo[a]pyrene 7,8-diol-9,10-epoxides in newborn mice: exceptional activity of (+)-7β,8α-dihydroxy-9α,10α-epoxy-7,8,9,10-tetrahydrobenzo[a]pyrene, *Proc. Natl. Acad. Sci. U.S.A.*, 75, 5358, 1978.
42. **Slaga, T. J., Bracken, W. J., Gleason, G., Levin, W., Yagi, H., Jerina, D. M., and Conney, A. H.,** Marked differences in the skin tumor-initiating activities of the optical enantiomers of the diastereomeric benzo(a)pyrene 7,8-diol-9,10-epoxides, *Cancer Res.*, 39, 67, 1979.
43. **Wood, A. W., Chang, R. L., Levin, W., Yagi, H., Thakker, D. R., Jerina, D. M., and Conney, A. H.,** Differences in mutagenicity of the optical enantiomers of the diastereomeric benzo[a]pyrene 7,8-diol-9,10-epoxides, *Biochem. Biophys. Res. Commun.*, 77, 1389, 1977.

44. **Levin, W., Buening, M. K., Wood, A. W., Chang, R. L., Kedzierski, B., Thakker, D. R., Boyd, D. R., Gadaginamath, G. S., Armstrong, R. N., Yagi, H., Karle, J. M., Slaga, T. J., Jerina, D. M., and Conney, A. H.,** An enantiomeric interaction in the metabolism and tumorigenicity of (+)- and (−)-benzo[a]pyrene 7,8-oxide, *J. Biol. Chem.*, 255, 9067, 1980.
45. **Jerina, D. M., and Daly, J. W.,** Oxidation at carbon, in *Drug Metabolism*, Parke, D. V. and Smith, R. L., Eds., Taylor and Frances Ltd., London, 1976, 15.
46. **Jerina, D. M., Lehr, R. E., Yagi, H., Hernandez, O., Dansette, P. M., Wislocki, P. G., Wood, A. W., Chang, R. L., Levin, W., and Conney, A. H.,** Mutagenicity of benzo(a)pyrene derivatives and the description of a quantum mechanical model which predicts the ease of carbonium ion formation from diol epoxides, in In Vitro *Metabolic Activation and Mutagenesis Testing*, deSerres, F. J., Fouts, J. R., Bend, J. R., and Philpot, R. M., Eds., Elsevier/North-Holland Biomedical, Amsterdam, 1976, 159.
47. **Swaisland, A. J., Hewer, A., Pal, K., Keysell, G. R., Booth, J., Grover, P. L., and Sims, P.,** Polycyclic hydrocarbon epoxides: the involvement of 8,9-dihydroxy-8,9-dihydrobenz[a]anthracene 10,11 epoxide in reactions with DNA of benz[a]anthracene-treated hamster embryo cells, *FEBS Lett.*, 47, 34, 1974.
48. **Malaveille, C., Bartsch, H., Grover, P. L., and Sims, P.,** Mutagenicity of non-K-region diols and diol-epoxides of benz[a]anthracene and benzo[a]pyrene in *S. typhimurium* TA 100, *Biochem. Biophys. Res. Commun.*, 66, 693, 1975.
49. **Wood, A. W., Levin, W., Lu, A. Y. H., Ryan, D., West, S. B., Lehr, R. E., Schaefer-Ridder, M., Jerina, D. M., and Conney, A. H.,** Mutagenicity of metabolically activated benzo[a]anthracene 3,4-dihydrodiol: evidence for bay region activation of carcinogenic polycyclic hydrocarbons, *Biochem. Biophys. Res. Commun.*, 72, 680, 1976.
50. **Slaga, T. J., Huberman, E., Selkirk, J. K., Harvey, R. G., and Bracken, W. M.,** Carcinogenicity and mutagenicity of benz(a)anthracene diols and diol-epoxides, *Cancer Res.*, 38, 1699, 1978.
51. **Wood, A. W., Levin, W., Chang, R. L., Lehr, R. E., Schaefer-Ridder, M., Karle, J. M., Jerina, D. M., and Conney, A. H.,** Tumorigenicity of five dihydrodiols of benz[a]anthracene on mouse skin: exceptional activity of benz[a]anthracene 3,4-dihydrodiol, *Proc. Natl. Acad. Sci. U.S.A.*, 74, 3176, 1977.
52. **Wislocki, P. G., Kapitulnik, J., Levin, W., Lehr, R., Schaefer-Ridder, M., Karle, J. M., Jerina, D. M., and Conney, A. H.,** Exceptional carcinogenic activity of benz[a]anthracene 3,4-dihydrodiol in the newborn mouse and the bay region theory, *Cancer Res.*, 38, 693, 1978.
53. **Wood, A. W., Chang, R. L., Levin, W., Lehr, R. E., Schaefer-Ridder, M., Karle, J. M., Jerina, D. M., and Conney, A. H.,** Mutagenicity and cytotoxicity of benz[a]anthracene diol epoxides and tetra-hydro-epoxides: exceptional activity of the bay region 1,2-epoxides, *Proc. Natl. Acad. Sci. U.S.A.*, 74, 2746, 1977.
54. **Levin, W., Thakker, D. R., Wood, A. W., Chang, R. L., Lehr, R. E., Jerina, D. M., and Conney, A. H.,** Evidence that benzo(a)anthracene 3,4-diol-1,2-epoxide is an ultimate carcinogen on mouse skin, *Cancer Res.*, 38, 1705, 1978.
55. **Wislocki, P. G., Buening, M. K., Levin, W., Lehr, R. E., Thakker, D. R., Jerina, D. M., and Conney, A. H.,** Tumorigenicity of the diastereomeric benz[a]anthracene 3,4-diol-1,2-epoxides and the (+)- and (−)-enantiomers of benz[a]anthracene 3,4-dihydrodiol in newborn mice, *J. Natl. Cancer Inst.*, 63, 201, 1979.
56. **Wood, A. W., Chang, R. L., Levin, W., Yagi, H., Thakker, D. R., van Bladeren, P. J., Jerina, D. M., and Conney, A. H.,** Mutagenicity of the enantiomers of the diastereomeric bay-region benz(a)anthracene 3,4-diol-1,2-epoxides in bacterial and mammalian cells, *Cancer Res.*, 43, 5821, 1983.
57. **Levin, W., Chang, R. L., Wood, A. W., Yagi, H., Thakker, D. R., Jerina, D. M., and Conney, A. H.,** High stereoselectivity among the optical isomers of the diastereomeric bay-region diol-epoxides of benz(a)anthracene in the expression of tumorigenic activity in murine tumor models, *Cancer Res.*, 44, 929, 1984.
58. **Wood, A. W., Levin, W., Ryan, D., Thomas, P. E., Yagi, H., Mah, H. D., Thakker, D. R., Jerina, D. M., and Conney, A. H.,** High mutagenicity of metabolically activated chrysene 1,2 dihydrodiol: evidence for bay region activation of chrysene, *Biochem. Biophys. Res. Commun.*, 78, 847, 1977.
59. **Wood, A. W., Chang, R. L., Levin, W., Ryan, D. E., Thomas, P. E., Mah, H. D., Karle, J. M., Yagi, H., Jerina, D. M., and Conney, A. H.,** Mutagenicity and tumorigenicity of phenanthrene and chrysene epoxides and diol epoxides, *Cancer Res.*, 39, 4069, 1979.
60. **Levin, W., Wood, A. W., Chang, R. L., Yagi, H., Mah, H. D., Jerina, D. M., and Conney, A. H.,** Evidence for bay region activation of chrysene 1,2-dihydrodiol to an ultimate carcinogen, *Cancer Res.*, 38, 1831, 1978.
61. **Buening, M. K., Levin, W., Karle, J. M., Yagi, H., Jerina, D. M., and Conney, A. H.,** Tumorigenicity of bay-region epoxides and other derivatives of chrysene and phenanthrene in newborn mice, *Cancer Res.*, 39, 5063, 1979.

62. **Chang, R. L., Levin, W., Wood, A. W., Yagi, H., Tada, M., Vyas, K. P., Jerina, D. M., and Conney, A. H.,** Tumorigenicity of enantiomers of chrysene 1,2-dihydrodiol and of diastereomeric bay-region chrysene 1,2-diol-3,4-epoxides on mouse skin and in newborn mice, *Cancer Res.*, 43, 192, 1983.
63. **Wood, A. W., Chang, R. L., Levin, W., Yagi, H., Tada, M., Vyas, K. P., Jerina, D. M., and Conney, A. H.,** Mutagenicity of the optical isomers of the diastereomeric bay-region chrysene 1,2-diol-3,4-epoxides in bacterial and mammalian cells, *Cancer Res.*, 42, 2972, 1982.
64. **Wood, A. W., Levin, W., Thakker, D. R., Yagi, H., Chang, R. L., Ryan, D. E., Thomas, P. E., Dansette, P. M., Whittaker, N., Turujman, S., Lehr, R. E., Kuman, S., Jerina, D. M., and Conney, A. H.,** Biological activity of benzo[e]pyrene: an assessment based on mutagenic activities and metabolic profiles of the polycyclic hydrocarbon and its derivatives, *J. Biol. Chem.*, 254, 4408, 1979.
65. **Wood, A. W., Chang, R. L., Huang, M.-T., Levin, W., Lehr, R. E., Kumar, S., Thakker, D. R., Yagi, H., Jerina, D. M., and Conney, A. H.,** Mutagenicity of benzo(e)pyrene and triphenylene tetrahydroepoxides and diol-epoxides in bacterial and mammalian cells, *Cancer Res.*, 40, 1985, 1980.
66. **Buening, M. K., Levin, W., Wood, A. W., Chang, R. L., Lehr, R. E., Taylor, C. W., Yagi, H., Jerina, D. M., and Conney, A. H.,** Tumorigenic activity of benzo(e)pyrene derivatives on mouse skin and in newborn mice, *Cancer Res.*, 40, 203, 1980.
67. **Chang, R. L., Levin, W., Wood, A. W., Lehr, R. E., Kumar, S., Yagi, H., Jerina, D. M., and Conney, A. H.,** Tumorigenicity of the diastereomeric bay-region benzo(e)pyrene 9,10-diol-11,12-epoxides in newborn mice, *Cancer Res.*, 41, 915, 1981.
68. **Wood, A. W., Chang, R. L., Levin, W., Ryan, D. E., Thomas, P. E., Croisy-Delcey, M., Ittah, Y., Yagi, H., Jerina, D. M., and Conney, A. H.,** Mutagenicity of the dihydrodiols and bay-region diol-epoxides of benzo(c)phenanthrene in bacterial and mammalian cells, *Cancer Res.*, 40, 2876, 1980.
69. **Levin, W., Wood, A. W., Chang, R. L., Ittah, Y., Croisy-Delcey, M., Yagi, H., Jerina, D. M., and Conney, A. H.,** Exceptionally high tumor-initiating activity of benzo(c)phenanthrene bay-region diol-epoxides on mouse skin, *Cancer Res.*, 40, 3910, 1980.
70. **Wood, A. W., Chang, R. L., Levin, W., Thakker, D. R., Yagi, H., Sayer, J. M., Jerina, D. M., and Conney, A. H.,** Mutagenicity of the enantiomers of the diastereomeric bay-region benzo(c)phenanthrene 3,4-diol-1,2-epoxides in bacterial and mammalian cells, *Cancer Res.*, 44, 2320, 1984.
71. **Levin, W., Chang, R. L., Wood, A. W., Thakker, D. R., Yagi, H., Jerina, D. M., and Conney, A. H.,** Tumorigenicity of optical isomers of the diastereomeric bay-region 3,4-diol-1,2-epoxides of benzo(c)phenanthrene in murine tumor models, *Cancer Res.*, 46, 2257, 1986.
72. **Wood, A. W., Chang, R. L., Levin, W., Ryan, D. E., Thomas, P. E., Lehr, R. E., Kumar, S., Sardella, D. J., Boger, E., Yagi, H., Sayer, J. M., Jerina, D. M., and Conney, A. H.,** Mutagenicity of the bay-region diol-epoxides and other benzo-ring derivatives of dibenzo(a,h)pyrene and dibenzo(a,i)pyrene, *Cancer Res.*, 41, 2589, 1981.
73. **Chang, R. L., Levin, W., Wood, A. W., Lehr, R. E., Kumar, S., Yagi, H., Jerina, D. M., and Conney, A. H.,** Tumorigenicity of bay-region diol-epoxides and other benzo-ring derivatives of dibenzo(a,h)pyrene and dibenzo(a,i)pyrene on mouse skin and in newborn mice, *Cancer Res.*, 42, 25, 1982.
74. **Wood, A. W., Levin, W., Thomas, P. E., Ryan, D., Karle, J. M., Yagi, H., Jerina, D. M., and Conney, A. H.,** Metabolic activation of dibenzo(a,h)anthracene and its dihydrodiols to bacterial mutagens, *Cancer Res.*, 38, 1967, 1978.
75. **Buening, M. K., Levin, W., Wood, A. W., Chang, R. L., Yagi, H., Karle, J. M., Jerina, D. M., and Conney, A. H.,** Tumorigenicity of the dihydrodiols of dibenzo(a,h)anthracene on mouse skin and in newborn mice, *Cancer Res.*, 39, 1310, 1979.
76. **Malaveille, C., Hautefeuille, A., Bartsch, H., MacNicoll, A. D., Grover, P. L., and Sims, P.,** Liver microsome-mediated mutagenicity of dihydrodiols derived from dibenz(a,c)anthracene in *S. typhimurium* TA100, *Carcinogenesis,* 1, 287, 1980.
77. **Chouroulinkov, I., Coulomb, H., MacNicoll, A. D., Grover, P. L., and Sims, P.,** Tumour-initiating activities of dihydrodiols of dibenz[a,c]anthracene, *Cancer Lett.*, 19, 21, 1983.
78. **Malaveille, C., Batsch, H., Marquardt, H., Baker, S., Tierney, B., Hewer, A., Grover, P. L., and Sims, P.,** Metabolic activation of 3-methylcholanthrene: mutagenic and transforming activities of the 9,10-dihydrodiol, *Biochem. Biophys. Res. Commun.*, 85, 1568, 1978.
79. **Wood, A. W., Chang, R. L., Levin, W., Thomas, P. E., Ryan, D., Stoming, T. A., Thakker, D. R., Jerina, D. M., and Conney, A. H.,** Metabolic activation of 3-methylcholanthrene and its metabolites to products mutagenic to bacterial and mammalian cells, *Cancer Res.*, 38, 3398, 1978.
80. **Chouroulinkov, I., Gentil, A., Tierney, B., Grover, P. L., and Sims, P.,** The initiation of tumours on mouse skin by dihydrodiols derived from 7,12-dimethylbenz(a)anthracene and 3-methylcholanthrene, *Int. J. Cancer,* 24, 455, 1979.
81. **Levin, W., Buening, M. K., Wood, A. W., Chang, R. L., Thakker, D. R., Jerina, D. M., and Conney, A. H.,** Tumorigenic activity of 3-methylcholanthrene metabolites on mouse skin and in newborn mice, *Cancer Res.*, 39, 3549, 1979.

82. **Malaveille, C., Bartsch, H., Tierney, B., Grover, P. L., and Sims, P.,** Microsome-mediated mutagenicities of the dihydrodiols of 7,12-dimethylbenz[a]anthracene: high mutagenic activity of the 3,4-dihydrodiol, *Biochem. Biophys. Res. Commun.*, 83, 1468, 1978.
83. **Marquardt, H., Baker, S., Tierney, B., Grover, P. L., and Sims, P.,** Introduction of malignant transformation and mutagenesis by dihydrodiols derived from 7,12-dimethylbenz[a]anthracene, *Biochem. Biophys. Res. Commun.*, 85, 357, 1978.
84. **Slaga, T. J., Gleason, G. L., DiGiovanni, J., Sukumaran, K. B., and Harvey, R. G.,** Potent tumor-initiating activity of the 3,4-dihydrodiol of 7,12-dimethylbenz(a)anthracene in mouse skin, *Cancer Res.*, 39, 1934, 1979.
85. **Wislocki, P. G., Gadek, K. M., Chou, M. W., Yang, S. K., and Lu, A. Y. H.,** Carcinogenicity and mutagenicity of the 3,4-dihydrodiols and other metabolites of 7,12-dimethylbenz(a)anthracene and its hydroxymethyl derivatives, *Cancer Res.*, 40, 3661, 1980.
86. **Wislocki, P. G., Juliana, M. M., MacDonald, J. S., Chou, M. W., Yang, S. K., and Lu, A. Y. H.,** Tumorigenicity of 7,12-dimethylbenz[a]anthracene, its hydroxymethylated derivatives and selected dihydrodiols in the newborn mouse, *Carcinogenesis*, 2, 511, 1981.
87. **Malaveille, C., Tierney, B., Grover, P. L., Sims, P., and Bartsch, H.,** High microsome-mediated mutagenicity of the 3,4-dihydrodiol of 7-methylbenz[a]anthracene in *S. typhimurium* TA 98, *Biochem. Biophys. Res. Commun.*, 75, 427, 1977.
88. **Marquardt, H., Baker, S., Tierney, B., Grover, P. L., and Sims, P.,** The metabolic activation of 7-methylbenz(a)anthracene: the induction of malignant transformation and mutation in mammalian cells by non-K-region dihydrodiols, *Int. J. Cancer*, 19, 828, 1977.
89. **Chouroulinkov, I. and Gentil, A.,** The metabolic activation of 7-methylbenz(a)anthracene in mouse skin: high tumour-initiating activity of the 3,4-dihydrodiol, *Cancer Lett.*, 3, 247, 1977.
90. **Wislocki, P. G., Fiorentini, K. M., Fu, P. P., Chou, M. W., Yang, S. K., and Lu, A. Y. H.,** Tumor-initiating activity of the dihydrodiols of 8-methylbenz[a]anthracene and 8-hydroxymethylbenz[a]anthracene, *Carcinogenesis*, 2, 507, 1981.
91. **Mushtaq, M., Fu, P. P., Miller, D. W., and Yang, S. K.,** Metabolism of 6-methylbenz[a]anthracene by rat liver microsomes and mutagenicity of metabolites, *Cancer Res.*, 45, 4006, 1985.
92. **Jerina, D. M., Sayer, J. M., Yagi, H., Croisy-Delcey, M., Ittah, Y., and Thakker, D. R.,** Highly tumorigenic bay-region diol epoxides from the weak carcinogen benzo[c]phenanthrene, in *Biological Reactive Intermediates*, Vol. 2, *Chemical Mechanisms and Biological Effects*, Snyder, R., Parke, D. V., Kocsis, J., Jollow, C. J., Gibson, G. G., Witmer, C. M., Eds., Plenum Press, New York, 1982, 501.
93. **Levin, W., Wood, A., Chang, R., Ryan, D., Thomas, P., Yagi, H., Thakker, D., Vyas, K., Boyd, C., Chu, S.-Y., Conney, A., and Jerina, D.,** Oxidative metabolism of polycyclic aromatic hydrocarbons to ultimate carcinogens, *Drug Metab. Rev.*, 13(4), 555, 1982.
94. **Yang, S. K., Chou, M. W., and Fu, P. P.,** Metabolic and structural requirements for the carcinogenic potencies of unsubstituted and methyl-substituted polycyclic aromatic hydrocarbons, in *Carcinogenesis: Fundamental Mechanisms and Environmental Effects*, Pullman, B., Ts'o, P. O. P., and Gelboin, H., Eds., D. Reidel, Dordrecht, Holland, 1980, 143.
95. **Slaga, T. J., Iyer, R. P., Lyga, W., Secrist, A., III, Daub, G. H., and Harvey, R. G.,** Comparison of the skin tumor-initiating activities of dihydrodiols, diol-epoxides, and methylated derivatives of various polycyclic aromatic hydrocarbons, in *Polynuclear Aromatic Hydrocarbons: Chemistry and Biological Effects*, Bjorseth, A. and Dennis, A. J., Eds., Battelle Press, Columbus, Ohio, 1980, 753.
96. **Jerina, D. M., Yagi, H., Thakker, D. R., Sayer, J. M., Van Bladeren, P. J., Lehr, R. E., Whalen, D. L., Levin, W., Chang, R. L., Wood, A. W., and Conney, A. H.,** Identification of the ultimate carcinogenic metabolites of the polycyclic aromatic hydrocarbons: bay-region (R,S)-diol-(S,R)-epoxides, *Foreign Compound Metabolism*, (Proc. Int. Symp.), Caldwell, J. and Paulson, G. D., Eds., Taylor and Francis, London, 257, 1984.
97. **Lehr, R. E., Wood, A. W., Levin, W., Conney, A. H., Thakker, D. R., Yagi, H., and Jerina, D. M.,** The bay-region theory: history and current perspectives, in *Polycyclic Aromatic Hydrocarbons, Phys. Biol. Chem. 6th Int. Symp.*, Cooke, M., Dennis, A. J., and Fisher, G. L., Eds., Battelle Press, Columbus, 1982, 21.
98. **Coles, B.,** Effects of modifying structure on electrophilic reactions with biological nucleophiles, *Drug Metab. Rev.*, 15(7), 1307, 1984—85.
99. **Flesher, J. W. and Sydnor, K. L.,** Possible role of 6-hydroxymethylbenzo[a]pyrene as a proximate carcinogen of benzo[a]pyrene and 6-methylbenzo[a]pyrene, *Int. J. Cancer*, 11, 433, 1973.
100. **Sydnor, K. L., Bergo, C. H., and Flesher, J. W.,** Effects of various substituents in the 6-position on the relative carcinogenic activity of a series of benzo[a]pyrene derivatives, *Chem. Biol. Interact.*, 29, 159, 1980.
101. **Watabe, T., Hiratsuka, A., Ogura, K., and Endoh, K.,** A reactive hydroxymethyl sulfate ester formed regioselectively from the carcinogen, 7,12-dihydroxymethylbenz[a]anthracene, by rat liver sulfotransferase, *Biochem. Biophys. Res. Commun.*, 131, 694, 1985.

102. **Watabe, T., Hakamata, Y., Hiratsuka, A., and Ogura, K.,** A 7-hydroxymethyl sulphate ester as an active metabolite of the carcinogen, 7-hydroxymethylbenz[a]anthracene, *Carcinogenesis,* 7, 207, 1986.
103. **Watabe, T., Ishizuka, T., Fujieda, T., Hiratsuka, A., and Ogura, K.,** Sulfate esters of hydroxymethyl-methyl-benz[a]anthracenes as active metabolites of 7,12-dimethylbenz[a]anthracene, *Jpn. J. Cancer Res.,* 76, 684, 1985.
104. **Rogan, E. G., Cavalieri, E. L., Walker, B. A., Balasubramanian, R., Wislocki, P. G., Roth, R. W., and Saugier, R. K.,** Mutagenicity of benzylic acetates, sulfates and bromides of polycyclic aromatic hydrocarbons, *Chem. Biol. Interact.,* 58, 253, 1986.
105. **Cavalieri, E., Roth, R., and Rogan, E.,** Hydroxylation and conjugation at the benzylic carbon atom: a possible mechanism of carcinogenic activation for some methyl-substituted aromatic hydrocarbons, in *Polynuclear Aromatic Hydrocarbons: 3rd Int. Symp. on Chemistry and Biology — Carcinogenesis and Mutagenesis,* Jones, P. W. and Leber, P., Eds., Ann Arbor Science Publishers, Ann Arbor, Mich., 1979, 517.
106. **Cavalieri, E. and Rogan, E.,** Role of radical cations in aromatic hydrocarbon carcinogenesis, *Environ. Health Perspect.,* 76, 69, 1985.

Chapter 2

BENZACRIDINES AND DIBENZACRIDINES: METABOLISM, MUTAGENICITY, AND CARCINOGENICITY

Roland E. Lehr, Alexander W. Wood, Wayne Levin, Allan H. Conney, and Donald M. Jerina

TABLE OF CONTENTS

I. INTRODUCTION

The cancer-causing properties of benzacridines and dibenzacridines have attracted research interest for more than fifty years. Very early, the potential benefit of using the nitrogen atom of the acridine nucleus as a probe of structure-activity relationships in polycyclic aromatic hydrocarbon (PAH) carcinogenesis was clearly recognized. More recently, research to determine the factors affecting aza-PAH carcinogenicity, including that of the benz- and dibenzacridines, has had added impetus from the growing realization that these compounds are environmental contaminants.

Most of the carcinogenicity data for benz- and dibenzacridines dates back several decades to the extensive work of Lacassagne and coworkers in the 1940s and 1950s at the Institut du Radium, University of Paris.[1] Tabulations and discussions of this and related early work have appeared in a number of publications.[1-5] Although no attempt will be made here to review these data, the most significant results will be summarized. More emphasis will be placed on recent experiments and quantum chemical calculations that have not been reviewed elsewhere. These studies have focused increasingly on the mode of formation and nature of the metabolically activated forms of these compounds.

II. CARCINOGENIC AND RELATED PROPERTIES OF BENZ- AND DIBENZACRIDINES

Structures and numbering schemes for the benz- and dibenzacridines and for benz[a]anthracene (BA) are shown in Figure 1, which also includes the abbreviations used in the present report.

A. Unsubstituted Benzacridines

BaACR and BbACR are generally regarded as noncarcinogenic. BcACR and BA, the PAH that is isosteric with BaACR and BcACR, are generally regarded as inactive or weakly active. In a recent comparison of tumor-initiating activity on mouse skin, BA and BcACR have been found to have weak, but significant activity.[6] BaACR was found to lack significant activity.

B. Monomethyl Benzacridines

In contrast to the unsubstituted benzacridines, selected methylated benzacridines (the present discussion will be confined to methyl substituents) exhibit moderate to strong activity. None of the monomethyl BaACRs exhibit activity, although studies have been limited to the 9, 10, and 12-methyl derivatives. The 12-methyl derivative has attracted the most attention, since its lack of activity contrasts with the carcinogenic activity associated with the corresponding 12-Me-BA. Recently, extensive attempts to induce tumors (primarily by subcutaneous injection) with 12-Me-BaACR were summarized.[7] Despite testing in more than five hundred mice, no tumors were found. In contrast, these workers found the corresponding BcACR derivative, 7-Me-BcACR, to be highly active under comparable conditions; all rodents subjected to the subcutaneous injection protocol developed tumors. In a recent study of tumorigenicity by initiation-promotion on mouse skin and in the newborn mouse model, 7-Me-BcACR was found to be three- to tenfold more active than BcACR.[8]

acridine
ACR

benz[a]anthracene

benz[a]acridine
BaACR

benz[c]acridine
BcACR

benz[b]acridine
BbACR

dibenz[a,h]acridine
DBahACR

dibenz[a,j]acridine
DBajACR

dibenz[c,h]acridine
DBchACR

FIGURE 1. Structures, numbering and abbreviations for acridine, benz[a]anthracene, benz-, and dibenzacridines.

The corresponding 7-Me-BA exhibits moderate or higher activity.[4] 7-Me-BcACR is the only monomethyl derivative of BcACR to exhibit tumorigenic activity. In limited testing, 8-, 9-, and 10-Me-BcACR have been found to be inactive.[1] On the other hand, 8-Me-BA has

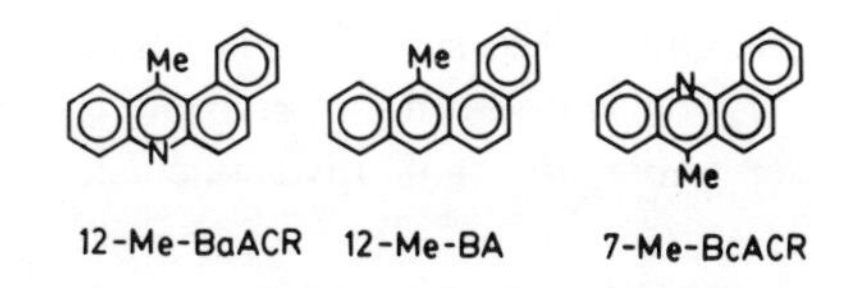

been judged to have at least moderate activity.[4] Benz[c]acridines methylated at positions 1 through 6 or 11 have not been tested.

C. Dimethyl Benzacridines

Although only a small portion of the possible dimethyl derivatives has been synthesized and tested (understandably, since there are 55 dimethyl derivatives for each compound!), the results further indicate higher activity for BcACRs relative to BaACR derivatives. Thus, 8,9-diMe-BaACR has been reported to be inactive on mouse skin and in subcutaneous injection experiments in mice.[1] The 8,12-, 9,12- and 10,12-benz[a]acridines are reported to be only weakly active.[2] In contrast, a number of dimethylbenz[c]acridines have exhibited moderate to high activity.[1] Thus, 7,9- and 7,10-diMe-BcACR have been reported to exhibit high activity, and 7,11-diMe-BcACR has moderate activity.[1] On the other hand, the 5,7- and 10,12-diMe-BcACR derivatives were inactive in limited testing.[1] Of identically substituted BA derivatives, only 7,11-diMe-BA has been tested, and it has moderate to high activity.[4] The 7,10-diMe-BA whose sarcomatogenic properties were cited by Lacassagne et al. (*vide supra*)[1] has subsequently been shown to be a 85:15 mixture of 7,10- and 7,8-diMe-

BcACR, respectively.[9] The pure compounds have been examined in a cell transformation assay and induce malignant transformation of hamster embryo cells.[9] The tumor-initiating

activity of BcACR with a 7,8-trimethylene bridge has recently been studied on mouse skin. The compound was only slightly more tumorigenic than BcACR.[6]

D. Tri- and Tetramethylated Benzacridines

A number of tri- and tetrasubstituted benzacridines have been tested. In the BaACR series, moderate activity is observed for the 8,10,12-trimethyl derivative (skin painting) and for the 9,10,12-derivative (injection).[1] The 3,8,12-trimethyl derivative was very slightly active (injection) and the 8,9,12-, 8,11,12-, and 9,11,12-derivatives were inactive both on skin and subcutaneously.[1] The 2,9,12-trimethyl, 2,8,10,12-, and 3,8,10,12-tetramethyl derivatives were inactive subcutaneously and were not tested on skin.[1] For the BcACR derivatives, the 5,7,11-, 7,8,11-, and 7,9,10-trimethyl derivatives exhibited slight to moderate activity on mouse skin, but were inactive subcutaneously. The 7,9,11-trimethyl and 7,8,9,11-tetramethyl derivatives had moderate to high activity on mouse skin, and slight to moderate activity subcutaneously.

The results for the methyl-substituted benzacridines support the general opinion that BcACRs are more active than BaACRs. The very limited number of carcinogenic BaACR derivatives and their potency is noteworthy, with the highest level, "moderate" activity, being confined to the trimethyl derivatives. For the BcACRs, there are several "moderate" or "high" activity methyl derivatives, which share as a common feature a methyl substituent at C-7. However, not all C-7 methyl-substituted BcACRs are active. Enthusiasm for pursuing quantitative correlations of carcinogenicity based upon these early results is diminished by the limited number of mice used in those studies and by the very limited range of dosages in the skin-painting experiments. Only 7-Me-BcACR and 12-Me-BaACR have been studied in any depth. Also, relatively few derivatives (none for BcACR) substituted at C1-C4 have been studied. Despite these reservations, the qualitative conclusions cited above seem secure.

E. Dibenz[a,h]- Dibenz[a,j]- and Dibenz[c,h]acridine

In the initial experiments of Lacassagne, DBajACR was found to be inactive both on mouse skin and when injected subcutaneously.[1] DBahACR had slight activity on mouse skin, and was not tested subcutaneously.[1] DBchACR had moderate to high activity on mouse skin, but was not active subcutaneously.[1] For the analogous PAH, dibenz[a,h]anthracene and dibenz[a,j]anthracene have been classed as "moderate" carcinogens,[4] but dibenz[a,h]anthracene is generally regarded as the more potent carcinogen. Among the derivatives of DBajACR studied, none showed activity subcutaneously, while 14-methyl-, 1-ethyl-, and 14-isobutyl DBajACR showed slight to moderate activity on mouse skin.[1] Of four DBahACR derivatives tested subcutaneously, 8-ethyl- and 14-n-butyl were weakly active, and 10-ethyl and 14-isobutyl-DBahACR were inactive. In later work, 14-methyl-DBahACR showed moderate activity subcutaneously.[10] Two DBahACR derivatives have been tested on mouse skin. The 3-methyl derivative was inactive, whereas 8-ethyl-DBahACR was moderately active.[1] For DBchACR, the 7-methyl derivative exhibited moderate activity upon subcutaneous injection.[10]

In a recent comparison of the ability of dibenz[a,h]anthracene, DBahACR, and dibenz[a,h]phenazine to induce sarcomas in rats by implantation of paraffin discs containing the compounds, the PAH was found to be considerably more active than the two aza analogues and the acridine was slightly more active than the phenazine.[11]

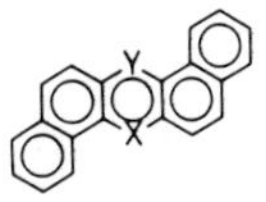

X = Y = CH: dibenz[a,h]anthracene
X = CH; Y = N: dibenz[a,h]acridine
X = Y = N: dibenz[a,h]phenazine

The carcinogenicities of ACR, BaACR, BcACR, DBahACR, and DBajACR were recently determined by injection/implant into lungs of female Osborne-Mendel rats.[12] ACR, BaACR, BcACR, and DBajACR were all judged noncarcinogenic at the doses tested (0.1 to 1.0 mg for BcACR, DBajACR, and DBahACR; higher doses for ACR and BaACR). DBahACR was carcinogenic in the 0.3 to 1.0 mg dose range, but was much less carcinogenic than BaP at the same dose levels.

III. MUTAGENICITY AND PHOTODYNAMIC ACTIVITY

Although benzacridines have been included in numerous experiments in which short-term mutagenesis assays have been examined for their ability to discriminate carcinogens and noncarcinogens, there are few published accounts that permit meaningful structure-activity conclusions. In one particularly thorough experiment, Glatt et al. have examined relationships between mutagenicity and carcinogenicity for a large number of heterocycles.[7] Included in the study was a detailed comparison of the mutagenicities of 7-Me-BcACR (moderate to strong carcinogen) with 12-Me-BaACR (noncarcinogen) when conditions of the in vitro tests were varied.[7] A comparison of mutagenicity in *Salmonella typhimurium* strains TA98, TA100, and TA1537 revealed that both compounds were metabolically activated to mutagens by microsomes from Aroclor-1254-treated rats. While the carcinogenic compound, 7-Me-BcACR, was activated to a greater extent (*circa* threefold), this difference in mutagenicity was much less than the difference in carcinogenicity. An approximate tenfold difference in mutagenicity was observed (7-Me-BcACR more active) in tests with a mammalian cell line (Balb 3T3 C1.A31-1 mouse fibroblasts).

In a recent structure-mutagenic activity study of PAH and aza-PAH relevant to synfuels, metabolic activation of ACR, BaACR, and DBajACR to compounds mutagenic to strain TA98 of *S. typhimurium* was studied. ACR and BaACR were not metabolically activated to mutagens, while DBajACR was metabolically activated to mutagenic products.[13]

The relative photodynamic activities of a number of benz- and dibenzacridines have been determined.[14,15] This property, which entails the destruction of organisms or other living matter by the combination of light and a compound, is likely mediated, at least in part, by photosensitized generation of singlet oxygen. Relationships between photodynamic activity and carcinogenicity have been postulated, and extensive data have been obtained for the photodynamic activity of compounds against *Paramecium caudatum* and against *Artemia salina* nauplii. In the latter case,[15] the relative photodynamic activities were DBajACR (2.2) > BcACR (1.0) > DBahACR (0.27) > DBacACR (0.08) > BaACR (0.06) > ACR (0.02). This relationship to carcinogenic activity is not very satisfactory for these compounds, since DBahACR is the most carcinogenic compound in the series and has lower photodynamic activity than the weakly carcinogenic BcACR and DBajACR. For photodynamic activity against *P. caudatum*,[14] activities observed were 7,9-diMe-BcACR (50.0) > BcACR (14.7)

> 12-Me-BaACR (8.3) > 5,6-diMeBcACR (6.2) > 7,10-diMeBcACR (5.6) > ACR (0.23) > DBacACR (0.19) > DBahACR (0.18) > 1.4-diMeBcACR (0.14) > DBajACR (0.11). While the more carcinogenic diMe-BcACRs show relatively high photodynamic activities, weakly carcinogenic BcACR has a high value, as does the noncarcinogenic 12-Me-BaACR. The very low photodynamic activity of DBahACR is also poorly related to its carcinogenicity.

IV. CALCULATIONS AND STRUCTURE-ACTIVITY RELATIONSHIPS FOR PARENT BENZACRIDINES

The generally higher carcinogenicity of BcACRs relative to BaACRs, and the demonstration of a wide variation in the carcinogenicity of methylated derivatives of the benzacridines, has prompted numerous attempts to find calculated or observed properties of these molecules which correlate with their carcinogenicities. Initially, calculations centered on the parent compounds. More recently, increased attention has been focused upon calculated properties of intermediates believed to be products of the metabolic activation of PAH and aza-PAH. These latter calculations will be described in a later section.

Early calculational approaches have been recently discussed by Dipple et al.[4] The most impressive correlations have involved calculations of various parameters at the K-region of these aza-PAH. It is found that higher calculated total charge at the K-region correlates well with higher observed carcinogenicities for the benzacridines. Higher electron density at the K-region has generally been taken as an indication of higher reactivity toward epoxidation or other reactions at the K-region. Subsequently, calculations of superdelocalizabilities at the K-region of benzacridines and their methylated derivatives have shown a similar good correlation with carcinogenicity.[16,17]

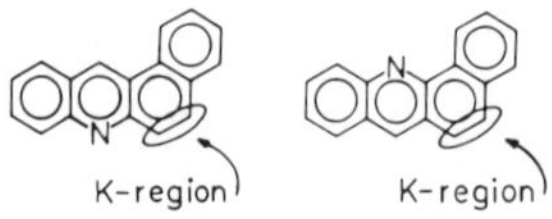

A recent study has revealed a generally good correlation between the rate of reaction of methyl-substituted BcACRs with OsO_4 (apparently at the K-region) with carcinogenicity and with the ability of the compounds to mutate *S. typhimurium* strain TA100 after metabolic activation.[18] Similarly, correlations have been found[19] between electrochemical oxidation and reduction potentials of a series of methylated BcACRs and carcinogenicity, as well as with quantum chemically calculated parameters.

V. BIOLOGICAL PROPERTIES OF METABOLITES AND RELATED DERIVATIVES OF BENZACRIDINES

Recently, both theoretical analysis and experimental studies of benzacridines have shifted to potential metabolites and related intermediates. Metabolism studies of selected benzacridines have now been reported, and others are in progress. Potential metabolites, and their analogues, have been synthesized, and their carcinogenicity and other biological properties have been examined. These studies will be reviewed in the following sections.

A. Theoretical Considerations

For the related PAH, oxidative metabolism leads to arene oxides, and subsequently to dihydrodiols, phenols, and diol-epoxides.[20] Studies of benzo[a]pyrene and, subsequently, numerous other PAH, have led to the discovery that bay-region diol-epoxides are the most

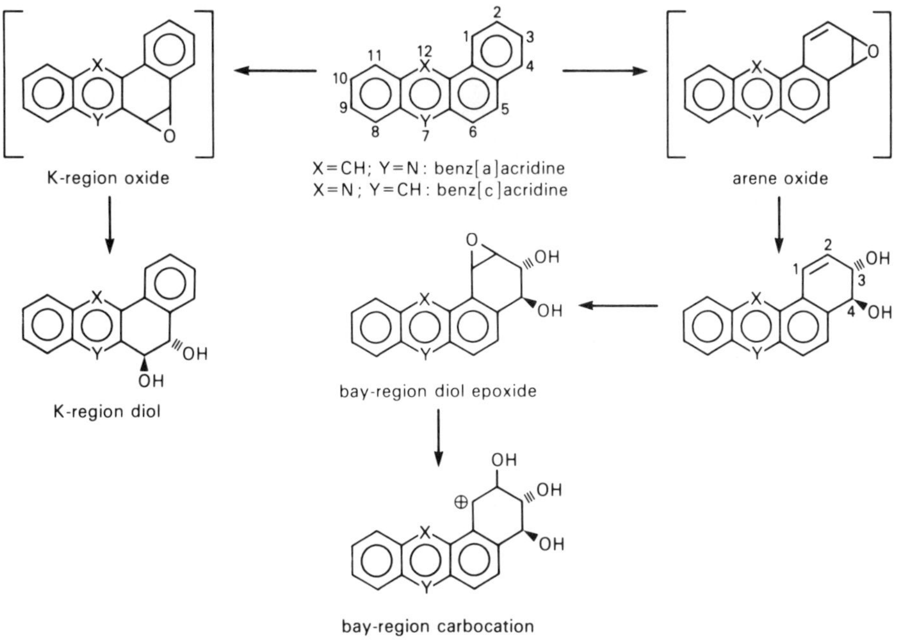

K-region oxide

X=CH; Y=N : benz[a]acridine
X=N ; Y=CH : benz[c]acridine

arene oxide

K-region diol

bay-region diol epoxide

bay-region carbocation

FIGURE 2. Metabolic activation of an arene to dihydrodiols and bay-region diol epoxide.

important "ultimate carcinogenic forms" of PAH. Furthermore, where more than one diol-epoxide can be formed from a carcinogenic PAH, the "bay-region" diol-epoxide has been found to be the only highly carcinogenic diol-epoxide. A part of the basis for the "bay-region theory,"[21,22] which predicted this observation, was the much higher reactivity calculated for those diol-epoxides in which the oxirane formed part of a bay region relative to those diol-epoxides in which the oxirane did not form part of a bay region. Experiments in the PAH field have revealed additional important determinants of biological activity, including conformational, steric, and absolute stereochemical factors.[23,24] Electronic considerations nonetheless remain one of the important components in predicting reactivity and biological activity of diol-epoxides, and benz- and dibenzacridines are particularly well suited for theoretical analysis.

Smith and Seybold[16] have compared quantum chemical estimates of the susceptibility of methylated BaACRs and BcACRs to reactions leading to bay-region diol-epoxides (Figure 2) with their carcinogenicities. The results were consistent with metabolic activation via bay-region diol-epoxides. However, extensive metabolism studies indicate that steric considerations are more important than electronic factors in determining the regiospecificity associated with the metabolism of PAH.[25] They also calculated charge densities and superdelocalizabilities for carbocations at the benzylic bay-region position of methylated BaACRs and BcACRs. Generally, carbocations derived from the more carcinogenic compounds were calculated to be more stable than the carbocations derived from the less or noncarcinogenic compounds.

Lowe and Silverman[26] and Lehr and Jerina[27] have used calculations to assess the effect of heteroatom substitution on the ease of carbocation formation from diol-epoxides. Lowe and Silverman examined the effect of heteroatom substitution upon ease of bay-region carbocation formation from BA. An examination of the charge distribution in the π-system of the carbocation at the bay-region position for BA, as calculated by simple Hückel theory, indicates that substitution of an electronegative nitrogen atom should be more destabilizing for the BaACR bay-region carbocation than for the BcACR bay-region carbocation (Figure 3). Lowe and Silverman supported these qualitative arguments with extended Hückel, INDO,

π-system for bay-region carbocation of BcACR

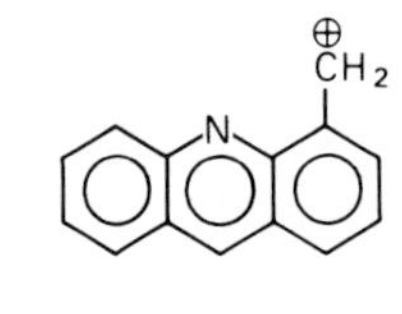

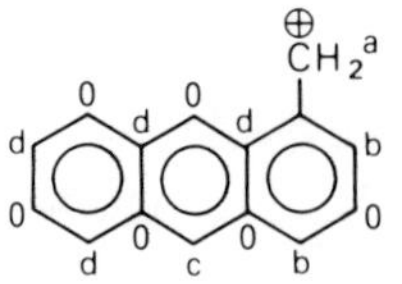

calculated charge distribution for π-system for the bay-region carbocation of BA
a=+.380; b=+.214;
c=+.095; d=+.024

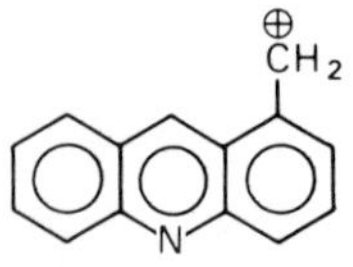

π-system for bay-region carbocation of BaACR

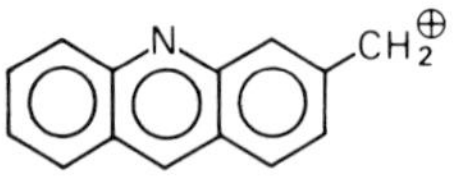

π-system for non-bay region carbocation of BcACR

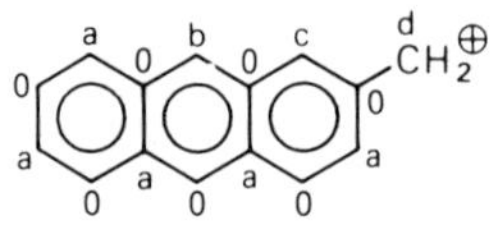

calculated charge distribution for π-system for the non-bay region carbocation of BA
a=+.03; b=+.12;
c=+.27; d=+.47

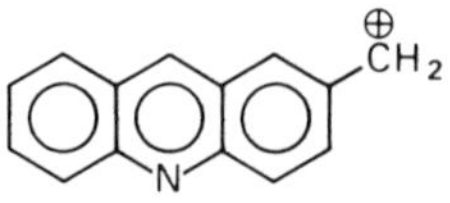

π-system for non-bay region carbocation of BaACR

FIGURE 3. Charge distribution in 1- and 2-anthracenylmethyl carbocations, and structures of analogous π-systems for the bay- and nonbay-region carbocations of benz[a]- and benz[c]acridine.

and Gaussian 70 calculations, and examined the effects of nitrogen atom substitution at other positions on the BA nucleus. Lehr and Jerina[27] used PMO and Hückel calculations to reach the same conclusion regarding the effects of nitrogen substitution upon ease of formation of bay-region carbocations from diol-epoxides of BaACR and BcACR. They also pointed out that the relative destabilization should be reversed for carbocation formation at the nonbay benzylic carbon atom on the angular ring of the benzacridines (Figure 3). In this case, the BcACR derivative should be more destabilized than the BaACR derivative, since the electronegative nitrogen atom occupies a position to which positive charge is delocalized in the former case, but not in the latter case.

B. Metabolism*

During the past few years, reports on the metabolism of acridine, BaACR, BcACR, and 7-Me-BcACR have appeared. Studies of the metabolism of DBchACR are in progress in our laboratories.

1. Acridine

There have been two recent reports concerning acridine metabolism.[28,29] McMurtrey and Knight[28] examined the metabolism of acridine by the 10,000 × *g* supernatant fractions (S10) of liver from control and Aroclor 1254-treated adult male Sprague-Dawley rats. A number

* The following abbreviations are used:

- "Dihydrodiol" refers to a trans-dihydroxydihydro derivative; thus, BcACR 3,4-dihydrodiol is trans-3,4-dihydroxy-3,4-dihydrobenz[c]acridine
- "Diol-epoxide-1" refers to the same benzo ring epoxide derivative of a dihydrodiol, in which the oxirane oxygen atom and the benzylic oxygen atom are cis (cf. Figure 5)
- "Diol-epoxide-2" refers to a stereoisomer of diol-epoxide-1 in which the oxirane oxygen atom and the benzylic hydroxyl group are trans (cf. Figure 5)
- "Tetrahydroepoxide" or "H_4-epoxide" refers to an epoxide derivative on a tetrahydrobenzo ring; thus, DBchACR-H_4-1,2-epoxide refers to 1,2-epoxy-1,2,3,4-tetrahydrodibenz[c,h]-acridine
- Absolute stereochemistry is indicated by the Cahn-Ingold-Prelog formalism, where appropriate
- Molecules lacking R,S designations at their chiral carbon atoms should be understood to be racemic (±) or of undetermined enantiomer composition

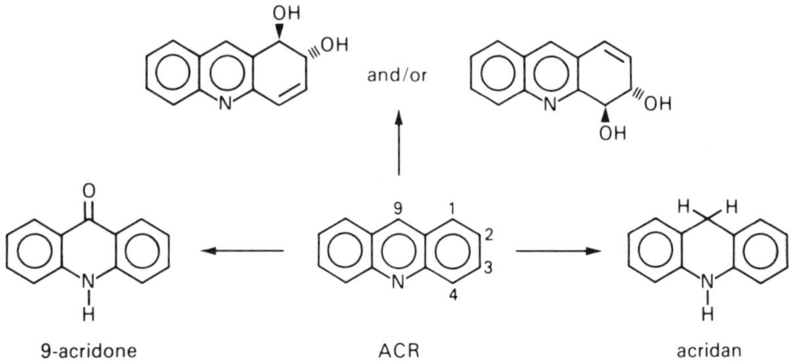

FIGURE 4. Metabolites of acridine.

of metabolites were produced in each case, as shown by HPLC analysis, but they were incompletely characterized, due to the lack of sufficient standards for comparison. The major metabolite with the S10 fraction from control rats was 9-acridone (Figure 4). A dihydrodiol, either the 1,2- or the 3,4-dihydrodiol, was also identified by GC/MS analysis of the trimethylsilyl derivatives. With the S10 fraction from Aroclor 1254-treated rats, the metabolic profile changed considerably. The dihydrodiol became the major metabolite, and very little 9-acridone was observed. Metabolism of acridine by partially purified aldehyde oxidase (EC 1.2.3.1) gave only 9-acridone as metabolite. The authors concluded that acridine was oxidized in vitro by aldehyde oxidase to 9-acridone, and by the cytochrome P-450 monooxygenase system and epoxide hydrolase to dihydrodiol.

Jacob et al.[29] have also briefly described the metabolism of acridine. With microsomes from lung tissue of untreated rats, a dihydrodiol and a phenol were detected as metabolites, the dihydrodiol being the major metabolite. With microsomes from liver of untreated rats, the same two metabolites were observed, with the dihydrodiol predominating to a greater extent. Small amounts of acridan and of a second dihydrodiol were also observed. With microsomes from liver of phenobarbital (PB)- and benzo[k]fluoranthene (BkF)-treated rats, enhanced formation of the major dihydrodiol metabolite was observed. No acridan was observed, but small amounts of the phenol and two additional metabolites were detected. Whether the phenol in this study is the 9-acridone found by McMurtrey and Knight is not clear from the limited experimental data that were provided.

2. *Benz[a]- and Benz[c]acridine*

Jacob et al.[29] have studied the metabolism of BaACR and BcACR by rat liver microsomes. Gas chromatography-mass spectrometry was used to identify and quantify metabolites which were converted to their trimethylsilyl derivatives. Only BaACR 3,4-dihydrodiol and BcACR 3,4-dihydrodiol were available as authentic standards. Other assignments were based on mass spectral analysis. For BaACR, only two dihydrodiols were observed. One was assigned as BaACR 5,6-dihydrodiol. The other dihydrodiol was not assigned a specific structure, but was not BaACR 3,4-dihydrodiol.

For BcACR, metabolism with liver microsomes of untreated rats led to formation of BcACR 5,6-dihydrodiol as the major metabolite. A non-K-region dihydrodiol which was not BcACR 3,4-dihydrodiol was observed as a minor metabolite.

The metabolism of BcACR was also studied with microsomes from PB-, 5,6-benzoflavone (BNF)-, and benzo[k]fluoranthene-treated rats. Three peaks in the GC-MS were assigned to non-K-region dihydrodiols. One of the peaks contained BcACR 3,4-dihydrodiol, but that peak was contaminated by another, unidentified dihydrodiol, which made it difficult to quantitate the level of BcACR 3,4-dihydrodiol. No direct observation of peaks corresponding to silylated derivatives of the bay-region diol-epoxide or its hydrolysis products was made.

However, GC-MS analysis of the silylated products of metabolism of BcACR 3,4-dihydrodiol led Jacob et al. to conclude that bay-region diol-epoxide is formed from the dihydrodiol, although other pathways of metabolism compete. The rigor of their conclusion is diminished by the lack of authentic diol-epoxide standard and of thorough identification of the metabolites. These metabolism studies suggest that bay-region diol-epoxides can be formed from BcACR, but that they are no more than very minor metabolites. These results are qualitatively similar to the metabolism results for BA.[30]

3. 7-Methylbenz[c]acridine

Holder and coworkers have extensively studied the metabolism of the potent carcinogen 7-methylbenz[c]acridine (7-Me-BcACR). They have been able to positively identify a large number of metabolites, due to the availability of synthetic standards that they prepared.[31,32] A number of metabolites have also been partially characterized by GC/MS and/or UV spectral properties. Holder et al. have studied the metabolism of 7-Me-BcACR by microsomes from lung and liver of normal and induced (PB, 3-MC, 7-Me-BcACR) rats,[33-36] by hepatocytes from normal and induced rats,[37] and by the intact rat.[38] Quantification was achieved by chromatographic separation of metabolites of radioactively labeled 7-Me-BcACR. Both 7-^{14}C-7-Me-BcACR and Me-^{3}H-7-Me-BcACR were used. The dihydrodiol with a bay-region double bond, 7-Me-BcACR-3,4-dihydrodiol, was found to be only a minor metabolite, present to the extent of about 3% of the ethyl acetate-extractable metabolites from lung and liver microsomes of untreated and induced rats.[36] It had eluded detection in earlier metabolism studies, due to its lability under the assay conditions.[33-35] The most dominant metabolites observed were 7-hydroxymethyl BcACR, 7-Me-BcACR-8,9-dihydrodiol, 7-Me-BcACR-5,6-oxide, and to a lesser extent, 7-Me-BcACR-5,6-dihydrodiol. Similarly, BA is extensively metabolized at the 5,6- and 8,9-double bonds.[30] Increased enzyme activity was observed for liver and lung microsomes from rats pretreated with MC, with the increased rate of formation of 7-MBAC-8,9-dihydrodiol being higher than the increased rate of formation of the other metabolites. The ability to observe the K-region oxides could well be due to slower enzymatic hydrolysis by epoxide hydrolase. Other minor metabolites were also identified, including 7-Me-BcACR-1,2-dihydrodiol, 7-Me-BcACR-3,4-dihydrodiol, and 7-hydroxymethyl-BcACR-5,6-oxide. Additional metabolites were also indicated by mass spectral analysis, but could not be assigned definite structures due to the absence of authentic standards. These included: (1) ''phenols,'' which were not identified, but which represented about one-fifth of the metabolites; (2) dihydrodiol and phenolic metabolites arising from metabolism of 7-OHMe-BcACR, and (3) an N-oxide of a dihydrodiol of 7-Me-BcACR.

Boux and Holder examined the metabolism of 7-Me-BcACR with hepatocytes from the untreated and induced rats.[37] The major metabolites were 7-OHMe-BcACR, 7-MeBcACR-5,6-dihydrodiol, and 7-Me-BcACR-8,9-dihydrodiol. They also examined the nature of the water-soluble metabolites and found that about 10% of the intracellular and about 25% of the extracellular water-soluble metabolites were glucuronides or sulfates.

Holder et al. have also determined the metabolism of 7-Me-BcACR by the intact rat.[38] They measured the ethyl acetate-extractable metabolites after β-glucuronidase treatment of bile. The major metabolites were BcACR-7-carboxylic acid and 7-OHMe-BcACR. Together these accounted for more than half the observed metabolites. Minor metabolites included 7-Me-BcACR-5,6-dihydrodiol, various phenols of 7-Me-BcACR, 7-Me-BcACR-8,9-dihydrodiol, 7-Me-BcACR-1,2-dihydrodiol, 7-OHMe-BcACR-5,6-oxide, and a dihydrodiol derivative of 7-OHMe-BcACR. This latter metabolite and 7-Me-BcACR-8,9-dihydrodiol increased significantly when MC-induced rats were used, whereas 7-OHMe-BcACR became almost undetectable. The structure of the dihydrodiol derivative of 7-OHMe-BcACR has not been determined, although Holder et al. have tentatively assigned the diol moiety to the 1,2- or 3,4-positions.

4. Dibenz[c,h]acridine

Our laboratories have been studying the metabolism, mutagenicity, and tumorigenicity of dibenz[c,h]acridine. A novel feature of this work, for the aza-PAH field, has been the preparation of and absolute configurational assignments for the enantiomeric forms of the bay-region[39] and K-region[40] dihydrodiols and the bay-region diol-epoxides.[39] This has enabled absolute stereochemical aspects of the metabolism of DBchACR to be examined, in addition to regiochemical aspects.

Metabolism of DBchACR[41] proceeds much faster with liver microsomes from MC-treated rats than with liver microsomes from control or PB-treated rats. In all cases, metabolism occurs primarily at the K-region, with DBchACR-5,6-oxide and DBchACR-5,6-dihydrodiol accounting for 80 to 86% of total metabolites. Of the K-region metabolites, DBchACR-5,6-oxide contributes 26 to 56%. The dihydrodiol with a bay-region double bond, DBchACR-3,4-dihydrodiol, accounts for 10 to 17% of total metabolites. The acridone, DBchACR-7-one, was not observed as a metabolite. Both DBchACR-3,4- and 5,6-dihydrodiols were formed with high enantiomeric purity, regardless of the source of microsomes. In the former case, (−)-(R,R)-DBchACR-3,4-dihydrodiol was the predominant (86%) enantiomer formed. For the K-region dihydrodiol, (+)-(R,R)-DBchACR-5,6-dihydrodiol was the predominant (88%) enantiomer formed. These results are in accord with predictions based on a steric model for the catalytic binding site for cytochrome P450.[25]

The metabolism of the enantiomeric forms of DBchACR-3,4-dihydrodiol has also been studied.[42] The major metabolites of (−)-(R,R)-DBchACR-3,4-dihydrodiol are a bis-dihydrodiol and a bay-region diol-epoxide for which the (1R,2S,3S,4R)-DBchACR-3,4-diol-1,2-epoxide stereoisomer predominates (Figure 5). A minor metabolite has been identified as the 8-(or 9-)-phenolic derivative of the 3,4-dihydrodiol. Metabolism of (+)-(S,S)-DBchACR-3,4-dihydrodiol also yields a 8-(or 9-)-phenolic derivative of the 3,4-dihydrodiol, a bis-dihydrodiol, and a diol-epoxide as metabolites (Figure 5). Due to (1) the trans-arrangement of the hydroxyl groups at the K-region, (2) the chiroptical properties of the tetraacetate derivatives of the bis-dihydrodiols, and (3) the diastereomeric relationship of the bis dihydrodiols, which are separable by HPLC, it was possible to assign the absolute configurations given in Figure 5. These results reveal that the absolute configuration of the K-region diol is (R,R), regardless of whether the configuration of the bay-region dihydrodiol from which it is formed is (R,R) or (S,S).

When racemic DBchACR-3,4-dihydrodiol was metabolized by microsomes from untreated, phenobarbital-, and 3-methylcholanthrene-treated rats, very similar distributions of metabolites were observed. The bis-dihydrodiols, obtained in roughly equal amounts, comprised 68 to 83% of the metabolites. The bay-region diol-epoxides comprised 15 to 23% of the metabolites, with the diol-epoxide-2, (1R,2S,3S,4R)-DBchACR-3,4-diol-1,2-epoxide-2, which is derived from the (R,R)-DBchACR-3,4-dihydrodiol, predominating (79 to 96%) over the diol-epoxide-1, (1R,2S,3R,4S)-DBchACR-3,4-diol-1,2-epoxide-1, derived from (S,S)-DBchACR-3,4-dihydrodiol (Figure 5). The phenolic dihydrodiols accounted for 2 to 9% of the total metabolites.

Metabolism by a highly purified monoxygenase system reconstituted with cytochrome P450c and epoxide hydrolase gave a metabolite profile very similar to that observed with liver microsomes from MC-treated rats. Greater than tenfold enhancement of the rate of metabolism was observed with microsomes from MC-treated rats relative to the rate of metabolism from control or PB-treated rats.

5. Summary of Metabolism Results

Based upon the results for the limited number of acridine derivatives that have been studied, it appears that their metabolism is very similar to that of the PAH. Major metabolites (arene oxides, dihydrodiols, diol-epoxides, phenols, bis-dihydrodiols, hydroxymethyl de-

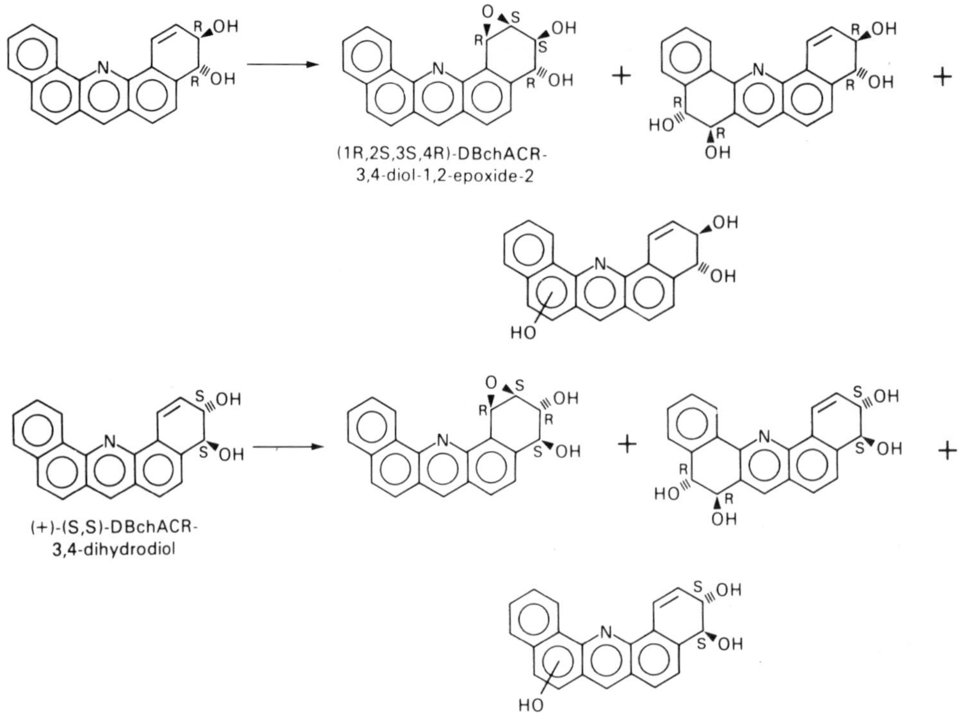

(1R,2S,3S,4R)-DBchACR-3,4-diol-1,2-epoxide-2

(+)-(S,S)-DBchACR-3,4-dihydrodiol

FIGURE 5. Metabolites of (−)-(R,R)- and of (+)-(S,S)-dibenz[c,h]acridine-3,4-dihydrodiol.

rivatives) are the metabolites typically observed for the PAH. There is little evidence that N-oxides are formed to more than a minor extent, and acridone formation has been found to be significant only for acridine.

The stereochemical results observed in the metabolism of DBchACR closely parallel those for the PAH. Thus, the observation of (R,R)-DBchACR-3,4-DHD as the major dihydrodiol enantiomer with a bay-region double bond is in accord with results for the PAH.[43] Similarly, the stereoselective metabolism of the (3R,4R)-dihydrodiol to bay-region diol-epoxide-2 of (1R,2S,3S,4R)-absolute configuration, and of the (3S,4S)-dihydrodiol to the bay-region diol-epoxide-1 of (1R,2S,3R,4S)-absolute configuration, is in agreement with results for bay-region dihydrodiols of PAH.[25] The absolute configurations of the bis-dihydrodiols obtained by metabolism of DBchACR-3,4-DHD are also consistent with a steric model for the catalytic binding site of cytochrome P450c that has been developed for PAH.[25]

For the benz- and dibenzacridines so far studied, metabolism to bay-region diol-epoxides is, at most, a minor pathway. DBchACR-3,4-dihydrodiol is a relatively minor metabolite for DBchACR (10 to 17%), and the analogous dihydrodiols with bay-region double bonds appear to be formed to a lower extent (*circa* 3 to 5%) from BcACR and 7-Me-BaACR, and possibly not at all from BaACR. Metabolism of the dihydrodiols with bay-region double bonds to bay-region diol-epoxides is indicated for both DBchACR and BcACR, although metabolism at the K-region appears to predominate in each case.

C. Mutagenicity of Dihydro- and Tetrahydro-Derivatives of Benz- and Dibenzacridines

1. K-Region Oxides

Kitahara et al.[44] have reported the synthesis of a variety of K-region oxides derived from aza-PAH, including DBahACR, DBajACR and DBchACRs. They also determined the intrinsic mutagenicities of the compounds to *S. typhimurium* strains TA98 and TA100, and their mutagenicities after metabolic activation with the 9000 × *g* supernatant fraction of rat

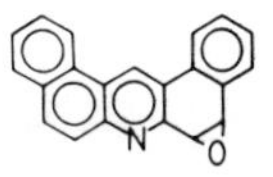

DBajACR-5,6-oxide

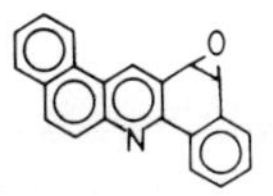

DBahACR-12,13-oxide

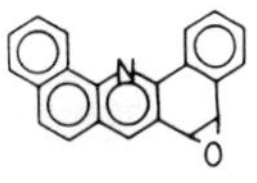

DBchACR-5,6-oxide

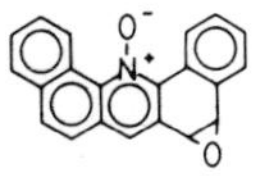

N-oxide of DBchACR-5,6-oxide

liver homogenates. None of the compounds shown exhibited measurable intrinsic mutagenicity in strain TA98 of *S. typhimurium*. DBchACR-12,13-oxide and the N-oxide of DBchACR-5,6-oxide were very weakly mutagenic to *S. typhimurium* TA100. The compounds could be metabolically activated to mutagens, but the parent aza-PAH were more highly activated.

In a separate study,[45] BcACR-5,6-oxide was found to be weakly mutagenic. This K-region oxide induced 2 and 15 *His*$^+$ revertants per nmole in *S. typhimurium* strains TA98 and TA100, respectively. These activities are 20- and 35-fold lower than the corresponding mutagenicities of BA-5,6-oxide. The very low mutagenicities of the K-region oxides make them poor candidates for activated forms of benz- and dibenzacridines.

2. *Benzo-Ring Derivatives of Benz[a]acridine*

A number of BaACR derivatives have been studied.[45] Neither BaACR nor BaACR-3,4-dihydrodiol were appreciably activated to mutagens by hepatic microsomes from Aroclor® 1254-treated rats. The bay-region diol-epoxides and angular ring tetrahydroepoxides were intrinsically mutagenic toward *S. typhimurium* strains TA98 and TA100, as well as toward Chinese hamster V79 cells (Figure 6). The tetrahydro (H_4) epoxides were much more mutagenic than the diol-epoxides. We have previously used H_4 epoxides as simplified model compounds for diol-epoxides of PAH, and have found that their mutagenicities more closely follow predictions based on calculated and observed reactivity than the mutagenicities of the more structurally complex diol-epoxides, wherein the hydroxyl groups lead to diastereomers and significant conformational differences.[23] The observation of higher mutagenic activity for the tetrahydroepoxides, relative to the diol-epoxides, is consistent with results for the PAH, although the difference in mutagenicity is greater than for PAH.

For all these systems, the bay-region diol-epoxide isomer-2, in which the oxirane oxygen atom is *trans* to the benzylic hydroxyl group, is slightly more mutagenic than isomer-1, in which the oxirane oxygen atom is *cis* to the benzylic hydroxyl group. The H_4-epoxides are exceptional, in that the bay-region (H_4-1,2) and nonbay-region (H_4-3,4) tetrahydroepoxides have similar mutagenicities in the systems in which they have been compared (TA98 and TA100). Usually, the bay-region tetrahydroepoxides are considerably more mutagenic, consistent with reactivity considerations inherent in the "bay-region theory".[21-23] For example, for BA, the bay-region H_4-epoxide is four- to fivefold more mutagenic than the nonbay H_4-epoxide in the same mutant bacterial strains. Here, the comparable mutagenic activity of the two epoxides appears to be due to diminished activity of the bay-region H_4 1,2-epoxide due to aza-substitution, rather than to enhanced activity of the nonbay-region isomer. This conclusion is supported by comparisons of mutagenic activity of other benzo-ring tetrahydroepoxides, which are presented in a later section, and is consistent with the predicted destabilizing effect of the nitrogen atom in BaACR upon a carbocation formed at C-1, the bay-region position.[26,27]

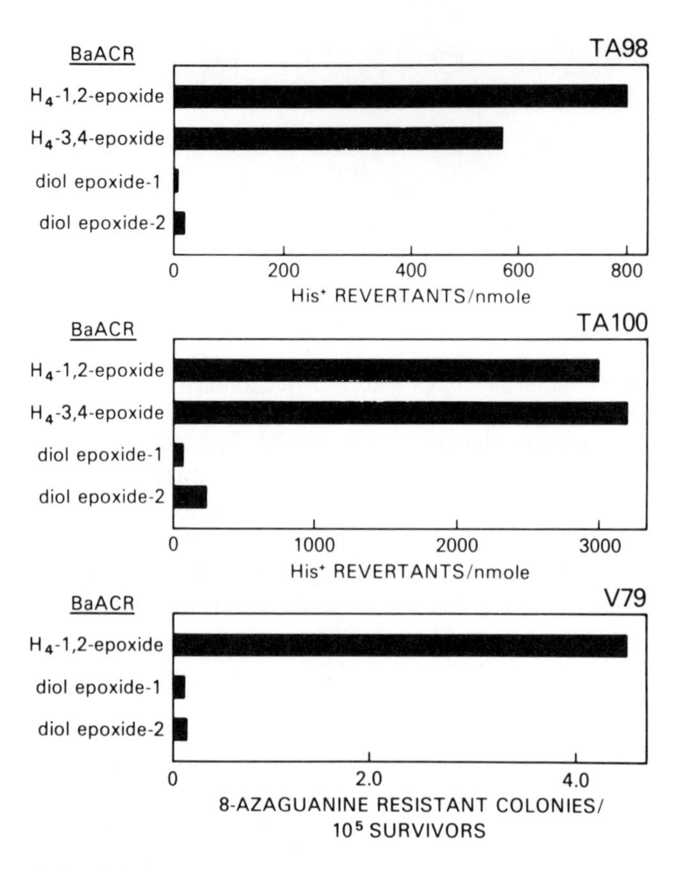

FIGURE 6. Intrinsic mutagenicities of H_4- and diol-epoxide derivatives of BaACR.

3. Benzo-Ring Derivatives of Benz[c]acridine

A large number of benz[c]acridine derivatives have been synthesized[46-49] and assayed for mutagenicity.[45] The ability of BcACR and all the metabolically possible dihydrodiols of BcACR, including the K-region dihydrodiol, to be activated to mutagens toward strain TA100 of *S. typhimurium*, by microsomes from Aroclor® 1254-treated rats, was examined.[45] Of these compounds, only BcACR-3,4-dihydrodiol, the dihydrodiol with an isolated double bond in the bay region of BcACR, was significantly activated to mutagens. The intrinsic mutagenic activities of the diol-epoxides that can potentially be formed by further metabolism of the dihydrodiols at the nonaromatic double bond of the dihydrodiol have been determined in bacteria and in Chinese hamster V79 cells (Figure 7). In all three test systems, only the bay-region diol-epoxides exhibit high inherent mutagenicity. The diol-epoxides on the nonangular ring, as well as the "reverse" diol-epoxide (BcACR 1,2-diol-3,4-epoxide) on the angular benzo ring, are either feebly active or inactive. The two diastereomeric series of bay-region diol-epoxides have comparable mutagenicities in *S. typhimurium* strain TA98. The bay-region diol-epoxide-1 diastereomer is more active in *S. typhimurium* strain TA100 than the bay-region diol-epoxide-2 diastereomer, whereas the reverse is true in Chinese hamster V79 cells.

The mutagenicities of the bay-region diol-epoxides are compared with the mutagenicities of the angular ring tetrahydro-epoxides in Figure 8. The bay-region tetrahydroepoxide is much more mutagenic than the bay-region diol-epoxides, in all test systems. The nonbay tetrahydroepoxide is the least mutagenic of the four compounds, and is virtually inactive in *S. typhimurium* strain TA100 and in Chinese hamster V79 cells. The much greater muta-

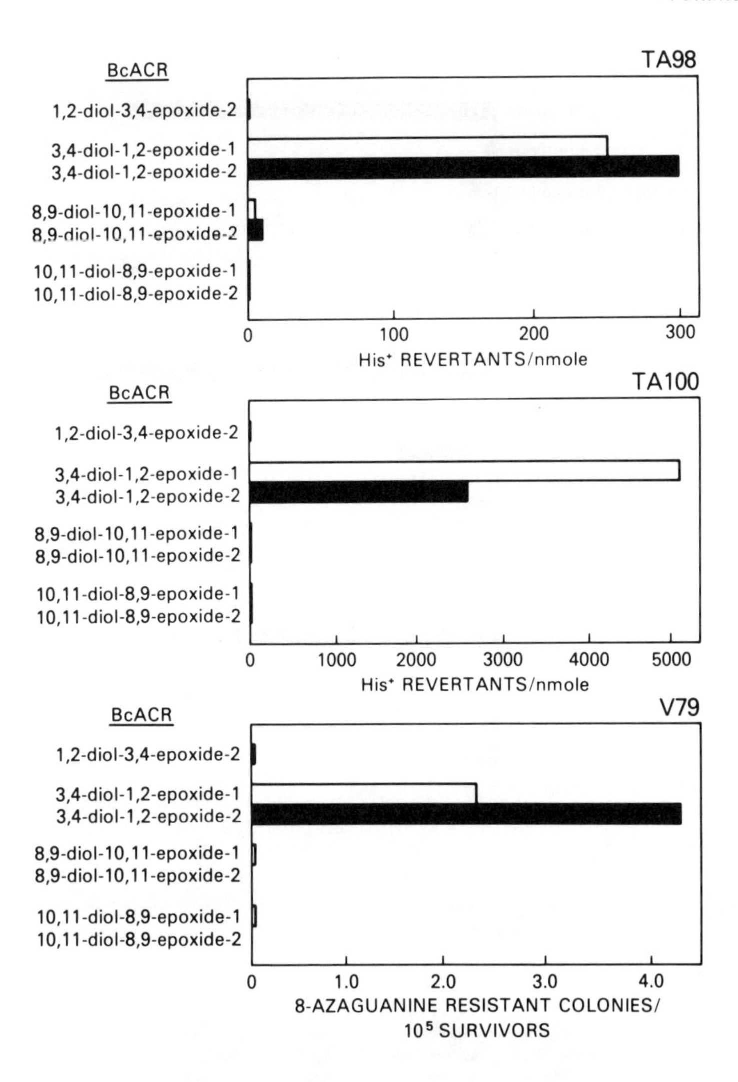

FIGURE 7. Intrinsic mutagenicities of BcACR diol-epoxides in bacteria and in mammalian cells.

genicity of the bay-region H_4-epoxide relative to the nonbay-region isomer is consistent with the predicted decrease in electrophilic character for the latter compound[26,27] due to the influence of the nitrogen atom in the acridine nucleus, and is in sharp contrast to the analogous BaACR tetrahydroepoxide derivatives (Figure 6).

4. Benzo-Ring Derivatives of 7-Methylbenz[c]acridine

Holder and coworkers have examined the ability of 7-Me-BcACR and its metabolically possible dihydrodiols, not including hydroxymethyl derivatives, to be activated to mutagens by microsomes from Aroclor®-treated rats or MC treated guinea pigs.[36] Of these compounds, the dihydrodiol with a bay-region double bond was by far the most metabolically activated toward *S. typhimurium* strains TA98 and TA100, and was the only compound that was activated to mutagens to a greater extent than 7-Me-BcACR.

Holder et al. also examined the metabolic activation of several dihydro derivatives of 7-Me-BcACR with liver microsomes from Mc-treated guinea pigs.[36] Only 3,4-H_2-7-Me-BcACR was metabolically activated highly mutagenic species. The other dihydro derivatives were

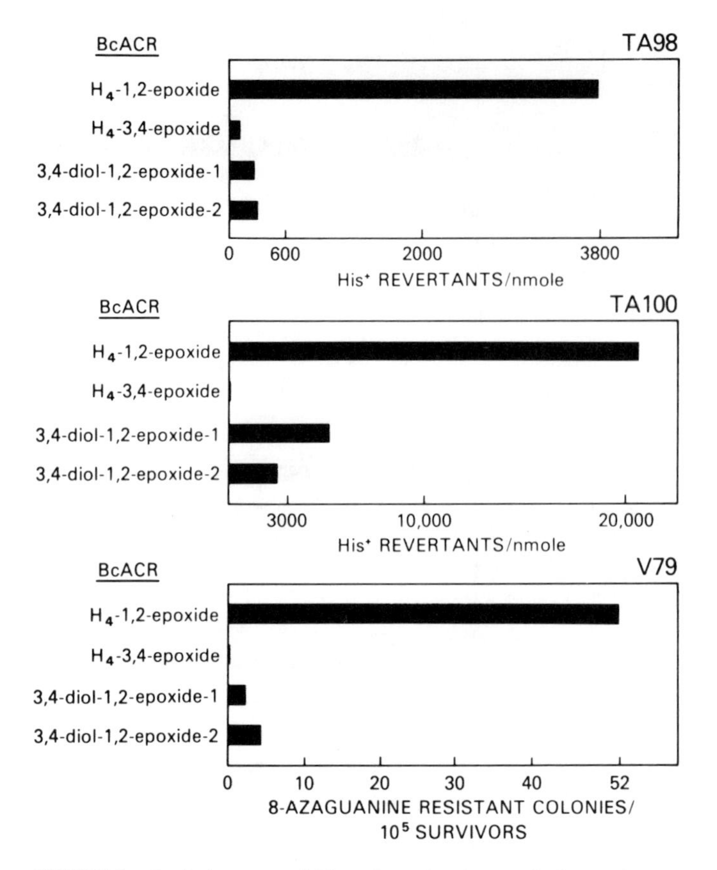

FIGURE 8. Intrinsic mutagenicities of angular ring tetrahydroepoxide and bay region diol-epoxide derivatives of BcACR.

1,2-H_2-7-Me-BcACR 3,4-H_2-7-Me-BcACR 8,9-H_2-7-Me-BcACR 10,11-H_2-7-Me-BcACR

activated to a lesser extent than 7-Me-BcACR. The tetrahydrobenzo-ring derivatives, 1,2,3,4-H_4- and 8,9,10,11-H_4-7-Me-BcACR, were also studied. Neither compound was metabolically activated to mutagens.

Holder et al. also examined the intrinsic mutagenicity of the racemic bay-region diol-epoxides-1 and -2 derived from 7-Me-BcACR.[36] The diol-epoxides were significantly mutagenic toward strains TA98 and TA100 of *S. typhimurium*. The diol-epoxide-2 isomer was

7-Me-BcACR-3,4-diol-1,2-epoxide-1 7-Me-BcACR-3,4-diol-1,2-epoxide-2

the more mutagenic in each strain, being 1.7 and 4.2 times as mutagenic as the diol-epoxide-1 isomer in strains TA98 and TA100, respectively. The latter result contrasts with mutagenicity data for BcACR bay-region diol-epoxides, wherein the diol-epoxide-1 isomer is the more mutagenic in *S. typhimurium* strain TA100. The diol-epoxide-2 isomer of 7-Me-BcACR was also observed to be severalfold more mutagenic than the diol-epoxide-1 isomer in Chinese hamster V79 cells.

In addition, 7-Me-BcACR and the five metabolically possible dihydrodiol derivatives were examined for their ability to induce sister chromatid exchange (SCE) in Don cells.[36] Only the dihydrodiol with a bay-region double bond showed activity greater than 7-Me-BcACR. It was two- to fourfold more active. A comparison of the ability of BcACR and 7-Me-BcACR to induce SCE revealed that 7-Me-BcACR was equal to or slightly more active than BcACR.

5. *Benzo-Ring Derivatives of Dibenz[c,h]acridine*

The mutagenicity of DBchACR and its derivatives has been studied extensively.[50] Of the racemic, metabolically possible 1,2-, 3,4-, and 5,6-dihydrodiols, DBchACR 3,4-dihydrodiol is most extensively metabolized to mutagens toward *S. typhimurium* strains TA98 and TA100 by hepatic microsomes from Aroclor®-1254-treated rats. The tetrahydrodiol, 3,4-dihydroxy-1,2,3,4-tetrahydro DBacACR, in which the bay-region double bond in the dihydrodiol has been saturated, was only minimally activated to mutagens. On the other hand, 3,4-dihydro-DBchACR was more extensively activated to mutagens than any of the other compounds

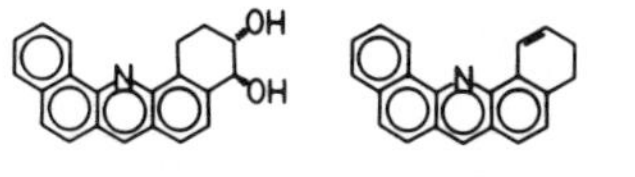

(±)-DBchACR-H_4-3,4-diol

3,4-H_2-DBchACR

studied. These results demonstrate the importance of the bay-region double bond for metabolic activation to mutagens. Of the optically active 3,4-dihydrodiols, the (3R,4R)-enantiomer was more extensively activated to mutagens than the (3S,4S)-enantiomer.

The intrinsic activities of tetrahydro- and diol-epoxides on the angular benzo ring were also examined, and the data are shown in Figure 9. For the H_4-epoxides, the bay-region isomers were considerably more mutagenic than their nonbay-region counterparts, which were virtually inactive in *S. typhimurium* strain TA100 and in Chinese hamster V79 cells. The enantiomeric forms of the bay-region DBchACR H_4-1,2-epoxides were of comparable mutagenicity in the bacteria, but the (1R,2S)-enantiomer was more than twice as mutagenic as the (1S,2R)-enantiomer in Chinese hamster V79 cells. Similarly, a comparison of the enantiomeric forms of each diol-epoxide within the diol-epoxide-1 or diol-epoxide-2 series reveals highly comparable mutagenicities in bacteria, with the enantiomers in the diol-epoxide-2 series being somewhat more mutagenic than the enantiomers in the diol-epoxide-1 series. In the Chinese hamster V79 cells, the (1R,2S,3S,4R)-stereoisomer is five- to sevenfold more mutagenic than the other stereoisomeric diol-epoxides, which were of comparable mutagenicity. In all cases, (1R)-absolute configuration at the benzylic epoxide carbon atoms of the tetrahydro and diol-epoxides results in higher mutagenic activity in Chinese hamster V79 cells.

The bay-region H_4-epoxides were much more mutagenic than the bay-region diol-epoxides in bacteria. Generally higher mutagenicities of the H_4-epoxides relative to diol-epoxides were also observed in Chinese hamster V79 cells, the only exception being the higher mutagenicity of the (1R,2S,3S,4R)-diol-epoxide-2 relative to the (1S,2R)-H_4-epoxide. If H_4- and diol-epoxides with identical configurations at the epoxide-bearing benzylic carbon atoms are compared [for example, the (1R,2S)-H_4-epoxide with the (1R,2S,3S,4R)-diol-epoxide], the tetrahydroepoxides are more mutagenic in all cases.

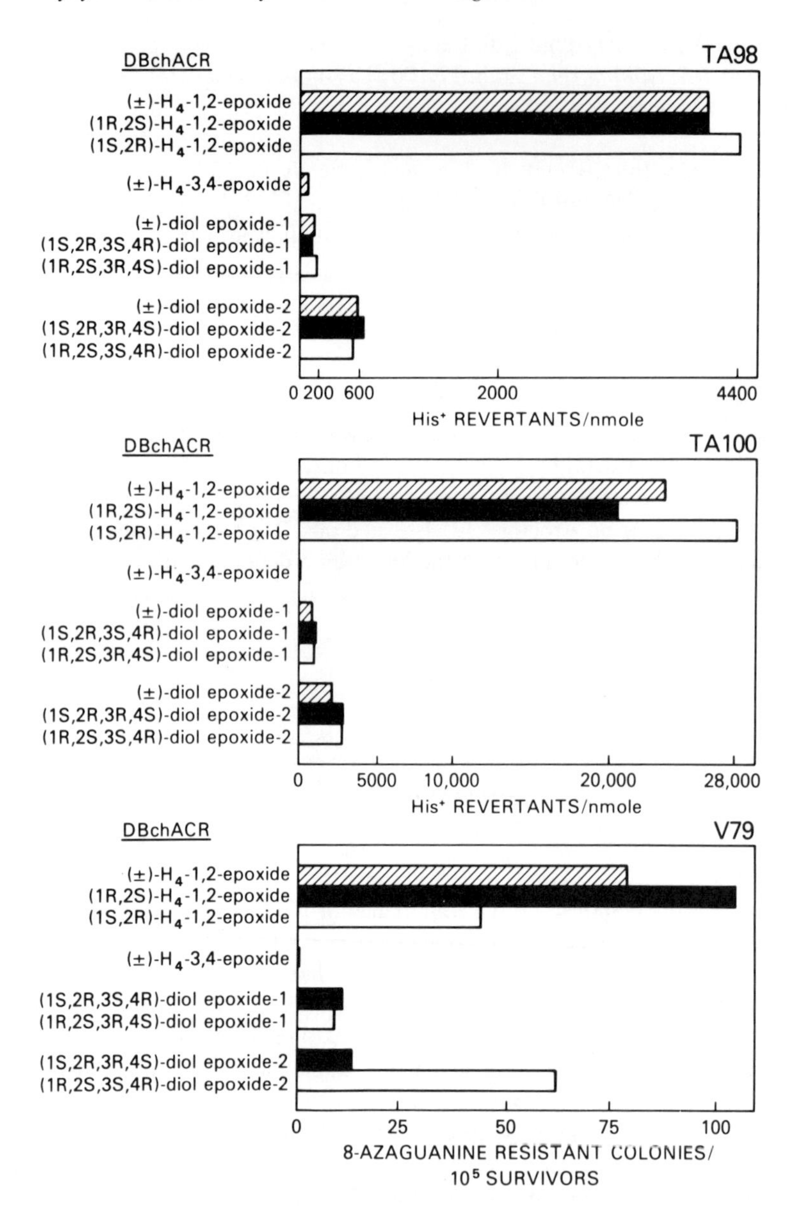

FIGURE 9. Intrinsic mutagenicities of tetrahydro epoxide and diol-epoxide derivatives of DBchACR in bacteria and in mammalian cells.

6. Comparison of Mutagenicities of Analogous Derivatives from Different aza-PAH and PAH

The mutagenicity data for BaACR, BcACR, and BA in the previous sections were obtained under conditions that permit confident comparison of their mutagenicities. The data for DBchACR were obtained under the same conditions used for the other compounds, and with the same strains of bacteria and of Chinese hamster V79 cells, but several months subsequent to the other measurements. Consequently, data for DBchACR have been included in the following comparisons, but should be of significant value only for qualitative or semi-quantitative comparisons.

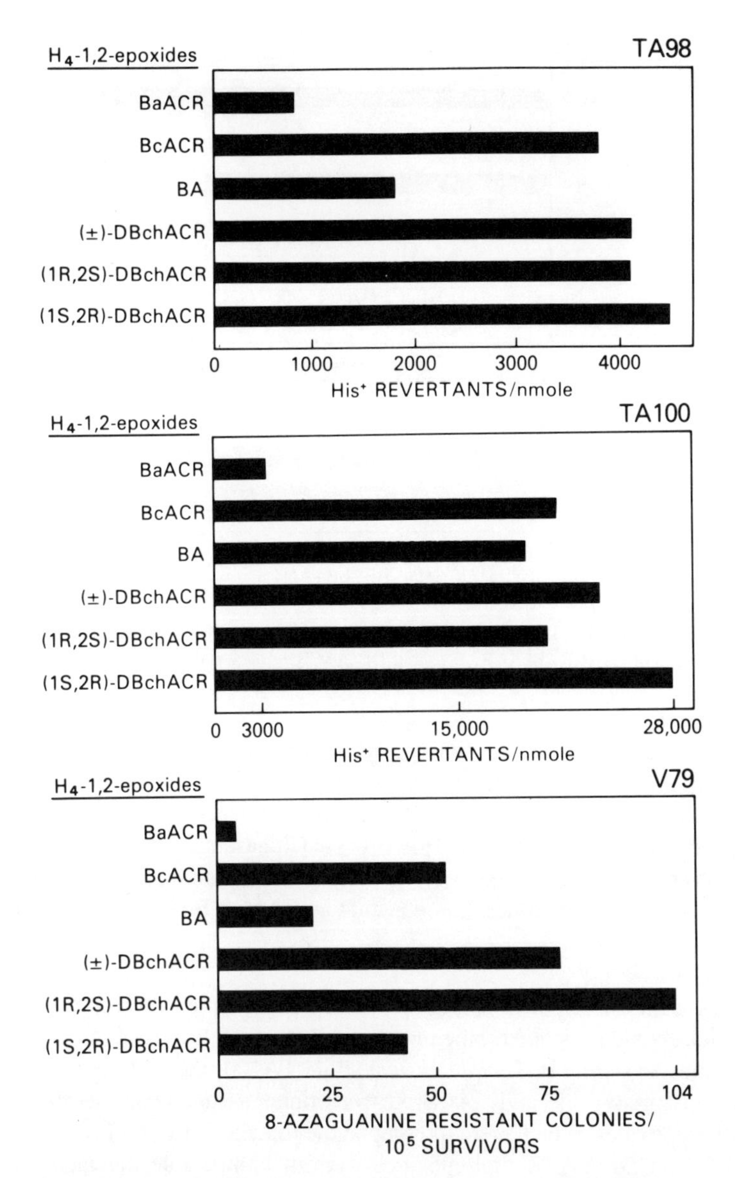

FIGURE 10. Intrinsic mutagenicities of bay-region tetrahydroepoxides of BaACR, BcACR, BA, and DBchACR in bacteria and in mammalian cells.

a. Bay-Region Tetrahydroepoxides

The relative mutagenicities of the bay-region H_4-epoxides of BaACR, BcACR, BA, and DBchACR are shown in Figure 10. The bay-region H_4-epoxide of BaACR is the least mutagenic compound and is severalfold less active than its aza-PAH counterparts. For BcACR and DBchACR, the bay-region H_4-epoxides are of comparable (TA100) or two- to fourfold greater (TA98 and V79) mutagenicity than the bay-region H_4-epoxide of BA.

b. Nonbay-Region Tetrahydroepoxides

Mutagenicity data in *S. typhimurium* strains TA98 and TA100 are shown in Figure 11 for the racemic nonbay-region H_4-epoxides of BaACR, BcACR, BA, and DBchACR. In contrast to the bay-region H_4-epoxides, it is the BcACR and DBchACR nonbay region H_4-epoxides that have low mutagenic activity, whereas the BaACR and BA nonbay region tetrahydroepoxides are of comparable, and significantly higher, mutagenicity.

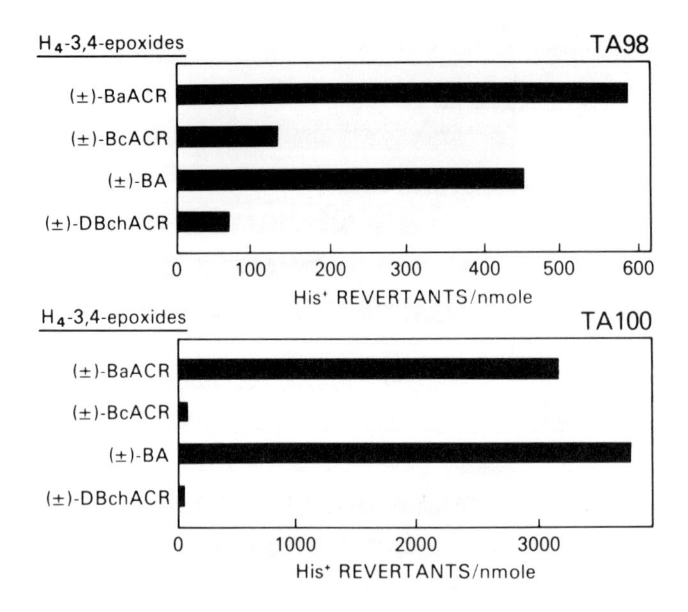

FIGURE 11. Intrinsic mutagenicities of nonbay-region tetrahydroepoxides of BaACR, BcACR, BA, and DBchACR in bacteria.

c. Diol-Epoxides-1

Mutagenicity data for bay-region diol-epoxides-1 are shown in Figure 12. In all systems, the BaACR bay-region diol-epoxide is by far the weakest mutagen. In bacteria, racemic bay-region diol-epoxide-1 derived from BA is more mutagenic than the analogous diol-epoxides from BcACR and DBchACR. However, in Chinese hamster V79 cells, the BcACR diol-epoxide has comparable mutagenicity to the BA diol-epoxide, and the DBchACR diol-epoxides are more than three times as mutagenic as BA and BcACR diol-epoxides.

d. Diol-Epoxides-2

Mutagenicity data for bay-region diol-epoxides-2 is presented in Figure 13. As was the case for the diol-epoxide-1 isomers, the most weakly mutagenic epoxide is the BaACR diol-epoxide, in all test systems. In *S. typhimurium* strain TA100, the diol-epoxide derived from BA was the most mutagenic and was about four times as mutagenic as the BcACR and DBchACR diol-epoxides, which were of comparable mutagenicity. In *S. typhimurium* strain TA98, the BA and DBchACR diol-epoxides were of comparable mutagenicity and were about twice as mutagenic as the BcACR diol-epoxide. In Chinese hamster V79 cells, the (1R,2S,3S,4R)-enantiomer of DBchACR was the most mutagenic diol-epoxide, by a considerable margin.

7. Summary of Mutagenicity Data

The available data strongly point toward bay-region diol-epoxides as ultimate mutagenic forms of those benz- and dibenzacridines that are metabolically activated to strong mutagens. The K-region oxides, in the four cases examined, have been very feebly mutagenic. In metabolic activation studies, only those derivatives (dihydrodiols and dihydro compounds) with a bay-region double bond have been activated to potent mutagens. The metabolism data for DBchACR shows that metabolism to a bay-region diol-epoxide occurs, and the data for BcACR and 7-Me-BcACR indicate that metabolism to bay-region diol-epoxides is likely, though minor, there also. Furthermore, the bay-region diol-epoxides and H_4-epoxides have, with the exception of BaACR diol-epoxides and H_4-epoxide, been potent mutagens and have exhibited the strongest intrinsic mutagenicity of the various epoxide derivatives of each aza-PAH studied.

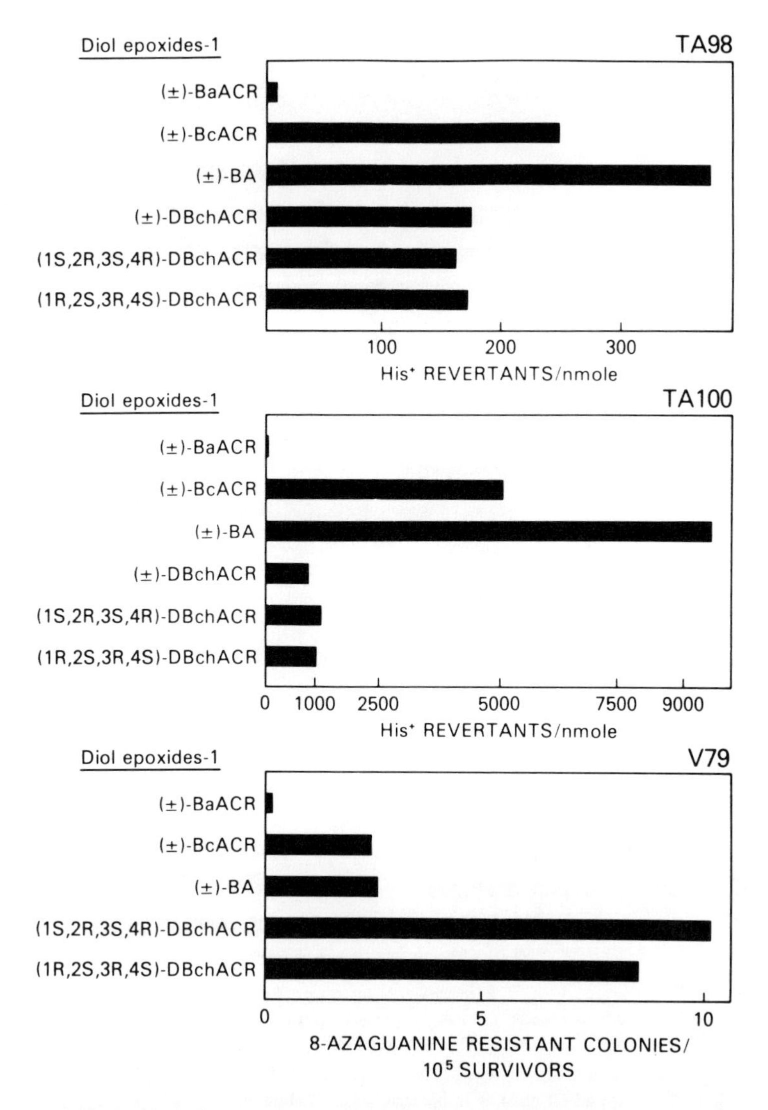

FIGURE 12. Intrinsic mutagenicities of bay-region diol-epoxides-1 of BaACR, BcACR, BA, and DBchACR in bacteria and in mammalian cells.

The diminished mutagenicities of the bay-region H_4- and diol-epoxides of BaACR point to the predicted significant attenuation of activity due to the nitrogen atom. Further evidence for this effect comes from the very low mutagenicities of the nonbay-region diol-epoxides and H_4-epoxides of BcACR and of the nonbay-tetrahydroepoxide of DBchACR. When the benzylic carbon atom cannot be formally conjugated with the nitrogen atom (for example, the bay-region H_4- and diol epoxides of BcACR, 7-Me-BcACR, and DBchACR), no clearly discernable effect of nitrogen atom substitution can be found. The H_4-epoxide derivatives appear to be either unaffected by nitrogen atom substitution or to have somewhat enhanced mutagenicities, as indicated by comparison of the mutagenicity of the nonbay-region H_4-epoxide of BaACR with the corresponding H_4-epoxide of BA (Figure 11) and of the BcACR bay-region H_4-epoxide with the bay-region H_4-epoxide of BA (Figure 10). Even though no destabilization due to resonance delocalization is expected in those cases, some reduction in the electrophilicity of the epoxides, due to pi-electron withdrawal by the nitrogen atom, might have been anticipated. Possibly, the electron density of the lone pair electrons on the nitrogen atom is able to stabilize positive charge at the bay-region benzylic carbon atom.

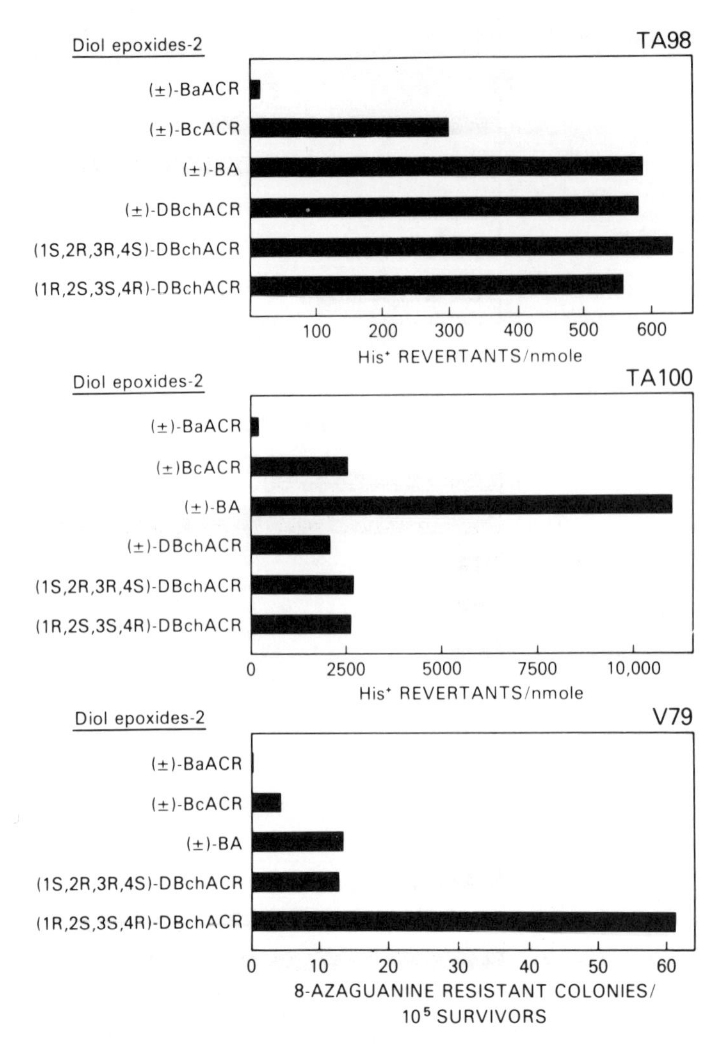

FIGURE 13. Intrinsic mutagenicities of bay-region diol-epoxides-2 of BaACR, BcACR, BA, and DBchACR in bacteria and in mammalian cells.

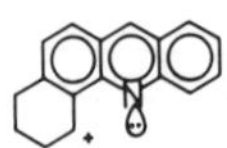

However, there is presently no basis for assessing the importance, if any, of such an effect.

For the diol-epoxides, the BcACR diol-epoxides range from slightly to considerably less (about fourfold in *S. typhimurium* strain TA100 for diol-epoxide-2) mutagenic than the BA diol-epoxides, whereas the DBchACR diol-epoxides range from considerably less mutagenic than the analogous BA diol-epoxides (*S. typhimurium* strain TA100) to considerably more mutagenic (Chinese hamster V79 cells).

The mutagenicities of the stereoisomeric DBchACR bay-region diol-epoxides are noteworthy. The enantiomeric forms of diol-epoxide-1 and of diol-epoxide-2 (Figure 9) have virtually identical mutagenicities in bacteria. In previous studies of the PAH chrysene,[51] benzo[a]pyrene,[52] BA,[53] and benzo[c]phenanthrene,[54] considerably greater differences in mutagenic activity have been noted for the enantiomers in bacteria, except for the enantiomeric benzo[c]phenanthrene diol-epoxides in *S. typhimurium* strain TA100. On the other hand, the most potent DBchACR diol-epoxide in Chinese hamster V79 cells is the (1R,2S,3S,4R)-diol-epoxide-2 enantiomer, which has been uniformly found to be the most potent stereoisomer for the PAH studied. Although this stereoisomer is about five to sevenfold more mutagenic than the other DBchACR diol-epoxide stereoisomers in Chinese hamster V79 cells, even these less active stereoisomers are roughly equal to or up to three times more mutagenic than the analogous racemic BA and BcACR diol-epoxides in the mammalian cell test system (Figures 12 and 13).

D. Tumorigenicity

For BaACR, only the parent aza-PAH has been studied, and it is nontumorigenic. However, tumor data have been obtained for potential metabolites and other derivatives of BcACR, 7-Me-BcACR, and DBchACR.

1. Benz[c]acridine and Derivatives

BcACR derivatives have been examined for tumorigenicity both on mouse skin[6] and in the newborn mouse.[55] All metabolically possible trans-dihydrodiols were tested. Of these, only the dihydrodiol with a bay-region double bond, BcACR 3,4-dihydrodiol, exhibited significant tumorigenicity. It induced tumors in 90% of the mice at a dose of 2.5 μmole, after 15 weeks of promotion, with 4.9 tumors per mouse. In contrast, BcACR at the same dosage induced tumors in 30% of the mice after 15 weeks, with 0.8 tumors per mouse, and tumors in 37% of the mice after 25 weeks, with 1.3 tumors per mouse. The K-region oxide, BcACR 5,6-oxide, was inactive at the 2.5 μmole dose, and the non-K-region diol-epoxides BcACR 8,9-diol-10,11-epoxide-2 and BcACR 10,11-diol-8,9-epoxide-2 lacked significant activity at a 1 μmole dose. In contrast, the bay-region diol-epoxide, BcACR-3,4-diol-1,2-epoxide-2, was as tumorigenic as the BcACR 3,4-dihydrodiol at the 1.0 μmole dose, and BcACR 3,4-diol-1,2-epoxide-1 had significant, but weak activity at the 1.0 μmole dose.[6] By far the most tumorigenic compound tested was 3,4-dihydro BcACR, which induced tumors in 97% of the mice after 15 weeks at a dose of 0.4 μmole, with 7.9 tumors per mouse.

Similar results were obtained in the newborn mouse.[55] At a total dose of 1.05 μmole, BcACR 3,4-dihydrodiol was the only dihydrodiol to induce significantly more pulmonary tumors than BcACR. For BcACR, 65% of the mice had tumors, with 2.5 tumors per mouse; for BcACR 3,4-dihydrodiol, 82% of the mice had tumors, with 5.1 tumors per mouse. At the same dose, BcACR 3,4-diol-1,2-epoxide-1 showed very weak activity, with 47% of the mice bearing tumors, and 0.55 tumors per mouse. However, BcACR 3,4-diol-1,2-epoxide-2 was highly active. At a dose of 0.5 μmole, it induced pulmonary tumors in 100% of the mice, with 33.4 tumors per mouse. A similar range of activity was observed for hepatic tumors induced by the compounds. 3,4-dihydro BcACR was not tested.

2. 7-Methylbenz[c]acridine and Derivatives

The metabolically possible dihydrodiols of 7-Me-BcACR and 7-Me-BcACR have been tested for tumorigenicity on mouse skin and in the newborn mouse.[8] At doses of 0.15 to 0.75 μmole on mouse skin, 7-Me-BcACR 3,4-dihydrodiol was found to be 4 to 6 times as tumorigenic as 7-Me-BcACR. The other dihydrodiols lacked tumorigenic activity.

In newborn mice, at a dose of 0.35 μmole, 7-Me-BcACR-3,4-dihydrodiol induced 7 times as many pulmonary tumors as 7-Me-BcACR, and several times as many hepatic tumors in male mice. The other dihydrodiols were nontumorigenic. A comparison of the relative tumorigenicity of 7-Me-BcACR and BcACR indicated that 7-Me-BcACR has three- to six times the tumorigenicity of BcACR in these tumor models.

3. Dibenz[c,h]acridine

An initiation-promotion study of the tumorigenicity of DBchACR and its derivatives on mouse skin has been completed.[56] Both the 1,2- and 5,6-dihydrodiols of DBchACR are inactive. At a dose of 0.05 μmole per mouse, DBchACR is tumorigenic (0.5 tumors per mouse) as is the dihydrodiol with a bay-region double bond (−)-(R,R)-DBchACR-3,4-dihydrodiol (0.6 tumors per mouse). The enantiomeric (+)-(S,S)-DBchACR-3,4-dihydrodiol had no significant activity at this dose (0.1 tumors per mouse). The most tumorigenic derivatives were 3,4-H_2-DBchACR (2.7 tumors per mouse) and BcACR H_4-1,2-epoxide (4.1 tumors per mouse). Three of the stereoisomeric bay-region 3,4-diol-1,2-epoxides had tumorigenic activity equal to that of DBchACR, and the (1R,2S,3S,4R)-DBchACR-3,4-diol-1,2-epoxide-2 was much more tumorigenic (3.0 tumors per mouse). The two enantiomers of DBchACR H_4-1,2-epoxide were of equal tumorigenicity. The observation that the (1R,2S,3S,4R)-diol-epoxide stereoisomer had the highest tumorigenicity of the DBchACR diol-epoxides is in accord with observations for PAH that indicate high tumorigenicity for bay-region (R,S)-diol-(S,R)-epoxides. The observation of tumorigenicity equal to or higher than that of the parent aromatic structure for the various stereoisomeric bay-region diol-epoxides of DBchACR is unusual, however. Of the many PAH previously studied, only benzo[c]phenanthrene has exhibited similar behavior.

VI. CONCLUSIONS

It seems likely that carcinogenic benz- and dibenzacridines are generally activated through metabolism to bay-region diol-epoxides. For the carcinogens BcACR, 7-Me-BcACR, and DBchACR, metabolism data have indicated that the dihydrodiol precursors of the bay-region diol epoxides are formed. For BcACR and DBchACR, evidence for further metabolism to diol-epoxides has been found. In all three cases, the dihydrodiols with bay-region double bonds have been shown to be tumorigenic, and for BcACR and DBchACR, diol-epoxide-2 isomers have been shown to be highly tumorigenic. Recently, evidence that binding of 7-Me-BcACR to DNA occurs on the angular ring of 7-Me-BcACR has been found,[57] and this is consistent with attack by DNA on a bay-region diol-epoxide of 7-Me-BcACR. At present, there are insufficient data to pinpoint a simple reason for the varying carcinogenic and mutagenic potencies of the compounds that have been studied, and these properties are likely to reflect a combination of both the extent of metabolism to the most carcinogenic or mutagenic bay-region diol-epoxide, and the intrinsic potency of the diol-epoxides. For the PAH, a variety of factors have been found to contribute to the mutagenic and carcinogenic activity of diol-epoxides, and these include electronic factors relating to the ease of formation of a benzylic-carbocation from the diol-epoxide, conformational, steric, and absolute stereochemical factors.[23,24] Absolute stereochemical factors have been indicated to be important for the DBchACR diol-epoxides, and the effects of nitrogen substitution on the mutagenicity of H_4- and diol-epoxides of BaACR, BcACR, and DBchACR indicate the importance of electronic effects for these aza-PAH also. Predictions of relative reactivities of bay-region or other diol-epoxides derived from different aza-PAH are problematic, in our view, since the results of calculations are heavily dependent upon the choice of parameters for the nitrogen atom in the calculations, and there is a general lack of solvolysis and other data for benzylic derivatives of aza-PAH against which to compare calculated results. If reactivity data, such as solvolysis rates, were available for a series of benzylic derivatives of aza-PAH, it might prove possible to develop Hammett constants for aza-aromatics, and to determine which parameters for nitrogen are most suitable in calculations.

One clear trend has emerged in these studies, however, and that is the consistent and significant deactivating effect of nitrogen atom substitution upon the mutagenicity of diol-epoxides and H_4-epoxides, when the nitrogen atom is positioned such that resonance delocalization of the benzylic carbocation that can be potentially formed by ring opening of the epoxide places positive charge on the nitrogen atom. This would seem to be a sufficient basis for the noncarcinogenicity of BaACR and most of its methylated derivatives, since the bay-region diol-epoxides of BaACR are only very weakly mutagenic in bacteria as well as in Chinese hamster V79 cells. While there are numerous examples of highly mutagenic epoxide derivatives of PAH that are not carcinogenic, there has been no instance where a carcinogenic epoxide derivative has not exhibited high mutagenicity.

However, another possibility should be considered. In metabolism studies,[29] the dihydrodiol with a bay-region double bond could not be detected, and it is possible, therefore, that metabolism to a bay-region diol-epoxide of BaACR does not occur. Engelhardt and Schaefer-Ridder[49] have found that peroxy acid oxidation of BaACR 3,4-dihydrodiol gives N-oxidation as well as diol-epoxide formation, whereas BcACR 3,4-dihydrodiol gives only diol-epoxide. They have suggested that detoxification of BaACRs through N-oxidation might be a basis for the lower carcinogenicity of those compounds. There is presently no support for this hypothesis of preferential N-oxidation of BaACRs relative to BcACRs from in vitro metabolism studies,[29] and it is only in metabolism studies of BcACR and 7-Me-BcACR that N-oxide derivatives have been inferred as minor metabolites[29,35] though insufficient spectral data was available to make reliable structural assignments. Nonetheless, the possibility of N-oxidation of BaACR cannot be excluded, since it was not demonstrated that BaACR N-oxides would have been detected, if present as metabolites, under the assay conditions used.

Schaefer-Ridder et al. have demonstrated that the BaACR diol-epoxides interact with DNA. In one study,[58] they examined the relative ability of BaP, BaACR, and BcACR bay-region diol-epoxides as well as phenanthrene 9,10-oxide to inactivate the thymidine kinase gene. Although higher concentrations of the BaACR diol-epoxides than of the BcACR diol-epoxides were required to cause detectable inactivation of the gene, the former compounds showed activity.

In another study,[59] Loquet, Engelhardt, and Schaefer-Ridder used an alkaline elution assay to determine the relative ability of BaACR and BcACR derivatives to induce DNA single-strand breaks in two rat hepatoma cell lines. For BcACR, the diol-epoxide-2 isomer was more active than diol-epoxide-1 in both cell lines, though both induced single-strand breaks. For BaACR, diol-epoxide-2 was much more effective in one cell line than was diol-epoxide-1, and the two isomers were of comparable activity in the other cell line. BcACR diol-epoxide-2 was more active than BaACR diol-epoxide-2 in one cell line, but BaACR diol-epoxide-2 was more active than BcACR diol-epoxide-2 in the other cell line. The dihydrodiols of BaACR and BcACR with bay-region double bonds were also tested in these cell lines, which are reported to have monoxygenase activity. The BaACR 3,4-dihydrodiol with a bay-region double bond failed to induce single-strand breaks, whereas the BcACR 3,4-dihydrodiol, also with a bay-region double bond, did to a limited extent. This latter observation is consistent with metabolism of the BcACR dihydrodiol to a bay-region diol-epoxide, and lack of metabolism of the BaACR dihydrodiol to a bay-region diol-epoxide.

These studies establish that the BaACR diol-epoxides can interact with DNA and can cause observable effects upon the DNA. Also, the argument that BaACR is not metabolized to bay-region diol-epoxides is somewhat strengthened by the observations in the rat hepatoma cell lines. Nonetheless, the very low mutagenicity of the BaACR bay-region diol-epoxides is a fact that is not easily dismissed. Schaefer-Ridder et al. speculate that the results in *S. typhimurium* may be due to bacteriostatic effects of the BaACR diol-epoxides. However, this argument does not apply to the results in Chinese hamster V79 cells, in which the mutagenicity of the BaACR diol epoxides is also very low (Figures 12 and 13). Since mutagenicity in Chinese hamster V79 cells has served as a good qualitative guide to the tumorigenicity of diol-epoxides, it seems highly unlikely that bay-region diol-epoxides of BaACR will be tumorigenic. However, this question can only be rigorously resolved by testing the compounds for carcinogenicity in an animal tumor model.

ACKNOWLEDGMENTS

Partial support of this research by grants from the National Cancer Institute to R. E. L. is gratefully acknowledged. We thank Professor Gerald M. Holder for sharing results for 7-methylbenz[c]acridine prior to publication.

REFERENCES

1. **Lacassagne, A., Buu-Hoï, N. P., Daudel, R., and Zajdela, F.,** The relation between carcinogenic activity and the physical and chemical properties of angular benzacridines, *Adv. Cancer Res.*, 4, 315, 1956.
2. **Badger, G. M.,** The carcinogenic hydrocarbons: chemical constitution and carcinogenic activity, *Br. J. Cancer,* 2, 22, 1948.
3. **Clayson, D. B.,** Carcinogenic and anticarcinogenic properties of acridines, in *Acridines,* Vol. 9, Acheson, R. M., Ed., Interscience, New York, 1973, 815.
4. **Dipple, A., Moschel, R. C., and Bigger, C. A. H.,** Polynuclear aromatic carcinogens, in *Chemical Carcinogens,* Vol. 1, 2nd ed., Searle, C. E., Ed., Monograph Series No. 182, American Chemical Society, Washington, D.C., 1984, 41.
5. **Arcos, J. C. and Argus, M. F.,** *Chemical Induction of Cancer,* Vol. IIA, Academic Press, New York, 1974, 103.
6. **Levin, W., Wood, A. W., Chang, R. L., Kumar, S., Yagi, H., Jerina, D. M., Lehr, R. E., and Conney, A. H.,** Tumor-initiating activity of benz[c]acridine and twelve of its derivatives on mouse skin, *Cancer Res.*, 43, 4625, 1983.
7. **Glatt, H. R., Schwind, H., Schechtman, L. M., Beard, S., Kouri, R. E., Zajdela, F., Croisy, A., Perin, A., Jacquignon, P. C., and Oesch, F.,** Mutagenicity of closely related carcinogenic and noncarcinogenic compounds using various metabolizing systems and target cells, in *Short-Term Test Systems for Detection of Carcinogens,* Norpoth, K. H. and Garner, R. C., Eds., Springer, Berlin, 1980, 103.
8. **Chang, R. L., Levin, W., Wood, A. W., Shirai, N., Jerina, D. M., Holder, G. M., and Conney, A. H.,** Tumorigenicity of the dihydrodiols of 7-methylbenz[c]acridine on mouse skin and in newborn mice, *Proc. Am. Assoc. Cancer Res.*, 26, 94, 1985.
9. **Papadopoulo, D., Levy, S., Poirier, V., Pene, C., Markovits, P., and Hubert-Habart, M.,** Effects of several dimethylbenzacridines on secondary hamster embryo cells: neoplastic transformation, *Eur. J. Cancer,* 17, 179, 1981.
10. **Lacassagne, A., Buü-Hoï, N. P., Zajdela, F., and Jacquignon, P.,** *Comptu rend., Acad. Sci., Paris,* 251, 1332, 1960.
11. **Bahns, L., Podamy, V., Benesova, M., Godal, A., and Vachalkova, A.,** Carcinogenicity and polarographic behavior of dibenz[a,h]anthracene, dibenz[a,h]acridine and dibenz[a,h]phenazine, *Neoplasma,* 25, 641, 1978.
12. **Deutsch-Wenzel, R. P., Brune, H., and Grimmer, G.,** Experimental studies of the carcinogenicity of five nitrogen containing polycyclic aromatic compounds directly injected into rat lungs, *Cancer Lett.*, 20, 97, 1983.
13. **Ho, C. H., Clark, B. R., Guerin, M. R., Barkenbus, B. D., Rao, T. K., and Epler, J. L.,** Analytical and biological analyses of test materials from the synthetic fuel technologies. IV. Studies of chemical structure-mutagenic activity relationships of aromatic nitrogen compounds relevant to synfuels, *Mutat. Res.*, 85, 335, 1981.
14. **Epstein, S. S., Small, M., Falk, H. L., and Mantel, N.,** On the association between photodynamic and carcinogenic activities in polycyclic compounds, *Cancer Res.*, 24, 855, 1964.
15. **Morgan, D. D. and Warshawsky, D.,** The photodynamic immobilization of *Artemia salina* nauplii by polycyclic aromatic hydrocarbons and its relationship to carcinogenic activity, *Photochem. Photobiol.*, 25, 39, 1977.
16. **Smith, I. A. and Seybold, P. G.,** Substituent effects in chemical carcinogenesis: methyl derivatives of the benzacridines, *J. Heterocyclic Chem.*, 16, 421, 1979.
17. **Memory, J. D.,** Electrophilic superdelocalizability and carcinogenesis by polycyclic aromatic hydrocarbons — Pullman Theory, *Int. J. Quantum Chem.*, 15, 363, 1979.
18. **Okano, T., Takashi, H., Koike, T., and Motohashi, N.,** Relationships of carcinogenicity, mutagenicity, and K-region reactivity in benz[c]acridines, *Gann,* 70, 749, 1979.
19. **Okano, T., Haga, M., Ujiie, K., and Motohashi, N.,** Half-wave potentials and LCAO-SCF-MO calculations for carcinogenic benz[c]acridines, *Chem. Pharm. Bull. Jpn.*, 26, 2855, 1978.

20. **Thakker, D. R., Yagi, H., Levin, W., Wood, A. W., Conney, A. H., and Jerina, D. M.,** Polycyclic aromatic hydrocarbons: metabolic activation to ultimate carcinogens, in *Bioactivation of Foreign Compounds,* Anders, M. W., Ed., Academic Press, New York, 1985, 177.
21. **Jerina, D. M., Lehr, R. E., Yagi, H., Hernandez, O., Dansette, P. M., Wislocki, P. G., Wood, A. W., Chang, R. L., Levin, W., and Conney, A. H.,** Mutagenicity of benzo[a]pyrene derivatives and the description of a quantum mechanical model which predicts the ease of carbonium ion formation from diol epoxides, in *In Vitro Activation in Mutagenesis Testing,* de Serres, F. J., Fouts, J. R., Bend, J. R., and Philpot, R. M., Eds., Elsevier/North-Holland, Biomedical Press, Amsterdam, 1976, 159.
22. **Jerina, D. M. and Lehr, R. E.,** The bay region theory: a quantum mechanical approach to aromatic hydrocarbon-induced carcinogenicity, in *Microsomes and Drug Oxidations,* Ullrich, V., Roots, I., Hildbrant, A. G., Estabrook, R. W., and Conney, A. H., Eds., Pergamon Press, Oxford, 1977, 709.
23. **Lehr, R. E., Kumar, S., Levin, W., Wood, A. W., Chang, R. L., Conney, A. H., Yagi, H., Sayer, J. M., and Jerina, D. M.,** The bay region theory of polycyclic aromatic hydrocarbon carcinogenesis, in *Polycyclic Hydrocarbons and Carcinogenesis,* Harvey, R. G., Ed., Symposium Series 283, American Chemical Society, Washington, D.C., 1985, 63.
24. **Jerina, D. M., Yagi, H., Thakker, D. R., Sayer, J. M., van Bladeren, P. J., Lehr, R. E., Whalen, D. L., Levin, W., Chang, R. L., Wood, A. W., and Conney, A. H.,** Identification of the ultimate carcinogenic metabolites of the polycyclic aromatic hydrocarbons: bay region (R,S)-diol-(S,R)-epoxides, in *Foreign Compound Metabolism,* Caldwell, J. and Paulson, G. D., Eds., Taylor and Francis Ltd., London, 1984, 257.
25. **Jerina, D. M., Sayer, J. M., Yagi, H., van Bladeren, P. J., Thakker, D. R., Levin, W., Chang, R. L., Wood, A. W., and Conney, A. H.,** Stereoselective metabolism of polycyclic aromatic hydrocarbons to carcinogenic metabolites, in *Microsomes and Drug Oxidations,* Boobis, A. R., Caldwell, J., DeMatteis, F., and Elcombe, C. R., Eds., Taylor and Francis, Ltd., London, 1985, 310.
26. **Lowe, J. P. and Silverman, B. D.,** Heteroatom effects in chemical carcinogenesis: effects of ring heteroatoms on ease of carbocation formation, *Cancer Biochem. Biophys.,* 7, 53, 1983.
27. **Lehr, R. E. and Jerina, D. M.,** Aza-polycyclic aromatic hydrocarbon carcinogenicity: predictions of reactivity of tetrahydrobenzo ring epoxide derivatives, *Tetrahedron Lett.,* 24, 27, 1983.
28. **McMurtrey, K. D. and Knight, T. J.,** Metabolism of acridine by rat-liver enzymes, *Mutat. Res.,* 140, 7, 1984.
29. **Jacob, J., Schmoldt, A., Kohbrok, W., Raab, G., and Grimmer, G.,** On the metabolic activation of benz[a]acridine and benz[c]acridine by rat liver and lung microsomes, *Cancer Lett.,* 16, 297, 1982.
30. **Thakker, D. R., Levin, W., Yagi, H., Karle, J. M., Lehr, R. E., Ryan, D., Thomas, P. E., Conney, A. H., and Jerina, D. M.,** Metabolism of benzo[a]anthracene to its tumorigenic 3,4-dihydrodiol, *Mol. Pharmacol.,* 15, 138, 1979.
31. **Boux, L. J., Cheung, H. T. A., Holder, G. M., and Moldovan, L.,** Potential metabolites of carcinogenic aza aromatic hydrocarbons. Synthesis of K-region oxide, phenol and dihydrodiols of 7-methylbenz[c]acridine, *Tetrahedron Lett.,* 21, 2923, 1980.
32. **Duke, C. C., Murphy, P. T., and Holder, G. M.,** Synthesis of the non-K-region dihydrodiols of 7-methylbenz[c]acridine, *J. Org. Chem.,* 49, 4446, 1984.
33. **Ireland, C. M., Holder, G. M., and Ryan, A. J.,** Studies in the metabolism of carcinogenic polycyclic heteroaromatic compounds. I. The hepatic microsomal metabolism of 7-methylbenz[c]acridine, *Biochem. Pharmacol.,* 30, 2685, 1981.
34. **Ireland, C. M., Cheung, H. T. A., Ryan, A. J., and Holder, G. M.,** Rat liver microsomal metabolites of 7-methylbenz[c]acridine, *Chem. Biol. Interact.,* 40, 305, 1982.
35. **Boux, L. J., Duke, C. C., Holder, G. M., Ireland, C. M., and Ryan, A. J.,** Metabolism of 7-methylbenz[c]acridine: comparison of rat liver and lung microsomal preparations and identification of some minor metabolites, *Carcinogenesis,* 4, 1429, 1983.
36. **Gill, J. H., Bonin, A. M., Podobna, E., Baker, R. S. V., Duke, C. C., Rosario, C. A., Ryan, A. J., and Holder, G. M.,** 7-Methylbenz[c]acridine: mutagenicity of some of its metabolites and derivatives, and the identification of trans-7-methylbenz[c]acridine 3,4-dihydrodiol as a microsomal metabolite, *Carcinogenesis,* 7, 23, 1986.
37. **Boux, L. J. and Holder, G. M.,** The metabolism of the carcinogen 7-methylbenz[c]acridine by hepatocytes isolated from untreated and induced rats, *Xenobiotica,* 15, 11, 1985.
38. **Wright, D. J., Robinson, H. K., Holder, G. M., and Ryan, A. J.,** The excretion and metabolism of the carcinogen, 7-methylbenz[c]acridine, in the rat, *Xenobiotica,* 15, 825, 1985.
39. **Lehr, R. E., Kumar, S., Shirai, N., and Jerina, D. M.,** Synthesis of enantiomerically pure bay-region 3,4-diol 1,2-epoxide diastereomers and other derivatives of the potent carcinogen dibenz[c,h]acridine, *J. Org. Chem.,* 50, 98, 1985.
40. **Balani, S. K., van Bladeren, P. J., Shirai, N., and Jerina, D. M.,** Resolution and absolute configuration of K-region trans dihydrodiols from polycyclic aromatic hydrocarbons, *J. Org. Chem.,* 51, 1773, 1986.

41. **Thakker, D. R., Shirai, N., Levin, W., Ryan, D. E., Thomas, P. E., Lehr, R. E., Conney, A. H., and Jerina, D. M.,** Metabolism of dibenz[c,h]acridine by rat liver microsomes and by cytochrome P450c with and without epoxide hydrolase, *Proc. Am. Assoc. Cancer Res.*, 26, 114, 1985.
42. **Adams, J. D., Sayer, J. M., Chadha, A., Shirai, N., Lehr, R. E., Kumar, S., Levin, W., and Jerina, D. M.,** Metabolism of *trans*-3,4-dihydroxy-3,4-dihydrodibenz[c,h]acridine to bay-region 3,4-diol-1,2-epoxides by hepatic microsomes and cytochrome P450c, in preparation.
43. **Jerina, D. M., Michaud, D. P., Feldmann, R. J., Armstrong, R. N., Vyas, K. P., Thakker, D. R., Yagi, H., Thomas, P. E., Ryan, D. E., and Levin, W.,** Stereochemical modeling of the catalytic site of cytochrome P450c, in *Microsomes, Drug Oxidations and Drug Toxicity*, Sato, R. and Kato, R., Eds., Scientific Societies Press, Tokyo, 1982, 195.
44. **Kitihara, Y., Okuda, H., Shudo, K., Okamoto, T., Nagao, M., Seino, Y., and Sugimura, T.,** Synthesis and mutagenicity of 10-azabenzo[a]pyrene-4,5-oxide and other pentacyclic aza-arene oxides, *Chem. Pharm. Bull. Jpn.*, 26, 1950, 1978.
45. **Wood, A. W., Chang, R. L., Levin, W., Ryan, D. E., Thomas, P. E., Lehr, R. E., Kumar, S., Schaefer-Ridder, M., Engelhardt, U., Yagi, H., Jerina, D. M., and Conney, A. H.,** Mutagenicity of diol-epoxides and tetrahydroepoxides of benz[a]acridine and benz[c]acridine in bacteria and in mammalian cells, *Cancer Res.*, 43, 1656, 1983.
46. **Lehr, R. E. and Kumar, S.,** Synthesis of dihydrodiol and other derivatives of benz[c]acridine, *J. Org. Chem.*, 46, 3675, 1981.
47. **Kumar, S. and Lehr, R. E.,** Tetrahydroepoxides and diol epoxides of benz[c]acridine, *Tetrahedron Lett.*, 23, 4523, 1982.
48. **Schaefer-Ridder, M. and Engelhardt, U.,** Synthesis of trans-3,4-dihydroxy-3,4-dihydrobenz[a]- and -[c]acridines, possible proximate carcinogenic metabolites of polycyclic azaarenes, *J. Org. Chem.*, 46, 2895, 1981.
49. **Engelhardt, U. and Schaefer-Ridder, M.,** N-oxidation *versus* epoxidation in polycyclic azaarenes, *Tetrahedron Lett.*, 22, 4687, 1981.
50. **Wood, A. W., Chang, R. L., Levin, W., Kumar, S., Shirai, N., Jerina, D. M., Lehr, R. E., and Conney, A. H.,** Bacterial and mammalian cell mutagenicity of four optically active bay-region 3,4-diol-1,2-epoxides and other derivatives of the nitrogen heterocycle, dibenz[c,h]acridine, *Cancer Res.*, 46, 2760, 1986.
51. **Wood, A. W., Chang, R. L., Levin, W., Yagi, H., Tada, M., Vyas, K., Jerina, D. M., and Conney, A. H.,** Mutagenicity of the optical isomers of the diastereomeric bay-region chrysene 1,2-diol-3,4-epoxides in bacterial and mammalian cells, *Cancer Res.*, 42, 2972, 1982.
52. **Wood, A. W., Chang, R. L., Levin, W., Yagi, H., Thakker, D. R., Jerina, D. M., and Conney, A. H.,** Differences in mutagenicity of the optical enantiomers of the diastereomeric benzo[a]pyrene 7,8-diol-9,10-epoxides, *Biochem. Biophys. Res. Commun.*, 77, 1389, 1977.
53. **Wood, A. W., Chang, R. L., Levin, W., Yagi, H., Thakker, D. R., van Bladeren, P. J., Jerina, D. M., and Conney, A. H.,** Mutagenicity of enantiomers of the diastereomeric bay-region benz[a]anthracene 3,4-diol-1,2-epoxides in bacterial and mammalian cells, *Cancer Res.*, 43, 5821, 1983.
54. **Wood, A. W., Chang, R. L., Levin, W., Thakker, D. R., Yagi, H., Sayer, J. M., Jerina, D. M., and Conney, A. H.,** Mutagenicity of the enantiomers of the diastereomeric bay-region benzo[c]phenanthrene 3,4-diol-1,2-epoxides in bacterial and mammalian cells, *Cancer Res.*, 44, 2320, 1984.
55. **Chang, R. L., Levin, W., Wood, A. W., Kumar, S., Yagi, H., Jerina, D. M., Lehr, R. E., and Conney, A. H.,** Tumorigenicity of dihydrodiols and diol-epoxides of benz[c]acridine in newborn mice, *Cancer Res.*, 44, 5161, 1984.
56. **Wood, A. W., Chang, R. L., Levin, W., Shirai, N., Jerina, D. M., Kumar, S., Lehr, R. E., and Conney, A. H.,** Mutagenicity and tumorigenicity of enantiomers of the diastereomeric bay-region dibenz[c,h]acridine 3,4-diol-1,2-epoxides, *Proc. Amer. Assoc. Cancer Res.*, 26, 95, 1985.
57. **Boux, L. J. and Holder, G. M.,** The activation and DNA binding of 7-methylbenz[c]acridine catalyzed by mouse liver microsomes, *Cancer Lett.*, 25, 333, 1985.
58. **Schaefer-Ridder, M., Moeroey, T., and Engelhardt, U.,** Inactivation of the thymidine kinase gene after *in vitro* modification with benzo[a]pyrene-diol-epoxide and transfer to LTK-cells as a eukaryotic test for carcinogens, *Cancer Res.*, 44, 5861, 1984.
59. **Loquet, C., Engelhardt, U., and Schaefer-Ridder, M.,** Differentiated genotoxic response of carcinogenic and noncarcinogenic benzacridines and metabolites in rat hepatoma cells, *Carcinogenesis*, 6, 455, 1985.

Chapter 3

CYCLOPENTA[a]PHENANTHRENES: CARCINOGENS RELATED STRUCTURALLY TO BOTH STEROIDS AND POLYCYCLIC AROMATIC HYDROCARBONS

Maurice M. Coombs

TABLE OF CONTENTS

I. INTRODUCTION

Of the six possible isomers obtained by fusion of a five-membered cyclopentane ring to phenanthrene, the [a] isomer, 16,17-dihydro-15*H*-cyclopenta[a]phenanthrene (3), holds a unique position because it possesses the same ring system as the steroids. Indeed, the nature of this hydrocarbon and related compounds became of interest during the elucidation of the structures of the sterols and bile acids in the 1930s. Selenium dehydrogenation of these natural products gave a hydrocarbon $C_{18}H_{16}$, later called Diels hydrocarbon and shown to be the 17-methyl derivative of (3), namely 16,17-dihydro-17-methyl-15*H*-cyclopenta[a]phenanthrene (5). Cyclopenta[a]phenanthrenes are now numbered like steroids as shown in Figure 1; the basic hydrocarbons bear one double bond in the five-membered ring D. Thus, hydrocarbon (1) is 15*H*-cyclopenta[a]phenanthrene and (2) is its 17*H*-isomer; compound (3) is therefore the 16,17-dihydro derivative of the former. In compounds such as the 17-ketone (4), the position of the "extra" hydrogen is fixed, and the correct name for this compound is 15,16-dihydrocyclopenta[a]phenanthren-17-one or 15,16-dihydro-17-oxocyclopenta[a]phenanthrene.

The isolation of Diels hydrocarbon as a sterol-dehydrogenation product initiated a phase of synthetic work which soon led to its synthesis by several different routes, together with a number of related compounds such as the 3-methoxy derivatives (6) and (7), which also result from dehydrogenation of the methyl ethers of the natural estrogens estrone and estradiol, respectively. By coincidence, at this time the structures of the carcinogenic polycyclic hydrocarbons present in coal tar were being established. Thus, in 1929, E. Clar[1] synthesised dibenz[a,h]anthracene (8), which now holds the distinction of being the first pure organic compound clearly to be demonstrated as carcinogenic in laboratory rodents.[2] This hydrocarbon (see Figure 2) was found to possess a fluorescence spectrum similar to that of the carcinogenic tars, and using this as a guide, Kennaway and his group at the Royal Cancer Hospital in London undertook the fractionation of two tons of gas-works pitch. In 1933, they finally isolated 7 g of yellow crystals consisting mainly of benzo[a]pyrene (9) which proved to be a potent carcinogen.[3] Both these hydrocarbons can be considered as benzo derivatives of phenanthrene, but when the phenanthrene-derived Diels hydrocarbon and other simple cyclopenta[a]phenanthrenes were tested for carcinogenicity, they were found to lack activity. The same year, two groups[4,5] independently discovered that deoxycholic acid (10) can be efficiently converted via pyrolysis of a simple intermediate into 3-methylcholanthrene (11), which turned out to be an even more powerful carcinogen than benzo[a]pyrene. This striking preparation of a potent carcinogen from a natural steroid had a remarkable psychological effect at that time. Thus, in his book *Chemistry of Natural Products Related to Phenanthrene*, Fieser[6] felt that " . . . many forms of cancer may originate in the metabolic production of methylcholanthrene or related substances from the bile acids, or perhaps from the sterols or sex hormones, of the body." Even twenty years later, Inhoffen[7] devoted the larger part of a chapter entitled "The relationship of natural steroids to carcinogenic aromatic compounds" to consideration of this idea. However, by that time considerable progress had been made in understanding the basic principles underlying structure/carcinogenicity relationships among polycyclic aromatic hydrocarbons. In particular, it was discovered that often specific methyl substitution appeared to be crucial. Thus, 5-methylchrysene (12) (Figure 3) is a strong carcinogen, while the parent hydrocarbon and the other methyl isomers are inactive or only weakly carcinogenic.[8] Of the methyl derivatives of benz[a]anthracene (13), itself essentially noncarcinogenic, carcinogenic activity decreases in the order 7-methyl > 8-methyl > 12-methyl; 7,12-dimethylbenz[a]anthracene (14) is one of the most potent carcinogens known.[9]

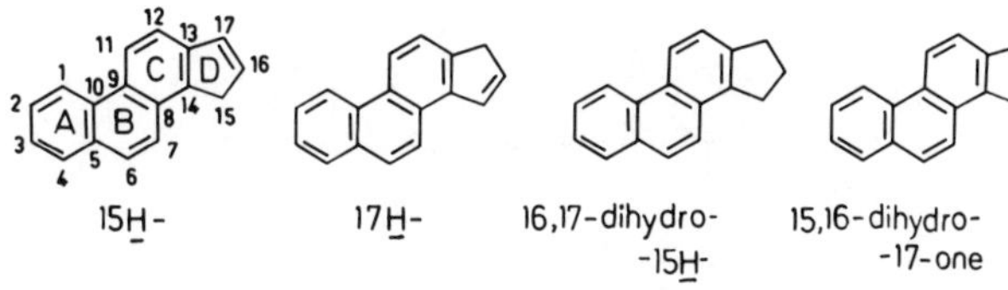

FIGURE 1. The structures of some cyclopenta[a]phenanthrenes, including Diels hydrocarbon (5).

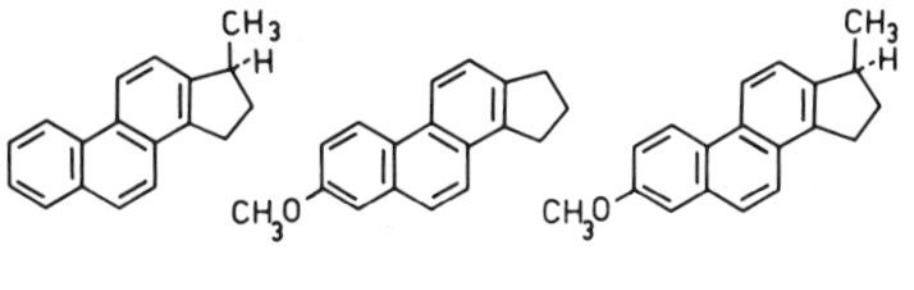

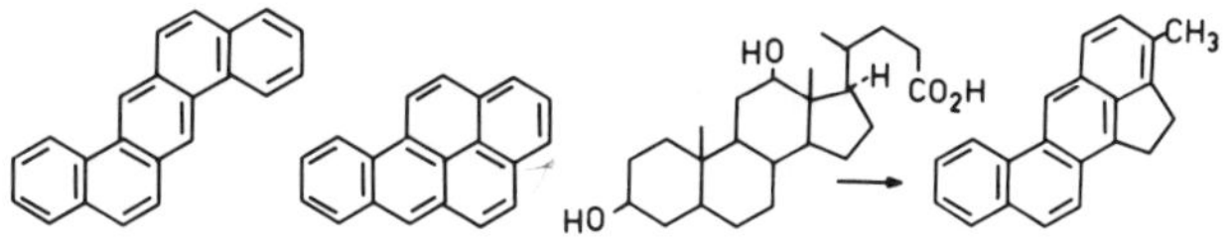

FIGURE 2. Classical polycyclic aromatic hydrocarbon carcinogens: dibenz[a,h]anthracene (8), benzo[a]pyrene (9), and 3-methylcholanthrene (11) obtained from deoxycholic acid (10).

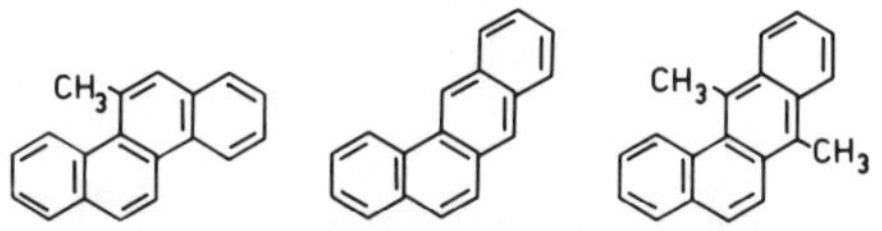

FIGURE 3. 5-Methylchrysene (12), benz[a]anthracene (13), and the potent carcinogen 7,12-dimethylbenz[a]anthracene (14).

II. CARCINOGENIC CYCLOPENTA[a]PHENANTHRENE HYDROCARBONS

With this idea of the enhancing effect of methyl substitution in mind, Butenandt,[10] who had isolated and established the structure of the androgen testosterone in 1933, recognized the need to synthesize the eight possible aryl-methyl derivatives of the parent cyclopenta[a]phenanthrene hydrocarbon (3). Starting from dehydroepiandrosterone (15), a simple transformation product of testosterone, Diels hydrocarbon (5) and the three 6-methyl hydrocarbon derivatives (16 to 18) were prepared, making use of the 5(6)-double bond to introduce a methyl group at C-6 via the epoxide and the tendency of the angular 18-methyl group to migrate to C-17 on selenium dehydrogenation to provide a methyl group at this position as shown in Figure 4. All the other methyl isomers were prepared by total synthesis during the next decade, using a variety of routes. Thus, judicious use of the intermediate

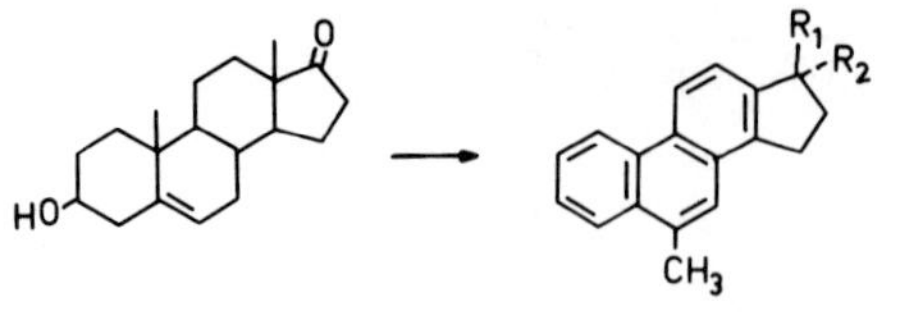

(15)

(16) $R_1 = R_2 = H$
(17) $R_1 = CH_3, R_2 = H$
(18) $R_1 = R_2 = CH_3$

FIGURE 4. Butenandt's preparation of the 6-methyl, 6,17-dimethyl, and 6,17,17-trimethyl derivatives of 16,17-dihydro-15*H*-cyclopenta[a]phenanthrene from the androgen (15).

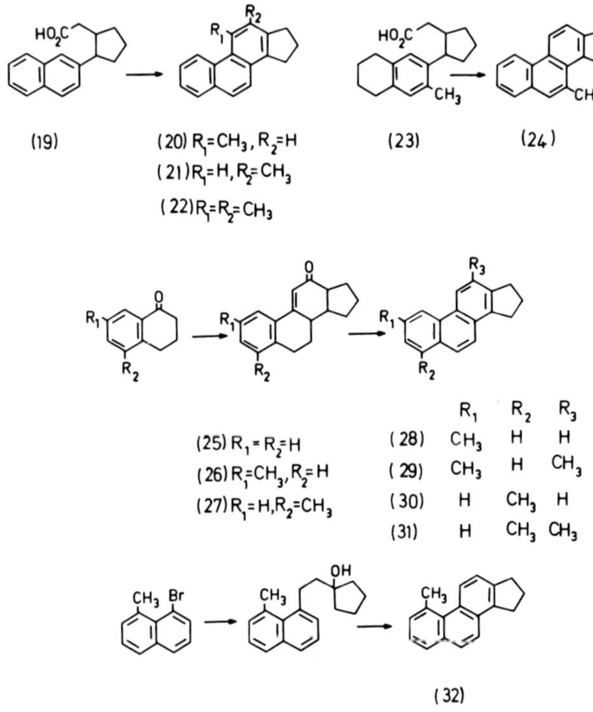

(19)

(20) $R_1 = CH_3, R_2 = H$
(21) $R_1 = H, R_2 = CH_3$
(22) $R_1 = R_2 = CH_3$

(23)

(24)

(25) $R_1 = R_2 = H$
(26) $R_1 = CH_3, R_2 = H$
(27) $R_1 = H, R_2 = CH_3$

	R_1	R_2	R_3
(28)	CH_3	H	H
(29)	CH_3	H	CH_3
(30)	H	CH_3	H
(31)	H	CH_3	CH_3

(32)

FIGURE 5. Syntheses of various isomeric methyl derivatives of 16,17-dihydro-15*H*-cyclopenta[a]phenanthrene.

(19) (Figure 5), first prepared by Robinson[11] during his early approaches to steroid synthesis, gave the 11-methyl (20), 12-methyl (21), and 11,12-dimethyl (22) hydrocarbons,[12] while the homologous 7-methyl acid (23) led to the 7-methyl compound (24).[13] In a similar way, another intermediate (25), first prepared by Robinson,[14] together with its 2- and 4-methyl homologues (26 and 27) were converted into the four cyclopenta[a]phenanthrene hydrocarbons (28 to 31).[15,16] The remaining 1-methyl hydrocarbon (32) was obtained[17] through a lengthy synthesis based on an earlier route to the parent hydrocarbon, as outlined in Figure 5.

All these methyl derivatives of 16,17-dihydro-15*H*-cyclopenta[a]phenanthrene were tested in mice for their capacity to induce skin tumors upon dorsal application by Butenandt and Dannenberg.[18] They applied two drops of a 0.4% benzene solution of these compounds

Table 1
SKIN-PAINTING EXPERIMENTS WITH METHYL HOMOLOGUES 16,17-DIHYDRO-15*H*-CYCLOPENTA[a]PHENANTHRENE (3)

Compound[a]	No. of mice used	Duration of expt (days)	Tumors at site of application		No. of mice with tumor/total no. of mice
			Papillomas	Carcinomas	
Unsubst. (3)	22	547	0	0	—
1-Methyl (32)	21	490	0	0	—
2-Methyl (28)	28	359	0	0	—
4-Methyl (30)	25	714	0	0	—
6-Methyl (16)	15	552	0	0	—
7-Methyl (24)	35	440	1	4	5/35
11-Methyl (20)	13	433	1	1	2/13
12-Methyl (21)	30	272	0	0	—
4,17-Dimethyl	10	270	0	0	—
6,17-Dimethyl (17)	10	339	0	0	—
2,12-Dimethyl (29)	25	506	0	0	—
4,12-Dimethyl (31)	14	550	0	0	—
6,7-Dimethyl	21	457	0	0	—
11,12-Dimethyl (22)	10	275	0	1	1/10
6,17,17-Trimethyl (18)	11	483	0	0	—

[a] Two drops of a 0.4% benzene solution twice weekly.[18]

twice weekly to groups of 10 to 35 mice of the B1.H1 strain and also mice of mixed strain obtained from a commercial source, with the result shown in Table 1. All the hydrocarbons were inactive, except those bearing methyl groups at C-7 and C-11. The 7-methyl (24) and 11-methyl (20) compounds were of similar activity (tumor incidence 14.3 and 15.4%, respectively), whereas the 11,12-dimethyl hydrocarbon (22) appeared to be somewhat less active (tumor incidence 10%). Latent periods for the production of these skin tumors were not quoted, but it is clear that these compounds are only weakly active. No local tumors were observed after injection (5 mg in oil). Dannenberg[19] also tested the three hydrocarbons (33-35) containing a double bond in ring D (i.e., 17-methyl-15*H*-, 17-*iso*propyl-15*H*-, and 15-methyl-17*H*-cyclopenta[a]phenanthrene, respectively; Figure 6). They were prepared from the corresponding 17- and 15-ketones by Grignard reactions followed by facile dehydration. These hydrocarbons appeared to be weakly carcinogenic both in skin-painting and injection experiments (Table 2). All three cyclopenta[a]phenanthrenes were active in the topical application test, although their activity was low as indicated by the long mean latent periods. The *iso*propyl derivative did not appear to be active by injection. Hydrocarbons of this type are formed from cholesterol on quinone dehydrogenation, and the author felt that their carcinogenicity might be explained by the reactive D-ring double bond acting as the "K-region".

III. 17-OXO-15,16-DIHYDROCYCLOPENTA[a]PHENANTHRENES

A. Synthesis

Steroids differ from cyclopenta[a]phenanthrenes by possessing C-18 and C-19 angular methyl groups which do not permit aromatization of the six-membered rings. However, there is one class of steroids, the estrogens, in which this has already partially occurred. Estrone (37) (Figure 7) is now known to be biosynthesized from the androgen androst-4-ene-3,17-dione (36) via oxidation of the angular methyl group, which is then eliminated together with hydrogen atom at C-1 as formaldehyde, the second hydrogen at C-2 being

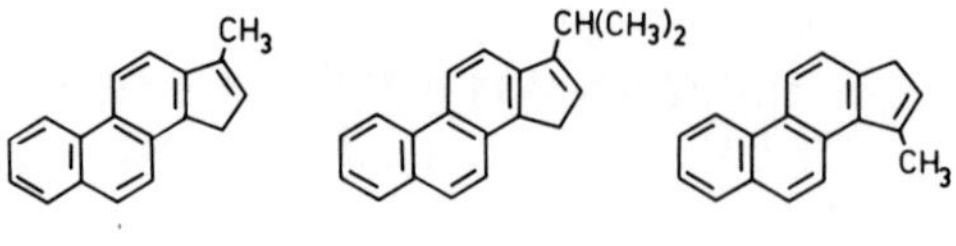

FIGURE 6. Weakly carcinogenic 15*H*- and 17*H*-cyclopenta[a]phenanthrenes tested by Dannenberg.

Table 2
SKIN-PAINTING AND INJECTION EXPERIMENTS WITH THE CYCLOPENTA[a]PHENANTHRENES (33)—(35)

Compound	No. of mice used (strain)[a]	Route of application	Tumors at site of application		No. of mice with tumor
			Papillomas	Carcinomas	
17-Methyl-15*H* (33)	15(S)	Topical[b]	0	1	1
	10(M)	Topical[b]	1	6	7[c]
17-Isopropyl-15*H* (34)	15(S)	Topical[b]	0	3	3
15-Methyl-17*H* (35)	15(S)	Topical[b]	1	8	9[d]
17-Methyl-15*H* (33)	10(S)	Injection[e]		3 Spindle cell sarcomas	
	10(M)	Injection[e]		0	
17-Isopropyl-15*H* (34)	10(S)	Injection[e]		0	
15-Methyl-17*H* (35)	10(S)	Injection[e]		1 Carcinoma	

[a] M, mice of mixed strain; S, Swiss mice with low spontaneous tumor incidence.
[b] Dose as used previously by Butenandt and Dannenberg.[18]
[c] Mean latent period, 73 weeks.
[d] Mean latent period, 83 weeks.
[e] 4 to 5 mg in 0.2 mℓ oil-injected s.c.

lost by enolization of the carbonyl group.[20] Estrogens are also known in which both rings A and B are aromatic, for example, the naphthol equilenin (38) isolated from mare's urine. Marrian, who had discovered another natural hormone estriol in 1930, isolated an unusual estrogen metabolite from human pregnancy urine by partition chromatography on celite.[21] This compound was shown to be 18-hydroxyestrone (39) by its ready loss of formaldehyde on treatment with dilute alkali, with the formation of 18-norestrone identical with a synthetic specimen. Soon after he became Director of Research at the Imperial Cancer Research Fund Laboratories, Marrian suggested that should this elimination occur in vivo, it would open the way to complete aromatization of the steroid to yield the phenol 15,16-dihydro-3-hydroxycyclopenta[a]phenanthren-17-one (40). It was therefore decided to synthesize this molecule in order to explore its biological properties. It was also proposed to investigate whether the two structural factors found by Dannenberg to lead to carcinogenicity in cyclopenta[a]phenanthrenes (viz. 7- or 11-methyl substitution, and an "extra" double bond in ring D), would reinforce one another to give higher potency.

Up to that time (1965), cyclopenta[a]phenanthrene synthesis had been directed towards preparation of the hydrocarbons of this series, and the methods used were not in the main readily adaptable to the synthesis of the 17-ketones, although several of the latter had already been synthesized in various ways. A reappraisal of these methods was therefore undertaken, and another of Robinson's steroid intermediates,[22] 11,12,13,14,15,16-hexahydrocyclopenta[a]phenanthren-11,17-dione (43), obtainable in five steps from 2-acetylnaphthalene as

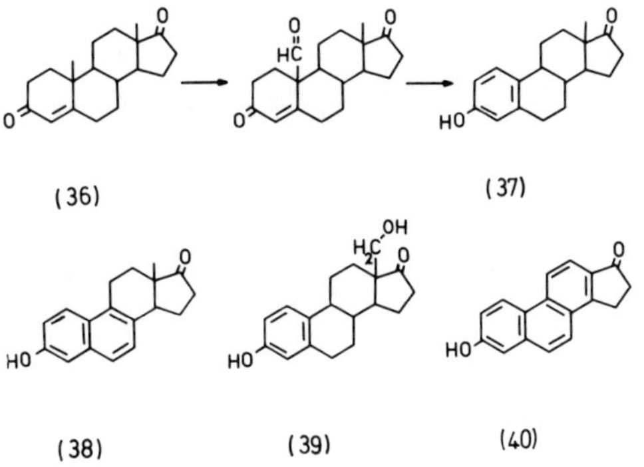

FIGURE 7. Estrogen biosynthesis: estrone (37) from androst-4-ene-3,17-dione (36), and the structures of equilenin (38), 18-hydroxy-oestrone (39), and the fully aromatic analogue 15,16-dihydro-3-hydroxy-cyclopenta[a]phenanthren-17-one (40).

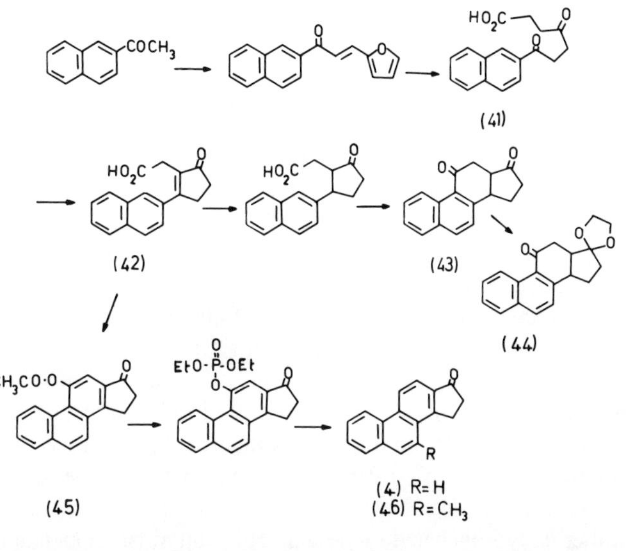

FIGURE 8. Synthesis of 17-oxocyclopenta[a]phenanthrenes utilizing Robinson's diketone (43).

shown in Figure 8, was selected as a starting material. As he had previously noted,[11] the two cyclizations (41) → (42) and (42) → (45) occur essentially quantitatively, and it was later discovered by Coombs and Jaitly[23] that the 11-oxygen function can be removed from (45), leaving the 17-carbonyl group intact to yield the parent ketone (4) by reduction of the derived diethyl phosphate with sodium in liquid ammonia containing tetrahydrofuran. This was found to be the best way of preparing the 7-methyl-17-ketone (46) from 2-acetyl-3-methylnaphthalene. The alternative route, from (42) to (43), suffers from difficulty in both reducing the double bond in the acid (42), and more particularly in the subsequent ring-closure. However, the diketone (43) is fairly available (we have made 50-g batches of it) and possesses a useful characteristic in that the 11-ketone function is relatively inert owing to steric hindrance. This allows the 17-ketone to be protected as the very useful 11-oxo-17-ketal (44) without difficulty.[24] The way in which this intermediate was used to prepare a number of 17-ketones is shown in Figure 9. The 11- and 12-alkyl derivatives (47 to 51), as well as the unsubstituted 17-ketone (4), were secured by generating 11,12-enes, either by reduction and dehydration, or by Grignard reductions followed by dehydration. The final

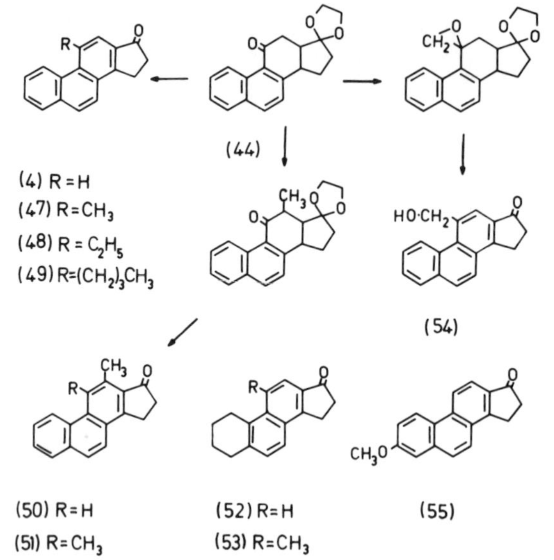

(44)

(4) R = H
(47) R = CH_3
(48) R = C_2H_5
(49) R = $(CH_2)_3CH_3$

(54)

(50) R = H
(51) R = CH_3

(52) R = H
(53) R = CH_3

(55)

FIGURE 9. Use of the oxoketal (44) for the preparation of various 17-oxocyclopenta[a]phenanthrenes.

13,14-double bond was inserted by heating these 11,12-enes with acetic and hydrochloric acids containing an excess of nitrobenzene, giving the 17-ketones in 70 to 90% yield. The 1,2,3,4,-tetrahydro-17-ketones (52 and 53) were obtained from the unsaturated A-ring analogue of the diketone (43) in the same way,[25] as was the 3-methoxy-17-ketone (55) from its 3-methoxy derivative.[24] The 11-hydroxymethyl-17-ketone (54) was prepared via the epoxide formed with the ylide from trimethylsulfonium iodide, followed by hydrolytic cleavage, dehydration, and quinone dehydrogenation.[26]

This synthetic scheme was not convenient for the preparation of other 17-ketones, mainly owing to the relative unavailability of the requisite 2-acetylnaphthalenes, and the method of Johnson and Peterson[27] was therefore further investigated. In this procedure a 1-oxo-1,2,3,4-tetrahydrophenanthrene is condensed with diethyl succinate (Stobbe reaction) and the half-ester obtained is cyclized and decarboxylated to give the 11,12,15,16-tetrahydro-17-ketone, which can be further dehydrogenated to yield a 15,16-dihydrocyclopenta[a]phenanthren-17-one as shown in Figure 10. Cyclization of the 13,14-ene occurs selectively in the direction shown, whereas cyclization of the corresponding phenanthrene occurs mainly in the opposite direction (i.e., at C-7).[28] This method is versatile and is limited chiefly by the availability of the tricyclic ketones. Thus, Riegel et al.[29] first prepared 12-alkyl hydrocarbons by way of the 12-alkyl-17-ketones (50,57,58). The method was subsequently employed extensively for the synthesis of the compounds (59 to 68) by Coombs et al.[26,30] and Ribeiro et al.[31] In each case the Stobbe reaction proceeded in fair to good yield, except in the case of the 7,11-dimethyl ketone (66) where steric hindrance by the *peri* 7-methyl group led to a poor yield of half ester. Cyclization with zinc chloride in acetic acid proceeded satisfactorily, provided strictly anhydrous conditions were maintained, and the final dehydrogenation step was readily achieved with high potential quinones, or by heating with palladium black in boiling *p*-cymene. Finally, the 11-alkoxy compounds (69 to 75) were readily prepared from the easily obtainable 11-acetoxy-17-ketone (45).[32]

B. Structure/Carcinogenic Activity Relationships

For the first carcinogenicity testing, the unsubstituted-17-ketone (4) and its 11-methyl (47), 12-methyl (50), 11,12-dimethyl (51), and 3-methoxy (55) and 11-methoxy (69) de-

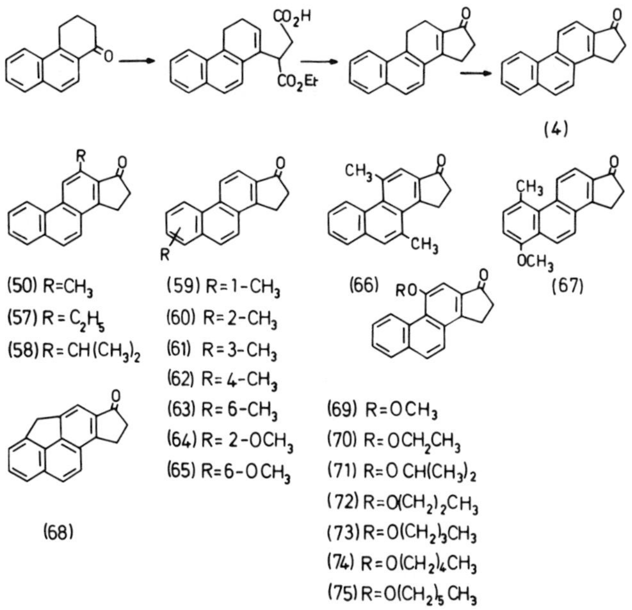

CO_2H CO_2Et

(4)

(50) R=CH_3
(57) R=C_2H_5
(58) R=$CH(CH_3)_2$

(59) R=1-CH_3
(60) R=2-CH_3
(61) R=3-CH_3
(62) R=4-CH_3
(63) R=6-CH_3
(64) R=2-OCH_3
(65) R=6-OCH_3

(66)

(67)

(68)

(69) R=OCH_3
(70) R=OCH_2CH_3
(71) R=$OCH(CH_3)_2$
(72) R=$O(CH_2)_2CH_3$
(73) R=$O(CH_2)_3CH_3$
(74) R=$O(CH_2)_4CH_3$
(75) R=$O(CH_2)_5CH_3$

FIGURE 10. 17-Oxocyclopenta[a]phenanthrenes prepared via the Stobbe reaction

rivatives were selected.[33] These were also converted[34] by means of the Grignard reaction into the 17-methyl-16(17)-enes (33 and 76 to 80) and thence to the 17-methyl hydrocarbons (5, 7, and 81 to 84) so that a comparison could be made with Dannenberg's testing results. In addition, the 11-acetate (45), the isomeric 15-ketone (85), and the three unsaturated D-ring hydrocarbons (2), (86), and (87) were also tested. The compounds involved in this initial test are shown in Figure 11. Swiss mice (T. O., Theiler's Original, outbred within a closed colony in the Imperial Cancer Research Fund for many years) were used in groups of 20 (ten male and ten female) for each compound. Compounds were applied to the shaved dorsal skin as toluene solutions (0.5% w/v, 1 drop = 6 μℓ = 30 μg or ~ 120 nmol) twice weekly for 52 weeks, and the animals were observed during a second year. Experiments were terminated at 2 years (about 730 days). Survival was good; 40% of the control animals treated with solvent alone were still alive at this time, and survival at 18 months was 90%. This has been found to be true also of the other similar skin-painting tests with T. O. mice discussed later. The results of this experiment are shown in Table 3, and Table 9 illustrates a small injection experiment with the three 17-ketones [unsubstituted (4); 11-methyl (47); and 11,12-dimethyl (51)] using the same strain of mice. Several points can be gleaned from Table 3:

1. The last column shows the Iball indexes calculated for these compounds. This index is defined as:

$$\text{Iball index} = \frac{\text{percentage of animals with tumors}}{\text{mean latent period in days}} \times 100$$

For this, "tumors" include both papillomas and carcinomas at the site of application, and the latent period is the time elapsed between the beginning of the experiment and the first appearance of the first skin tumor on each mouse. The mean latent period is the average latent period for all the animals in that group. This index is a better

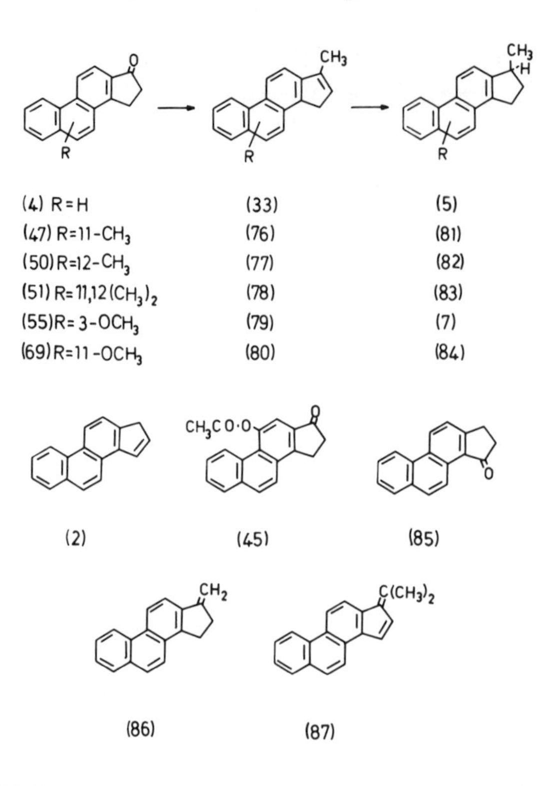

FIGURE 11. Hydrocarbons and 17-ketones tested for carcinogenicity in mice by skin painting.

indicator of tumorigenicity than tumor incidence alone, since it also takes the latent period into account, although it neglects other important parameters such as the number of skin tumors per animal, the carcinoma/papilloma ratio, and the rate of tumor growth. In agreement with Dannenberg, it is clear that the hydrocarbons with a saturated D ring are weakly carcinogenic when they bear an 11-methyl group (compounds 81 and 83). A methoxy group at this ring position (compound 84) also leads to very weak activity. In agreement with others, the methyl group at C-17 has no effect, as shown by the lack of carcinogenicity of Diels hydrocarbon (5).

2. Also in agreement with this author, introduction of a 16(17)-double bond into Diels hydrocarbon to give 17-methyl-15*H*-cyclopenta[a]phenanthrene (33) brings with it weak activity; the tumor incidence for this compound is 30%, but the latent period is very long, thus resulting in the low Iball index. The 17-methylene hydrocarbon (86) is also of similar activity, although the meaning of this is a little unclear because this compound is known to isomerize to its 16(17)-isomer (33) very readily. However, substitution of a methyl group at C-11 augments the activity as before, increasing the tumor incidence and decreasing the mean latent periods, so that the 11,17-dimethyl-16-ene (76) and its 11,12,17-trimethyl homologue (78) are moderately strong carcinogens (Iball indexes 27 and 23, respectively). Thus a D-ring double bond and 11-methyl substituent reinforce one another, whereas 3-oxygen substitution diminishes carcinogenicity.
3. The most interesting fact to emerge is that the 11-methyl-17-ketone (47) (15,16-dihydro-11-methylcyclopenta[a]phenanthren-17-one) is the most active compound to be tested. The 11,12-dimethyl homologue (51) and even the 11-methoxy-17-ketone (69) are also fairly carcinogenic, while the unsubstituted-17-ketone (4) and its 12-methyl (50) and 3-methoxy (55) derivatives are completely inert. This series therefore

Table 3
SKIN-PAINTING EXPERIMENT WITH THE CYCLOPENTA[a]PHENANTHRENE DERIVATIVES SHOWN IN FIGURE 11

Compound[a]	Tumors at site of application		Tumor incidence (%)	Iball index
	Papillomas	Carcinomas		
15,16-Dihydrocyclopenta[a]phenanthren-17-ones				
(4) Unsubst.	0	0	0	0
(47) 11-Methyl	1	13	70	36
(50) 12-Methyl	0	0	0	0
(51) 11,12-Dimethyl	2	11	65	30
(55) 3-Methoxy	0	0	0	0
(69) 11-Methoxy	2	9	55	25
17-Methyl-15*H*-cyclopenta[a]phenanthrenes				
(33) Unsubst.	3	3	30	8
(76) 11-Methyl	1	16	85	27
(77) 12-Methyl	1	2	15	7
(78) 11,12-Dimethyl	1	14	75	23
(79) 3-Methoxy	0	1	5	2
(80) 11-Methoxy	1	4	25	11
16,17-Dihydro-17-methyl-15*H*-cyclopenta[a]phenanthrenes				
(5) Unsubst.	0	0	0	0
(81) 11-Methyl	0	2	10	3
(82) 12-Methyl	0	0	0	0
(83) 11,12-Dimethyl	0	6	30	12
(7) 3-Methoxy	0	0	0	0
(84) 11-Methoxy	0	0	0	0
Miscellaneous cyclopenta[a]phenanthrenes				
(85) 16,17-Dihydrocyclopenta[a]phenanthren-15-one	0	0	0	0
(2) 17*H*-cyclopenta[a]phenanthrene	0	0	0	0
(86) 15,16-Dihydro-17-methylene-cyclopenta[a]phenanthrene	1	4	25	5
(87) 17-Isopropylene-cyclopenta[a]phenanthrene	0	0	0	0
Solvent control	0	0	0	0

[a] 120 nmol in toluene (6 $\mu\ell$) twice weekly.

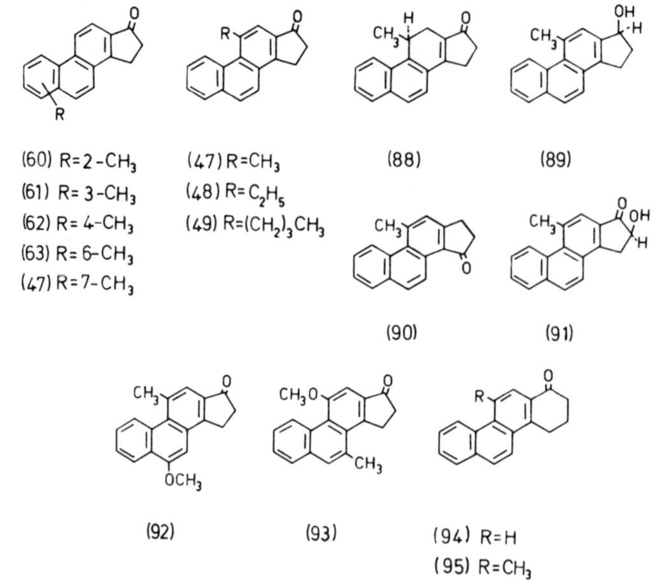

(60) R=2-CH_3
(61) R=3-CH_3
(62) R=4-CH_3
(63) R=6-CH_3
(47) R=7-CH_3

(47) R=CH_3
(48) R=C_2H_5
(49) R=$(CH_2)_3CH_3$

(88) (89) (90) (91) (92) (93)

(94) R=H
(95) R=CH_3

FIGURE 12. Further 17-oxocyclopenta[a]phenanthrenes and two chrysenes (94,95) tested for carcinogenicity in mice by skin painting.

differs from the 16(17)-ene series in this respect, and also in the higher activity of the 11-methoxy compound. It is interesting that although as a general rule introduction of oxygen functions into polycyclic aromatic hydrocarbon carcinogens diminishes their activity, here addition of a 17-carbonyl function considerably augments the effect of both methyl and methoxyl groups at C-11.

In the second skin-painting experiment,[35] carried out as before, the 11-methyl-17-ketone was tested together with its 2-, 3-, 4-, 6-, and 7-methyl isomers and the ten other related compounds shown in Figure 12; the results of this test appear in Table 4. The 11-methyl-17-ketone (47) was again found to be a strong carcinogen on repeated application to T.O. mice, whereas the 2-,3-,4-, and 6-methyl isomers were without activity, but the 7-methyl-17-ketone (46) was moderately carcinogenic. These structure/activity results therefore resemble those found for the analogous hydrocarbons by Dannenberg, except that the 11-methyl ketone was much more active than its 7-methyl isomer. The 11-ethyl homologue (48) was much less active than the 11-methyl ketone, and the 11-butyl compound (49) was inactive. The 11-methyl-11,12-dihydro derivative (88) was also inactive, showing the need for a complete phenanthrene ring system, and transposition of the ketone function in the 11-methyl compound to C-15 (compound 90) drastically reduced activity; in fact, this ketone is of about the same activity as the corresponding 11,17-dimethyl hydrocarbon (81). The activating effect of an 11-methyl group is clearly seen in the ring D-homo derivatives (94) and (95), 1,2,3,4-tetrahydrochrysen-1-one and its 11-methyl homologue; the latter was as carcinogenic as the corresponding 11-methyl cyclopenta[a]phenanthren-17-one (47), whereas the unsubstituted chrysenone was inactive, like the unsubstituted ketone (4). Introduction of a methoxy group at C-6 (compound 92) also reduced activity in the 11-methyl-17-ketone, whereas introduction of the same group at C-11 in the 7-methyl-17-ketone to give compound (93) led to an increase in activity.

The high potency of the 11-methylchrysene ketone (95) was further investigated by converting it into the three 1,11-dimethylchrysene derivatives, as shown in Figure 13. These chrysenes were then tested for carcinogenicity by the method already described (Table 5).[36]

Table 4
SKIN-PAINTING EXPERIMENT WITH COMPOUNDS RELATED TO THE CARCINOGEN 15,16-DIHYDRO-11-METHYLCYCLOPENTA[a]PHENANTHREN-17-ONE SHOWN IN FIGURE 12

Compound[a]	Tumor incidence[b] (%)	Iball index
(60) 2-Methyl-17-one	0	0
(61) 3-Methyl-17-one	0	0
(62) 4-Methyl-17-one	0	0
(63) 6-Methyl-17-one	0	0
(46) 7-Methyl-17-one	30	10
(47) 11-Methyl-17-one	90	46
(48) 11-Ethyl-17-one	30	8
(49) 11-*n*-Butyl-17-one	0	0
(88) 11-Methyl-11,12-dihydro-17-one	0	0
(89) 11-Methyl-17-ol	60	23
(90) 11-Methyl-15-one	20	5
(91) 11-Methyl-16-hydroxy-17-one	25	11
(92) 11-Methyl-6-methoxy-17-one	55	14
(93) 7-Methyl-11-methoxy-17-one	50	17
(94) 1,2,3,4-Tetrahydrochrysen-1-one	0	0
(95) 11-Methyl-1,2,3,4-tetrahydrochrysen-1-one	90	45
Toluene control	0	0

[a] Doses as for Table 3.
[b] Skin tumors at site of application.

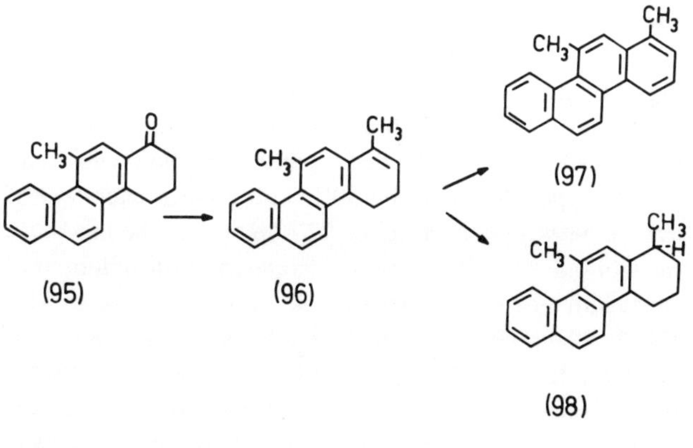

FIGURE 13. Chrysene derivatives prepared for comparison of carcinogenicity in mice by skin painting.

Of the four, the 11-methyl-1-ketone (95) was by far the most active; next came 1,11-dimethylchrysene (97) followed by its 3,4-dihydro derivative (96). The tetrahydro hydrocarbon (98) was the least active, similar to the corresponding 11,17-dimethylcyclopenta[a]phenanthrene hydrocarbon (5). Hence, again the ketone function four carbon atoms away from the 11-methyl group leads to remarkably high potency, higher than that of the fully aromatic 1,11-dimethylchrysene (97).

Continuing the structure/carcinogenicity investigation in the cyclopenta[a]phenanthrene series, the additional thirteen 17-ketones shown in Figure 14 have now been screened in

Table 5
SKIN-PAINTING RESULTS WITH THE FOUR CHRYSENE DERIVATIVES SHOWN IN FIGURE 13

Compound[a]	Tumor incidence[b] (%)	Iball index
(95) 11-Methyl-1-oxo-1,2,3,4-tetrahydrochrysene	85	44
(96) 1,11-Dimethyl-3,4-dihydrochrysene	40	10
(97) 1,11-Dimethylchrysene	75	22
(98) 1,11-Dimethyl-1,2,3,4-tetrahydrochrysene	15	4

[a] Doses as for Table 3.
[b] Skin tumors at site of application.

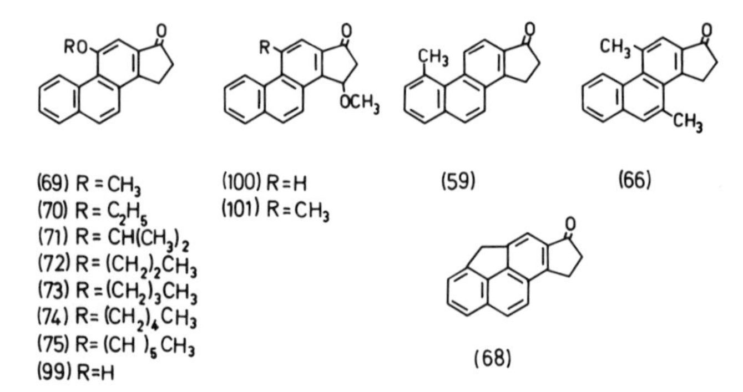

FIGURE 14. Additional 17- oxocyclopenta[a]phenanthrenes tested for carcinogenicity in mice.

several experiments. In the first,[37] it was established that the 11-methyl-17-ketone (47) was also active in the two-stage system. With a single topical initiating dose of 400 μg (1600 nmoles) in toluene (40 $\mu\ell$) followed by twice weekly promotion with croton oil, skin tumor incidence at the site of application reached 90% with a mean latent period very similar to that achieved by twice weekly application of 120 nmoles of the carcinogen. However, as has been observed with other carcinogens, the carcinoma to papilloma ratio was markedly in favor of papillomas in the two-stage experiment, but was the reverse when the carcinogen was applied repeatedly. This two-stage method, of course, measures initiating potential, whereas the repeated application method measures ''complete carcinogenicity''. Initiating activity of the alkoxy-17-ketones (69 to 75, 100, and 101) was investigated[32] using this two-stage method (Table 6). As in the 11-alkyl series, activity decreased with lengthening of the ether chain at C-11. The 11-methoxy compound (69) was the most active, its ethyl homologue (70) was less so, and the straight-chain propyl, pentyl, and hexyl ethers were not tumor initiators. The *iso*propyl (72) and *n*-butyl (73) ethers appeared to be very weakly active; however, occasionally croton oil itself induces a papilloma in this system so it would be necessary to retest these two compounds using a much larger number of mice to be certain of their weak activity. The 15-methoxy-17-ketone (100) was essentially inactive, but initiating activity was again associated with its 11-methyl homologue (101). The 11-phenol (99) could not be tested in this way owing to its insolubility in toluene (and most other organic solvents). It was therefore suspended in olive oil (10 mg in 0.2 mℓ) and injected subcutaneously into mice whose dorsal skin remote from the injection site was subsequently promoted as usual by repeated applications of croton oil. The same dose of the 11-methyl-17-ketone

Table 6
INITIATING ACTIVITY OF 15,16-DIHYDRO-11-ALKOXY CYCLOPENTA[a]PHENANTHREN-17-ONES AND TWO 15-METHOXY-17-KETONES

	Tumors at site of application			
Compound[a]	**Papillomas**	**Carcinomas**	**Tumor incidence**	**Iball index**
(69) 11-Methoxy	9	2	55	32
(70) 11-Ethoxy	6	2	40	23
(71) 11-n-Propoxy	0	0	0	0
(72) 11-iso-propoxy	2	0	10	4
(73) 11-n-butoxy	4	0	20	5
(74) 11-n-Pentoxy	0	0	0	0
(75) 11-n-Hexoxy	0	0	0	0
(100) 15-Methoxy	1	0	5	1
(101) 11-Methyl-15-methoxy	3	2	25	15
Croton oil control	0	0	0	0

[a] 1600 nmole, once, followed by croton oil twice weekly.

[b] This index cannot, of course, be compared directly with that in the other tables where tumors were induced by repeated application of the carcinogens.

Table 7
SKIN TUMORS INDUCED IN T.O. MICE WITH COMPOUNDS (59) AND (66)

		Total No.		
Compound	**Tumor incidence (%)**	**Papillomas**	**Carcinomas and sarcomas**	**Mean latent period (weeks)**
	Tumors induced by repeated application[a]			
(59) 1-Methyl-17-ketone	5	1	0	71
(66) 7,11-Dimethyl-17-ketone	65	18	12	19[b]
	Tumors induced in two-stage experiments[c]			
(59) 1-Methyl-17-ketone	10	2	0	45
(66) 7,11-Dimethyl-17-ketone	80	22	6	17
(47) 11-Methyl-17-ketone (positive control)	80	19	2	20
Croton oil (negative control)	5	1	0	22

[a] 200 nmol in toluene (10 $\mu\ell$) twice weekly.

[b] Owing to lack of material, skin painting was continued twice weekly for only 10 weeks (50 weeks for 1-methyl-17-ketone, as usual).

[c] 1600 nmol in toluene (40 $\mu\ell$) once, then croton oil twice weekly.

Table 8
SKIN TUMORS INDUCED BY 15,16-DIHYDRO-1,11-METHANO-CYCLOPENTA[a]PHENANTHREN-17-ONE (68)

Treatment	Tumor incidence	No. of mice with		Mean latent period (weeks)
		Papillomas	Carcinomas	
1,600 nmol once, croton	60	7	5	29
oil twice weekly	70	8	6	27
200 nmol twice weekly	40	6	2	39
	55	8	3	45

(47), employed as a positive control, was very active as a systemic initiator under these conditions, giving a 95% skin tumor incidence with a short mean latent period of 21 weeks. The 11-phenol was found to be moderately active; in two separate but identical experiments the tumor incidences were 60%, 25 weeks, and 65%, 29 weeks. The conclusion is that provided that they are small, electron-releasing groups at C-11 are capable of inducing carcinogenicity in this series.

The two methyl derivatives (59 and 66) were tested[38] both by repeated application (200 nmol twice weekly) and by the two-stage method (1600 nmol once topically, followed by repeated applications of croton oil) using groups of 20 T.O. mice as before (Table 7). As anticipated, the 1-methyl-17-ketone (59) was essentially inactive in both tests, while the 7,11-dimethyl-17-ketone (66) was a powerful carcinogen, apparently more active than the 11-methyl-17-ketone (47) (used as a positive control) on the basis of both the total number of tumors induced and their shorter mean latent period in the two-stage experiment. It was also very active as a complete carcinogen, giving a total of 30 skin tumors, over a third of which were malignant, in 13/20 mice. Most probably both the tumor incidence and the total number of tumors would have increased further if treatment had been continued for 50 weeks as usual; unfortunately, it was discontinued at ten weeks owing to lack of the compound, the preparation of which had proved to be difficult.

The 1,11-methano-17-ketone (68) was synthesized in the expectation that the methylene bridge would block metabolism at C-1,2 in the bay region (known by 1979 to be involved in the biological activation of the carcinogenic 11-methyl analogue) without sacrificing the small electron-releasing substituent at C-11.[39] When this compound began to show initiating activity in the first two-stage experiment, a second similar experiment was begun, and it was also tested in two separate repeated-application tests (Table 8). It is clear that contrary to anticipation, this compound is substantially active as a skin tumor initiator, but appears to be less active as a complete carcinogen.

IV. THE CARCINOGENICITY OF 15,16-DIHYDRO-11-METHYLCYCLOPENTA[a]PHENANTHREN-17-ONE

We have already seen that 15,16-dihydro-11-methylcyclopenta[a]phenanthren-17-one (47) is a strong carcinogen both by repeated topical application and by the initiation/promotion regime in T.O. mice. It is also a potent skin carcinogen in Balb C mice[39a] and in C57 B1 mice, but not in DBA/2 mice by both methods.[40] The reason for resistance of mice of the latter strain is unknown, because Abbott found that the total DNA adducts in the treated skin, as well as the pattern of adducts disclosed by high pressure liquid chromatography of the derived nucleosides, were similar for T.O., C57B1, and DBA/2 mice. Moreover, the persistence of these DNA adducts in vivo also appeared to be similar for mice of all three

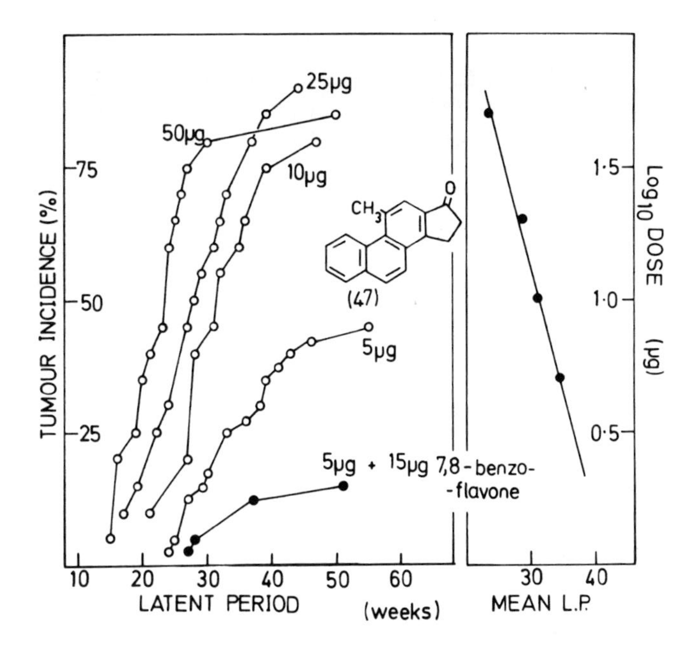

FIGURE 15. First appearance of the first skin tumor for each mouse painted with the carcinogen 15,16-dihydro-11-methylcyclopenta[a]phenanthren-17-one (47) at 4 doses (50,25,10, and 5μg) twice weekly for 1 year. Note the substantial drop in tumor yield when the aryl hydrocarbon hydroxylase inhibitor 7,8-benzoflavone is applied together with the carcinogen. The graph on the right illustrates the linear relationship between mean latent period and the logarithm of the repeated dose.

strains. In T.O. mice the mean latent period for skim tumor induction is related to dose over a tenfold dose range (5 to 50 μg, twice-weekly) by the well-known equation:

$$L = a - b(\log_{10} d + c)$$

where L = mean latent period, d = dose, and a, b, and c are constants, equal to 29.4, 10.8, and 1.2 in this instance[41] (Figure 15). Benzo[a]pyrene painted at the lowest dose (5 μg, twice weekly) had virtually the same activity, in terms of both tumor incidence and latent period, as the 11-methyl-17-ketone (47) at the same dose, showing that these two compounds are of comparable activity with regard to skin-tumor induction in this strain of mice. When the aryl hydrocarbon hydroxylase inhibitor 7,8-benzoflavone was painted together with this carcinogen at three times the dose of the latter, skin tumor appearance was markedly suppressed, indicating that the carcinogenicity of this ketone was probably dependent upon metabolic oxygenation of the aromatic ring(s). This idea was supported by enhancement of its carcinogenicity by co-administration of the epoxide-hydratase inhibitor 1,1,1-trichloro-propane oxide.[41]

The 11-methyl-17-ketone (47) was also carcinogenic after subcutaneous injection without deliberate subsequent promotion. Thus, the first skin-painting experiment was supported by the small injection experiment[33] summarized in Table 9. Of the three compounds tested, the 11-methyl-17-ketone was the only one to give local sarcomas at the injection site and carcinomas on skin remote from it, although the dose required was high. These remote tumors occurred mainly on the ventral skin of these animals; both this compound and its 11,12-dimethyl analogue also gave rise to papillomas and sebaceous adenomas in this area. The unsubstituted compound (15,16-dihydrocyclopenta[a]phenanthren-17-one) gave no skin

Table 9
SUBCUTANEOUS INJECTION EXPERIMENT WITH 15,16-DIHYDROCYCLOPENTA[a]PHENANTHREN-17-ONE AND ITS 11-METHYL AND 11,12-DIMETHYL HOMOLOGUES

				No. of mice with remote skin tumors	
Compound	Dose (mg)	No. of mice treated	No. of mice with local spindle cell sarcoma	Squamous papillomas and sebaceous adenomas	Squamous carcinomas
(4) Unsubst.-17-ketone	50	18	0	0	0
(47) 11-Methyl-17-ketone	50	18	3	5	5
(47) 11-Methyl-17-ketone	8	21	3	8	0
(51) 11,12-Dimethyl-17-ketone	8	21	0	9	0
Olive oil control	—	21	0	0	0

tumors even at the high dose (50 mg). Systemic initiation of the dorsal skin by subcutaneous injection of the 11-methyl-17-ketone (10 mg per mouse) has already been noted, but this was disclosed only by promotion of this area with croton oil. The ventral skin tumors appeared without apparent promotion; ventral skin is thicker and contains much more fat than dorsal skin, possibly retarding the loss of this lipophilic compound and thereby allowing it to act over a prolonged period at this site. Dorsal skin was efficiently initiated by this carcinogen at a lower subcutaneous dose (3 mg), subsequent promotion giving a 65% skin tumor incidence, with a mean latent period of 33 weeks. These animals, whether their dorsal skin was promoted or not, also developed a number of other skin tumors as shown in Table 10. The eyelids were particularly affected, and tumors on the head and ventral surfaces were again observed. When dorsal promotion was delayed for six months following injection, and then begun as usual, dorsal skin tumor incidence was still 45%. The mean latent period was less, 24 weeks, when it was taken from the start of promotion rather than the date of injection, and Iball indexes for these two experiments, calculated in this way, were virtually identical (28 for immediate promotion and 27 for delayed promotion). This proves that initiation with this carcinogen is an irreversible and permanent state. Unlike a typical polycyclic hydrocarbon carcinogen such as benzo[a]pyrene, this ketone (47) does not give rise to local sarcomas at the injection site at this dose, and this useful property has been utilized in transplantation experiments involving initiated skin in Balb C mice.[39a] It has also made possible experiments which have proved that silica fiber isolated from seeds of the grass *Phalaris canariensis* can act as tumor promoters in T.O. mice.[42] In these experiments, ventral skin tumors were again conspicuous.

Gastric instillation of the carcinogen into female Sprague-Dawley rats (30 mg per rat) led to mammary adenocarcinomas as shown in Table 11. This experiment establishes that 15,16-dihydro-11-methylcyclopenta[a]phenanthren-17-one (47) is active systemically in animals of more than one species.[41]

V. MUTAGENICITY OF CYCLOPENTA[a]PHENANTHRENES

It is now well established that the great majority of chemical carcinogens are mutagenic, although most first require conversion to a reactive form, usually by metabolism. No cyclopenta[a]phenanthrenes are mutagenic per se, and this is therefore a good indication that metabolism is probably involved when they act as carcinogens in animals. Our interest in this area arose out of the desire to establish a rapid test for biological activity that could be

Table 10
TUMORS OTHER THAN THOSE APPEARING ON PROMOTED SKIN AFTER SUBCUTANEOUS INJECTION OF THE 11-METHYL-17-KETONE (47) INTO MICE AT 3 mg/MOUSE

	Site of tumors[a]				
No. of mice injected	Eyelid	Ear	Head	Ventral surfaces	Lung adenomas
60	3 Carc. 9 Pap. 1 Sebaceous adenoma	2 Carc.	2 Pap.	2 Carc. 2 Pap.	19

[a] No tumors at these sites were found in 20 control mice injected with olive oil. Carc., squamous carcinoma. Pap., squamous papilloma.

Table 11
MAMMARY ADENOCARCINOMAS INDUCED IN FEMALE RATS AFTER INTRAGASTRIC INSTILLATION OF THE 11-METHYL-17-KETONE (47) AT 30 mg/RAT

		No. of rats					
		At 20 weeks		At 30 weeks		At 50 weeks	
Adenocarcinomas	At time of treatment	With tumor	Alive	With tumor	Alive	With tumor	Alive
In treated rats	27	2	26	4	26	6	20
In untreated rats	96	0	94	0	92	1	25
p	—	0.455		0.0019		0.0001	

used to distinguish the active metabolites of the carcinogen 15,16-dihydro-11-methylcyclopenta[a]phenanthren-17-one (47) among the many separable by high-pressure liquid chromatography (HPLC). As a preliminary to this, it seemed advisable to explore the correlation between mutagenicity and carcinogenicity among these closely related compounds. Ames histidine deficient *Salmonella typhimurium* TA100 were found to be suitable for this purpose, and 36 cyclopenta[a]phenanthrenes and two related chrysen-1-ones were screened for mutagenicity in this way by a standard plate assay.[32,39,43] In Table 12, mutagenicity is expressed as the number of revertant colonies per nmol (after substracting the number growing on control plates from the total on the test plates). Carcinogenicity is expressed as the Iball index as previously defined; two series are necessary: A refers to repeated application, and B to the two-stage method, as already outlined. It is evident that there is a remarkably good correlation in that with two possible exceptions all the carcinogens are mutagens in the Ames test. The two exceptions (compounds 72 and 73) may be very weak carcinogens; however, low figures under B (two-stage system) are questionable, because occasionally the promoting agent, croton oil alone, causes the appearance of a skin tumor. The reverse relationship is not true, however, because there are a number of noncarcinogens which are mutagenic (compounds 4, 71, 94, 85, 2, and 5). There is also little evidence for a quantitative correlation between mutagenicity and carcinogenicity in this series. Where compounds have been tested by both repeated application (A) and initiation/promotion (B) (compounds 47, 68, and 69), it is clear that the Iball index is higher with the latter procedure which is usually associated with shorter latent periods (see for example, Table 8). Despite these shortcomings, it was

Table 12
MUTAGENICITY OF CYCLOPENTA[a]PHENANTHRENES IN AMES TEST COMPARED WITH THEIR CARCINOGENICITY

Compound	Mutagenicity[a] (No. of revertant colonies/nmol)	Carcinogenicity[b] (Iball Index)	
		A	B
15,16-Dihydrocyclopenta[a]phenanthren-17-ones			
(4) Unsubstituted	9.5	<1	<1
(59) 1-Methyl	<0.2	1	3
(60) 2-Methyl	<0.2	<1	—
(61) 3-Methyl	<0.2	<1	<1
(62) 4-Methyl	<0.2	<1	—
(63) 6-Methyl	<0.2	<1	—
(46) 7-Methyl	12.3	10	—
(47) 11-Methyl	21.7	46	57
(50) 12-Methyl	<0.2	<1	—
(51) 11,12-Dimethyl	1.1	30	—
(68) 1,11-Methano	1.9	16[c]	33
(57) 11-Ethyl	1.8	8	—
(58) 11-*n*-Butyl	<0.2	<1	—
(55) 3-Methoxy	0.2	<1	—
(69) 11-Methoxy	3.1	25	38
(70) 11-Ethoxy	1.8	—	28
(71) 11-*n*-Propoxy	0.3	—	<1
(72) 11-iso-Propoxy	1.1	—	4
(73) 11-*n*-Butoxy	<0.2	—	5
(74) 11-*n*-Pentoxy	<0.2	—	<1
(93) 7-Methyl-11-methoxy	34.5	17	—
(92) 11-Methyl-6-methoxy	1.7	14	—
(88) 11-Methyl-11,12,15,16-tetrahydro-17-one	<0.2	<1	—
(91) 11-Methyl-16-hydroxy	18.3	11	—
1,2,3,4-Tetrahydrochrysen-1-ones			
(94) Unsubstituted	7.2	<1	—
(95) 11-Methyl	7.1	45	—
16,17-dihydrocyclopenta[a]phenanthren-15-ones			
(85) Unsubstituted	6.7	<1	—
(90) 11-Methyl	3.7	5	—
Hydrocarbons, etc.			
(33) 17-Methyl-15*H*-cyclopenta[a]phenanthrene	1.6	6	—
(76) 11,17-Dimethyl-15*H*-	1.0	27	—
(77) 12,17-Dimethyl-15*H*-	1.0	7	—
(78) 11,12,17-Trimethyl-15*H*-	1.2	23	—
(79) 3-Methoxy-17-methyl-15*H*-	0.7	2	—
(80) 11-Methoxy-17-methyl-15*H*-	0.9	11	—
(2) 17*H*-	4.0	<1	—
(5) 17-Methyl-15,16-dihydro-	0.7	<1	—
(81) 11,17-Dimethyl-15,16-dihydro-	1.5	3	—
(7) 3-Methoxy-17-methyl-15,16-dihydro-	<0.2	<1	—

Note: Compounds marked <1 are classed as noncarcinogenic, i.e., tumor incidence is <5% (less than 1 animal in 20) in 700 days. Spontaneous dorsal skin tumors have never been observed in control mice painted with the solvent (toluene); however, occasionally they are observed in uninitiated mice painted with croton oil repeatedly. Low Iball indexes B are therefore open to some doubt.

Table 12 (continued)
MUTAGENICITY OF CYCLOPENTA[a]PHENANTHRENES IN AMES TEST COMPARED WITH THEIR CARCINOGENICITY

[a] Compounds marked <0.2 are classed as non-mutagenic; this value is approximately equal to the spontaneous mutation rate observed in the control plates.
[b] Skin tumor induction in T.O. mice: A, 120 nmol of compound in toluene twice weekly for 1 year, mice observed for up to 2 years; B, 400 nmol of compound in toluene applied once, followed by croton oil twice weekly for up to 2 years.
[c] As A but 200 nmol twice weekly.

Table 13
COMPARISON OF IBALL INDEXES OF 17-METHYL-16,17-ENES AND 17-KETONES

Substitution	17-Methyl-16,17-ene		17-ketone	
Unsubstituted	(33)	6	(4)	<1
11-Methyl	(76)	27	(47)	46
12-Methyl	(78)	7	(50)	<1
11,12-Dimethyl	(77)	23	(51)	30
11-Methoxy	(80)	11	(69)	25
3-Methoxy	(79)	2	(55)	<1

felt that the Ames test would provide a valid method for identifying potentially carcinogenic metabolites among the range of derivatives formed by in vitro metabolism of these carcinogens, and it has indeed proved useful for this purpose.

VI. METABOLIC ACTIVATION OF CYCLOPENTA[a]PHENANTHRENES

It is clear from the foregoing that cyclopenta[a]phenanthrenes, like most polycyclic compounds, require correct metabolism to render them biologically active. Dannenberg[19] found that introduction of a 16,17-double bond into Diels hydrocarbon (5, giving 33) led to weak carcinogenic activity (see Tables 1 and 2), and noted that this double bond was the most reactive in the molecule, yielding for example the *cis*-16,17-diol with osmium tetroxide in preference to the K-region 6,7-double bond. He suggested that the carcinogenicity of these hydrocarbons, such as (33) and its 15-methyl-17*H* isomer (35), might well be due to the reactivity of this double bond in the five-membered D ring. This does not seem to have been followed up, although structure/activity relationships established since do indicate differences between this series and the series of 17-ketones. Thus, comparing the Iball indexes of several similarly substituted 17-methyl-16,17-enes and 17-ketones (Table 13), it appears that carcinogenicity among the latter is confined to 11-substituted derivatives, whereas all the hydrocarbons are active to some extent. The active ketones are more carcinogenic than the corresponding hydrocarbon, and this is true in particular of the 11-methoxy derivatives.

Over the last few years, the metabolism of several 17-ketones, especially the strongly mutagenic and carcinogenic 11-methyl derivative (47), has been studied in detail. This has been greatly facilitated by: (1) the use of high pressure liquid chromatography (HPLC) for the efficient separation of small amounts of complex mixtures of metabolites; (2) the Ames test to identify biologically active fractions; (3) the use of sophisticated modern analytical techniques such as mass spectrometry and Fourier-transform high-field nuclear magnetic resonance spectroscopy (NMR) to establish their structures.

Ultraviolet spectroscopy (UV) has proved to be invaluable in this work because of its sensitivity, and the fact that mild *in situ* reduction of the conjugated 17-carbonyl group in these molecules with sodium borohydride leads to chromophores very similar to those of the corresponding phenanthrenes, most of which are known. Although rat-liver homogenate is employed in the Ames test to cause biological activation, a more defined system consisting of isolated rat-liver microsomes and an excess of NADPH in 0.1 molar tris buffer in the presence of air, is all that is required for efficient in vitro metabolism. Use of isolated microsomes eliminates most of the conjugating enzymes, thus preserving the products of primary metabolism, so that simple extraction of the medium at the end of the incubation with ethyl acetate routinely leads to recovery of more than 80% of the material incubated.[45] Use of substrate containing a radioactive label, either tritium or carbon-14, permits easy quantitation, and experiments with 1 to 10 μmol or more are conveniently carried out in this manner.

Figure 16 shows a typical HPLC separation of in vitro metabolites obtained from the carcinogenic 11-methyl-17-ketone (47) by these means. A reverse phase column was employed with a gradient of aqueous methanol increasing in methanol concentration, so that the more polar materials elute first, and the last peak to emerge is the unchanged ketone. The profile shown was monitored by its UV absorption at 254 nm; the profile of eluted radioactivity was very similar. The peak marked M was of microsomal origin because it was not radioactive; the structures of the metabolites (a through g), established by methods outlined above, were as follows: (a) 1,2,15-triol; (b) 3,4,16-triol; (c) 3,4,15-triol; (d) 15-hydroxy-11-hydroxymethyl derivative; (e) 3,4-diol, (f) 15-ol; (g) 16-ol. The principle metabolic reactions leading to these products are shown in Figure 17.[44-46] Thus, the 17-ketone function remains unchanged while oxygen is introduced into the molecule at both terminal rings. There is no evidence of attack at the 6,7-double bond, unlike the result of chemical oxidation with chromium trioxide or osmium tetroxide. The absolute stereochemistry of these 1,2- and 3,4-diols has been deduced from comparisons of the circular dichroism curves of the derived 17-alcohols with those of the phenanthrene diols of established absolute stereochemistry.[47,48]

When equimolar amounts of these main metabolites (a) to (f) were tested for mutagenicity in the Ames test, none were mutagenic unless a microsomal metabolizing system was supplied. Under these conditions, metabolite (e) was a powerful mutagen, about three times more active than the original carcinogen; (b), (c), (f), and (g) were less active, and (a) and (d) were not mutagens. Thus, the 3,4-diols or compounds that could give rise to 3,4-diols were active, while the 1,2-diols were not. The carcinogenicity of metabolite (e) was proved in a direct manner; enough of it was prepared by large-scale incubations and HPLC separations to carry out a skin tumor induction experiment with twenty T.O. mice.[49] Each received a single initiating dose of 100 nmol of this metabolite, followed by twice weekly promotion with croton oil. Other groups painted with the same quantity of metabolites (a) to (d) or the original carcinogen were treated in the same way. The result of this experiment is shown in Table 14, from which is easily appreciated that the 3,4-diol (e) is by far the most carcinogenic, much more active than the original carcinogen both in terms of tumor incidence and mean latent period. The other 3,4-diols (b) and (c) were weakly active, but the 1,2,15-triol (a) was not carcinogenic, in line with its lack of mutagenicity. Metabolites (g) and (f) were not tested in this way because not enough of (f) was obtained, and it is known that peak (g) consists the 15-ol together with a small amount of the 11-hydroxy-methyl-17-ketone (h). Instead, these three synthetic hydroxy derivatives were tested in the conventional way by the two-stage method, using single topical doses of 1,600 nmol; acetone/toluene (1:1 v/v) was used as the solvent as for the metabolites (see Table 14). Under these conditions (Table 15), the 16-ol (g) appeared to be almost as active as the original carcinogen (47), whereas the 15-ol (f) and the 11-hydroxy-methyl-17-ketone (h) were much less carcinogenic. By comparison of Tables 14 and 15, it is seen that the 3,4-diol (e) is about fifteen times more active than the original carcinogen (47) as a skin tumor initiator.

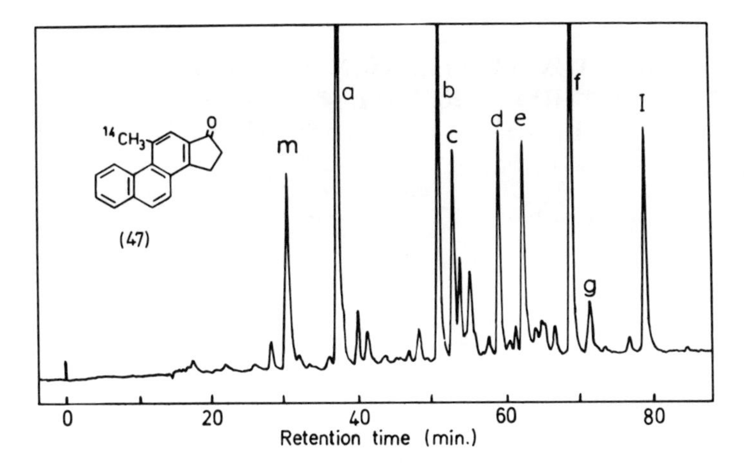

FIGURE 16. Separation of the metabolites of the carcinogenic 11-methyl-17-ketone (47) by HPLC using a reverse phase column with an aqueous gradient increasing in methanol content; the profile is a UV trace monitored at 254 nm, and the structures of the metabolites are discussed in the text. The last peak (I) is the unchanged compound (47).

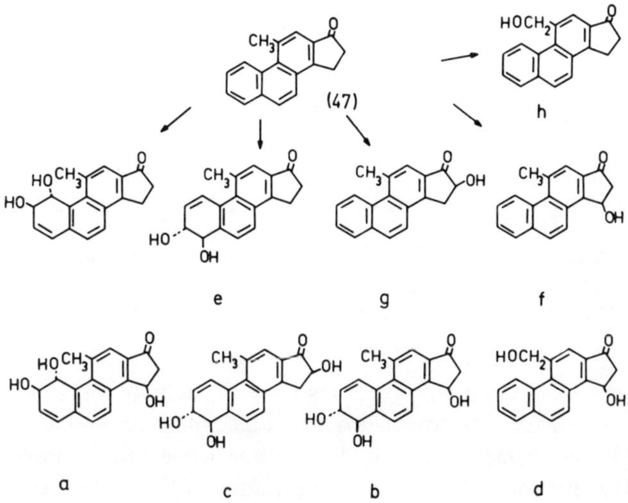

FIGURE 17. Structures of the in vitro metabolites a through h of the carcinogenic [^{14}C]-11-methyl-17-ketone (47) separated by HPLC as shown in Figure 16.

Since further metabolism of the 3,4-diol (e) was necessary to demonstrate its mutagenic potential in the Ames test, it seemed probable that a reactive derivative of this diol was the ultimate carcinogen. Further, since DNA damage is involved in both mutagenicity and carcinogenicity, it was logical to examine the reaction of the diol with DNA. To this end, the original carcinogen (47) was incubated with calf thymus DNA and the usual microsomal metabolizing system; the DNA was then subsequently recovered, purified, and hydrolyzed enzymatically to its constituent nucleosides which were separated by chromatography on a column of Sephadex® LH20.[50] Using carbon-14 labeled carcinogen, two radioactive adduct peaks were eluted from the column following the nonradioactive normal nucleosides which were located by their UV absorption. When DNA isolated from the skin of mice treated

Table 14
SKIN TUMOR INITIATION WITH THE 11-METHYL-17-KETONE (47) AND ITS MICROSOMAL METABOLITES

Compound[a]	No. of mice with skin tumors[b]	Mean latent period (weeks)
(a) 1,2,15-Triol	0	—
(b) 3,4,16-Triol	5	53.6
(c) 3,4,15-Triol	2	66
(d) 11-CH_2OH,15-ol	1	(38)
(e) 3,4-Diol	18	24
(47) Original carcinogen	12	33.8

[a] Each mouse received a single topical application of 100 nmol in acetone/toluene (1:1 v/v), then repeated applications of croton oil.
[b] Groups of 20 mice were employed.

Table 15
SKIN TUMOR INDUCTION WITH MONOHYDROXY DERIVATIVES OF THE 11-METHYL-17-KETONE (47)

Compound[a]	No. of mice with skin tumors[b]	Mean latent period (weeks)
(47) 11-Methyl-17-ketone	19	21.4
(g) 15-Ol	8	45
(f) 16-Ol	20	27
(h) 11-Hydroxymethyl-17-ketone	10	41

[a] Single topical dose of 1600 nmol in acetone/toluene (1:1 v/v) followed by repeated applications of croton oil.
[b] Groups of 20 mice were employed.

with the carcinogen labeled with tritium at high-specific activity was processed in this way, the same pattern was seen. The two radioactive adduct peaks from the Sephadex® column were resolved into six distinct adducts by HPLC, one accounting for more than 80% of the whole (Figure 18). By using the metabolites (a) through (f) in place of the carcinogen in the in vitro incubation with DNA, it was established that this major adduct was derived from the 3,4-diol (e), whereas the other minor adducts were mainly derived from the triols (b) and (c). Also, by using DNA specifically labeled with [^{3}H]-deoxyguanosine and [^{3}H]-deoxyadenosine, it was shown that the major adduct involved the former, but not the latter nucleoside. The nature of this major adduct was further investigated by mass spectrometry which indicated that the deoxyguanosine was attached at C-1 through its amino group to a 2,3,4-trihydroxy derivative of the carcinogen (47),[51] as shown in Figure 19. The stereochemistry at C-1 and C-2 in the adduct, and thence in the proposed intermediary diol-epoxide, was suggested by the behavior of the adduct on chromatography on a borate column. Under these conditions, the adduct was eluted appreciably earlier than from a column of tris buffer of the same pH and molarity, showing that the molecule contains a pair of vicinal *cis* hydroxyl groups that can form a comparatively nonpolar complex with the borate.[46] If the stereochemistry at C-1,2 had been the reverse (i.e., a *syn* diol-epoxide), all three hydroxyl groups in the adduct would have been mutually *trans*, and no complexation with borate would have occurred. This carcinogen therefore appears to be activated in the same way as

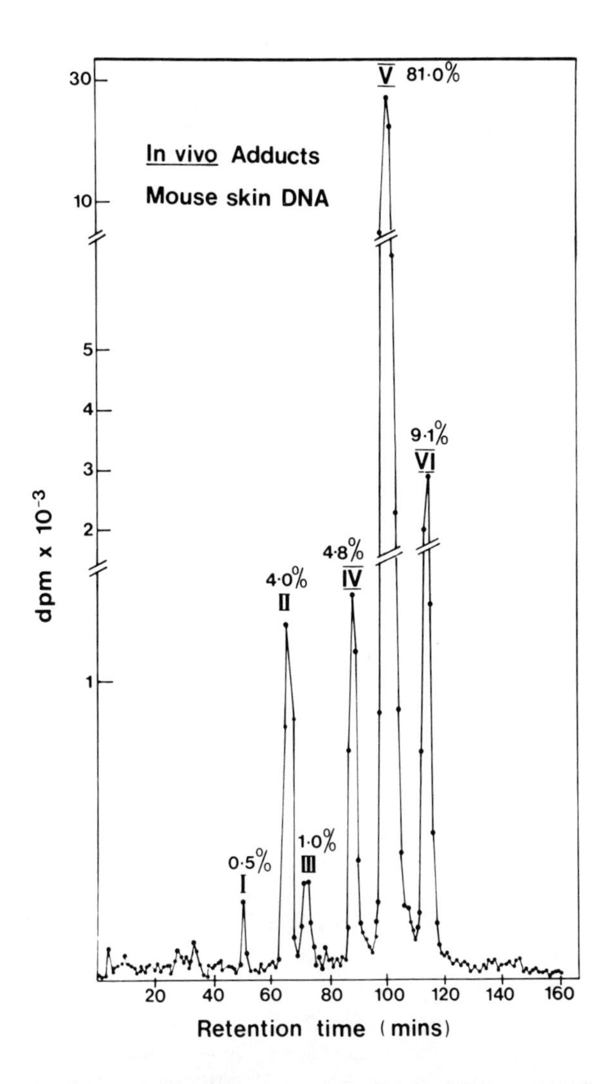

FIGURE 18. Nucleoside adducts isolated from mouse skin treated with the [^{3}H]-11-methyl-17-ketone (47) by topical application in vivo. The skin was subsequently degraded enzymatically and nucleosides were isolated by Sephadex®LH20 chromatography; the radioactive nucleoside fraction was then reseparated by HPLC on a reverse phase column using an aqueous-methanolic gradient to give the six nucleoside adducts shown. All were deoxyguanosine adducts with the exception of VI, which was derived from deoxyadenosine.

benzo[a]pyrene, giving a diol-epoxide completely analogous to the most carcinogenic diol-epoxide formed from this hydrocarbon.[52]

VII. STRUCTURE/ACTIVITY RELATIONSHIPS IN RELATION TO METABOLISM AND COVALENT BINDING TO DNA

The fascinating structure/activity relationships uncovered in this series of cyclopenta[a]phenanthrenes undoubtedly owe their origins to the way these various closely related compounds are metabolized and bind to DNA. This has now been studied with eight of these compounds shown in Figure 20, and some conclusions can be tentatively reached.

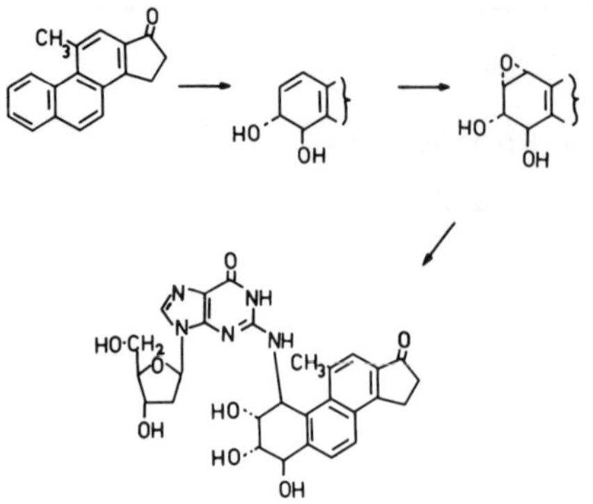

FIGURE 19. Biological oxidation of the carcinogenic 11-methyl ketone leads to its A-ring diol-epoxide which reacts with the exocyclic amino group of deoxyguanosine to give the nucleoside adduct shown. Note the pair of vicinal *cis*-hydroxyl groups at C-2 and C-3.

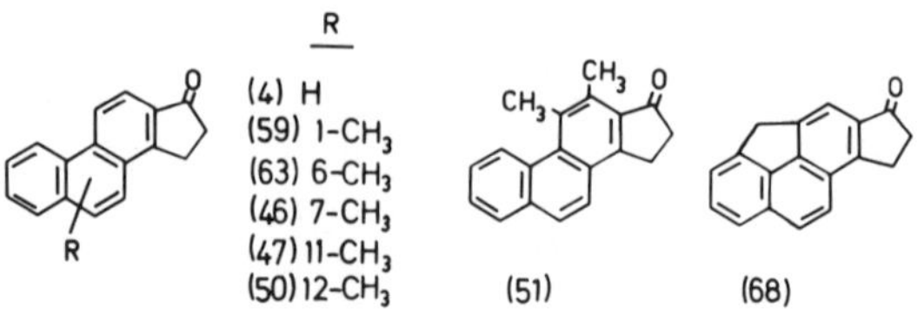

FIGURE 20. 17-Ketones on which in vitro metabolic studies have been conducted.

In vitro metabolism of the unsubstituted 17-ketone (4) and its 12-methyl derivative (50) (both noncarcinogens), and the 11-methyl (47) and 11,12-dimethyl-17-ketones (51) (carcinogens), follow very similar courses[48] giving 1,2- and 3,4-diols, 15- and 16-ols, and triols derived from the various combinations of A- and D-ring oxidation. In particular, all give similar amounts of 3,4-diols, and where they have been studied by circular dichroism,[47] they are of the same absolute stereochemistry (i.e., 3R, 4R, or 3α, 4β). They are all *trans*, and *quasi* diequatorial, as shown by their NMR coupling constants for H-3 and H-4. The 6-methyl-17-ketone (63), a noncarcinogen, also yields a 3,4-diol as a major metabolite; however, it is more polar than the others because it is diaxial, as disclosed by the lack of magnetic coupling between H-3 and H-4.[53] It is probably not further metabolized to a 1,2-oxide because it does not react with DNA even in vitro in the presence of the microsomal system. Others have noted this inhibiting effect of a *peri* substituent, and in fact, introduction of a 6-methoxy group into the 11-methoxy-17-ketone (47) to give the ketone (92) considerably diminishes its carcinogenicity (Iball indexes 46 and 14, respectively). The 7-methyl-17-ketone (46), a moderately active carcinogen (Iball index 17), gives a large number of metabolites, including a small amount of *trans*-diequatorial-3,4-diol, and binds to DNA in vitro much less extensively than its 11-methyl isomer (47), thus accounting for its lower activity as a carcinogen. The 1,11-methano-17-ketone (68) also yields a prominent *trans*-3,4-diol, although its H-3 and H-4 coupling constants reveal that its hydroxyl groups are approximately halfway between being axial and equatorial. Surprisingly, it is also metabolized to a 1,2-diol despite the carbon substituent at C-1, and probably yields a diol-epoxide since it binds to DNA in vitro and in vivo and is mutagenic and carcinogenic.[54] By contrast, the inactive 1-methyl-17-ketone (59) gives little evidence of a 3,4-diol, being converted instead to its 4-phenol and derivatives of the latter. Presumably, the initially formed 3,4-

oxide rapidly isomerizes to this phenol before it can be enzymatically opened to give the *trans*-diol.[48] Thus, consideration of the metabolism of the 11-methyl, 11-12-dimethyl, 7-methyl, and 1,11-methano-17-ketones accounts reasonably for their carcinogenicity, and of the 1-methyl and 6-methyl-17-ketones for their lack of this property.

Of the compounds so far studied, this leaves the unsubstituted parent ketone (4) and its 12-methyl homologue (50), both of which are inactive although they yield *trans*-diequatorial-3,4-diols on metabolism. They both also bind to DNA after microsomal activation in vitro to give nucleoside adducts similar to those given by the strongly carcinogenic 11-methyl isomer (47).[55] However, whereas on borate chromatography the 12-methyl adduct elutes early like the 11-methyl adduct, suggesting that it is derived from an *anti*-diol-epoxide, the retention time of the nucleoside adduct derived from the parent ketone is unchanged. This implies that it is derived from a *syn*-diol epoxide (i.e., a 3α,4β-diol 1β,2β-oxide). This difference is unexpected and difficult to account for, but again, by analogy with benzo[a]pyrene, a *syn* diol-epoxide would be expected to be more chemically reactive and less biologically active than its *anti* isomer.[56,57] In keeping with this expectation, the unsubstituted ketone fails to bind to the DNA of mouse skin after topical application, thus accounting for its lack of carcinogenicity in this species. It is about half as potent a mutagen as its 11-methyl derivative in *S. typhimurium*, but much less active in mammalian V79 Chinese hamster cells.[58] Also, it fails to increase the frequency of sister chromatid exchanges in human lymphocytes, while the 11-methyl-17-ketone is active in this test.[59]

The 12-methyl-17-ketone (50) binds to DNA in vitro to about 15% of the extent of its 11-methyl isomer (47); in vivo (mouse skin DNA after topical application of 1,000 nmol) the figure is higher, about 30%, although the 12-methyl compound is not carcinogenic. Topical application of the 11-methyl isomer at a dose (340 nmol), which would lead to the *same* amount of binding to the DNA of the treated skin as given by the 12-methyl-17-ketone at 1,000 nmol, yields a skin tumor incidence of over 50%. The answer to this paradox may lie in the persistence of these adducts in the DNA, for while the 12-methyl adducts appear to possess a halflife in skin of around 3.5 days, loss of the 11-methyl adducts cannot be measured above the normal rate of DNA turnover in the skin (halflife about 7 days). Moreover, following systemic initiation by injection of this carcinogen, DNA adducts are lost rapidly from the liver (2.5 days), which is not a target for this carcinogen, whereas adduct loss from the lung, which is a target, is similar to that from the dorsal skin after topical application.[60]

Thus, at least three lines of defense against mutation induced by carcinogens binding to DNA in mouse skin have so far been identified:

1. Correct metabolism of the original compound, including correct stereochemical configurations in intermediate metabolites
2. Sufficient stability of the reactive metabolite so that it can survive transport to the nuclear DNA in the mammalian cell and react covalently with it there
3. Persistence of the DNA lesion for at least one round of DNA replication in order to fix the mutation by causing a base change in the newly formed DNA

Undoubtedly other defenses will come to light.

VIII. PHYSICAL AND CHEMICAL PROPERTIES OF CYCLOPENTA[a]PHENANTHRENES AND THEIR INFLUENCE ON BIOLOGICAL ACTIVITY

While the metabolism and DNA interactions of cyclopenta[a]phenanthrenes discussed in the previous section are undoubtedly crucial in determining their mutagenicity and carcin-

Table 16
MELTING-POINTS AND ULTRAVIOLET SPECTRAL MAXIMA ALONGSIDE CARCINOGENICITY AND MUTAGENICITY OF THE EIGHT MONOMETHYL DERIVATIVES OF 15,16-DIHYDROCYCLOPENTA[a]PHENANTHREN-17-ONE (4)

17-Ketone	Melting point (°C)	λmax	Carcinogenicity	Mutagenicity
(4) Unsubstituted	200—201	265(4.89)	<1	9.5
(59) 1-Methyl	189—190	266(4.84)	1	<0.2
(60) 2-Methyl	221—222	267(5.05)	<1	<0.2
(61) 3-Methyl	203—204	267(5.02)	<1	<0.2
(62) 4-Methyl	265—266	266(4.98)	<1	<0.2
(63) 6-Methyl	211—212	266(5.06)	<1	<0.2
(46) 7-Methyl	198—199	268(4.79)	10	12
(47) 11-Methyl	171—172	264(4.83)	46	22
(50) 12-Methyl	233	268(4.85)	<1	<0.2

Note: Spectral maxima expressed in nm with $\log_{10}$ (molar extinction coefficient) shown in parentheses. Carcinogenicity expressed in terms of the Iball index A shown in Table 11. Mutagenicity expressed in revertant colonies per nmole.

ogenicity, these biochemical properties must themselves be determined by the structures of the original molecules. It is therefore tempting to compare the chemical, physico-chemical, and physical attributes of these molecules with a view to illuminating the marked biological differences observed. If our attention is confined to an isomeric series such as the positional methyl isomers of 15,16-dihydrocyclopenta[a]phenanthren-17-one, a number of complications are eliminated because, as expected, these compounds are extremely similar. They are all crystalline solids with melting points in the range 170 to 260°, and all possess similar spectral properties; Table 16 illustrates this alongside the remarkable diversity in biological properties associated with these molecules. Unlike polycyclic hydrocarbons, these polycyclic ketones are not appreciably soluble in hexane; they are, however, moderately soluble in aromatic hydrocarbon solvents and in alcohols, and readily soluble in dipolar solvents such as chloroform and dimethyl sulfoxide. Their solubility in water is of the order of 0.5 μg/mℓ, but is somewhat enhanced by the presence of protein as in tissue culture media. No quantitative comparison of their solubilities in organic solvents has been made, but from the virtual identity of their chromatographic properties in a number of different systems, it is clear that these are closely similar for the isomeric members of this series. It therefore seems likely that the tissue distribution of these isomers will not be dissimilar. The carcinogenic 11-methyl isomer resembles benzo[a]pyrene in its ability to initiate skin tumors in mice, but unlike the latter does not frequently give rise to local sarcomas after injection.[41] Also, in an ongoing experiment with Sprague-Dawley rats, injection of this ketone has given no local sarcomas, but a number of animals have developed leukaemia and skin tumors on the face and trunk.[61] This compound is less lipophilic than hydrocarbons such as benzo[a]pyrene, and it is therefore probably cleared from the injection site more quickly, especially when oil is employed as the vehicle.

The presence of the 11-methyl group in this molecule (47) does not appear to have any marked effect on its chemical reactions, as demonstrated by several careful comparison of it with its unsubstituted parent 17-ketone (4) (Figure 21). On oxidation[62] with chromium trioxide in acetic acid, both are largely converted into the 6,7-quinones (102a,b); the *cis*-6,7-diols (103a,b) are produced with osmium tetroxide, and undergo acid-catalyzed dehy-

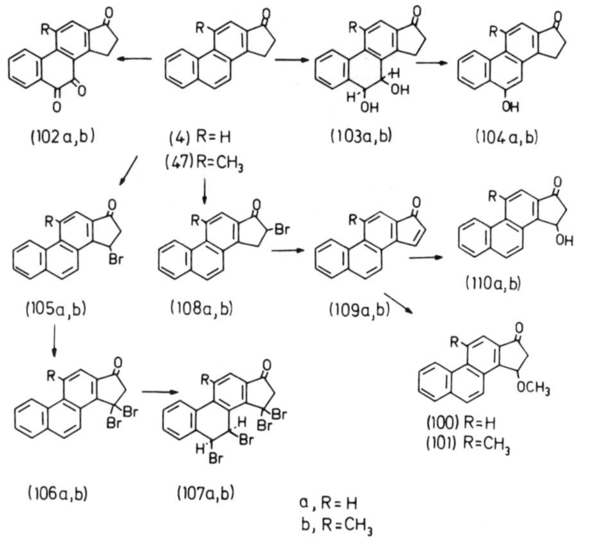

FIGURE 21. Some chemical reactions of 15,16-dihydrocyclopenta[a]phenanthren-17-one (4) and its carcinogenic 11-methyl derivative (47). In all these reactions only minor differences in reaction rates and product yields were noted between the two compounds.

dration to give mostly the 6-phenols (104a,b). By contrast, electrophilic bromination[63] with bromine in acetic acid or chloroform leads exclusively to attack at the benzylic methylene group in ring D to form the 15-bromo (105a,b) and the 15,15-dibromo (106a,b) derivatives. Further bromination then occurs by *addition* to the 6,7-double bond to yield the 6,7-dihydro-6,7,15,15-tetrabromides (107a,b).[64] Free-radical bromination with N-bromo-succinimide in the presence of visible light occurs exclusively at C-16, and treatment of the 16-bromides (108a,b) with trimethylamine causes facile dehydrobromination to yield the 15,16-enes (109a,b).[63,26] Acid-catalyzed addition of water or methanol to the latter furnishes the 15-hydroxy (110a,b) and 15-methoxy (100,101) derivatives.[32,26] In all these reactions only minor differences in reaction rates and product yields were noted between the parent unsubstituted-17-ketone (4) and its 11-methyl homologue (47), emphasizing the considerable similarity between them.

Bigger differences are seen when the stability of the hydrogen atoms at C-16 adjacent to the 17-carbonyl group in these compounds is examined. This can be achieved by labeling this position with tritium by exchange with tritiated water, and then measuring the rate of hydroxide-catalyzed detritiation.[65] Some second-order rate constants ($k^{T}_{OH^-}$) are shown in Table 17, alongside the nuclear magnetic resonance signals for H-15 in these 17-ketones. The effect of the methyl group in lowering the rate of detritiation decreases as expected with its distance from ring D. The rates for the parent ketone (4) and its 1-methyl derivative are essentially the same, some 20% higher than the 3-, 4-, and 11-methyl isomers; the effect is more pronounced for the 6-methyl compound and is at its maximum for the 12-methyl isomer, where the rate is nearly one-third of that of the parent ketone. The rate for the 1,11-methano-17-ketone is also low. The 7-methyl isomer is, however, anomalous, the detritiation rate being approximately doubled compared with the parent compound. This is reflected in the NMR chemical shift of the C-15 protons which are deshielded 0.42 to 0.46 ppm in this compound in comparison with the other methyl isomers.

The reason for these anomalies becomes clear when the three-dimensional structures of these compounds, as disclosed by X-ray crystallography, are considered.[38,66] A total of 12 compounds, shown in Figure 22, have now been studied in this way, and an interesting

Table 17
SECOND-ORDER RATE CONSTANT ($10^2k^T_{OH^-}$ - $dm^3mol^{-1}sec^{-1}$) FOR HYDROXIDE-CATALYZED DETRITIATION AT C-16 IN [16-^{3}H]-17-KETONES IN 90:10 (v/v) WATER-DIOXANE AT 298.2°K, AND NMR H-15 RESONANCE SIGNALS (δ) MEASURED IN $CDCl_3$ IN ppm FROM TETRAMETHYLSILANE

17-Ketone	$k^T_{OH^-}$	δ
(4) Unsubstituted	1.83 ± 0.14	3.28
(59) 1-Methyl	1.85 ± 0.05	3.25
(61) 3-Methyl	1.42 ± 0.02	3.25
(62) 4-Methyl	1.47 ± 0.19	3.26
(63) 6-Methyl	1.15 ± 0.14	3.27
(46) 7-Methyl	3.17 ± 0.08	3.70
(47) 11-Methyl	1.47 ± 0.06	3.24
(50) 12-Methyl	0.67 ± 0.07	3.28
(68) 1,11-Methano	0.87 ± 0.07	3.25

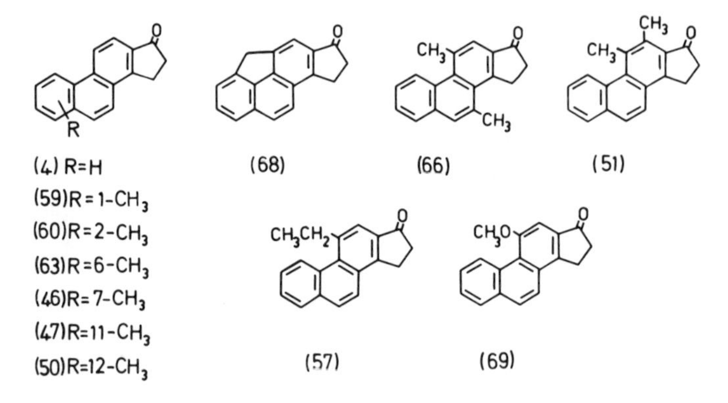

FIGURE 22. 17-Oxocyclopenta[a]phenanthrenes studied by x-ray crystallography.

correlation with carcinogenicity has become apparent. The original publications should be consulted for full tables of bond lengths and angles, and other crystallographic details; only the relevant ring A/C dihedral angles and exocyclic bay-region bond angles are shown here in Table 18. Compounds lacking a substituent at C-11 are essentially planar (A/C dihedral angle <5°), as expected. Introduction of a methyl group at this position causes this angle to increase to 12.5° in compound (47), to relieve the resulting steric strain in the bay region, and this compound is a strong carcinogen. The A/C angle in the carcinogenic 11,12-dimethyl-17-ketone (50) is similar (13.6°), and this angle reaches a maximum in the 7,11 dimethyl isomer (66), which is also the most potent carcinogen so far prepared in this series. Its A/C dihedral angle (20.6°) is not far short of that (23°) for 7,12-dimethylbenz[a]anthracene,[67] which is an extremely potent carcinogen. The A/C angle for the weakly carcinogenic 11-ethyl-17-ketone (57) is less (7.3°), and the weak to moderately active 7-methyl (46), 1,11-

Table 18
RINGS A/C DIHEDRAL ANGLES AND SUM OF THE EXOCYCLIC BAY-REGION BOND ANGLES [C(1)–C(10)–C(9) PLUS C(10)–C(9)–C(11)] (°) COMPARED WITH THE IBALL INDEXES (A, TABLE 12) FOR 15,16-DIHYDROCYCLO PENTA[a]PHENANTHREN-17-ONE(4) AND SOME OF ITS DERIVATIVES

17-Ketone	A/C Dihedral angle	Sum of exocyclic bay-region bond angles	Iball index
(4) Unsubstituted	2.6	245.5	<1
(59) 1-Methyl	4.9	249.5	1
(60) 2-Methyl	4.2	244.9	<1
(63) 6-Methyl	1.5	244.0	<1
(46) 7-Methyl	3.9	242.5	10
(47) 11-Methyl	12.5	249.6	46
(50) 12-Methyl	2.6	245.1	<1
(68) 1,11-Methano	3.9	220.6	16
(66) 7,11-Dimethyl	20.6	246.4	>50[a]
(51) 11,12-Dimethyl	13.6	249.0	30
(57) 11-Ethyl	7.3	249.2	8
(69) 11-Methoxy	1.9	249.2	25

[a] Incomplete experiment, see Table 7.

methano (68), and 11-methoxy (69) 17-ketones are essentially flat. Steric strain in the two former compounds is shown, however, in reduction of their exocyclic bay-region bond angles [(C-1)–C(10)–C(9) plus C(10)–C(9)–C(11)]. The 1,11-methylene bridge in (68) causes the whole molecule to bow in towards the bridge, so that each bay-region angle is about 10° less than that ideal for sp^2 hybridization, and 26.2° less than the average sum for the other eleven compounds. This molecule also has an exceptionally short C(9)–C(10) bond and long C(6)–C(7) bond. In the 7-methyl-17-ketone, the sum of these bond angles is 4.2° less than the average, resulting in a nonbonding distance of 1.94 Å for H(1)- - - H(11), the shortest observed. The contact distances between the hydrogens at C-15 and those of the methyl group are very short (1.82 and 1.84 Å), and cannot be increased by out-of-plane deformation because the former are on either side of the molecular plane containing the methyl group, and movement in either direction would further shorten one of them. These structural constraints are sufficiently large to alter the electronic distribution in the 1,11-methano-17-ketone, so that its ultra-violet spectrum, λ_{max} 266.5 (4.77), 277(4.76), is appreciably different from related compounds lacking the bay-region bridge. Molecular strain in the 7-methyl-17-ketone was suggested by the difficulties in its synthesis, and gives rise to the enhancement in reactivity at C-16 and the NMR deshielding at C-15 already observed.

There therefore seems to be a rough correlation between molecular strain and carcinogenicity in this series; Newman of course has also noted a similar relationship in the benzanthracene series.[68] However, the way in which these two properties are interrelated is quite obscure at the present time. There is one exception to this general rule among the compounds shown in Figure 22; the 11-methoxy-17-ketone (69) is a fairly strong carcinogen, but is planar and its bay-region bond angles are normal. The ether oxygen in the molecular plane of this compound is only 1.98 Å from the hydrogen atom at C-1, suggesting that this contact is attractive. This is supported by the NMR spectrum of this compound, in which H-1 resonates at δ 0.2 compared with δ 1.2 for the 11-methyl-17-ketone (47). Biologically, there are marked differences between the 11-alkyl and 11-alkoxy series. Thus, whereas the

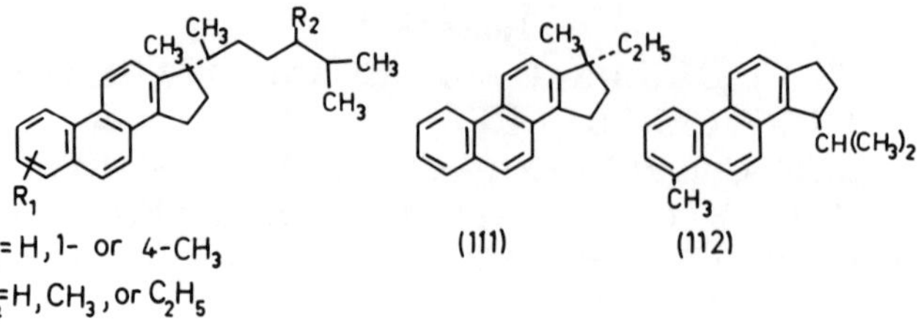

R_1 = H, 1- or 4-CH_3
R_2 = H, CH_3, or C_2H_5

(111) (112)

FIGURE 23. Some cyclopenta[a]phenanthrenes recently identified in natural sources.

11-ethyl-17-ketone (57) is much less active than its 11-methyl isomer, there is less difference between the 11-methoxy (69) and 11-ethoxy-17-ketones (70) (see Table 6). Moreover, introduction of a methyl group at C-7 into the 11-methyl-17-ketone augments its activity, while introduction of this group into the 11-methoxy-17-ketone (to give compound 93, Table 4) has the reverse effect. It therefore seems possible that 11-alkyl and 11-alkoxy groups may mediate their activating effect in quite different manners.

IX. CONCLUSION

Interest in cyclopenta[a]phenanthrenes was revived about 20 years ago by the theoretical possibility of their endogenous formation by aberrant estrogen metabolism. Substantial progress has since been made both in establishing the chemistry and metabolic transformation of compounds of this type, and in investigating some of their biological properties. Structure/activity relationships have been rationalized to some extent, but much more remains still to be done in this area. In the intervening years there have, however, been no reports of the occurrence of cyclopenta[a]phenanthrenes in mammalian tissues of excretion products, although it is improbable that sensitive, modern techniques have yet been applied to this end. Of course Diels hydrocarbon (5) and other simple cyclopenta[a]phenanthrenes have long been known[69] among the pyrolysis products of steroids and sterols. For example, this hydrocarbon occurs together with the phenanthrene derivative of cholesterol in edible oils that have been overheated, where they are formed by aromatization of the natural sterols, making the oil fluorescent. Steroid aromatization can also be brought about by vigorous dehydrogenation with high potential quinones,[70] but again, elevated temperatures are required. It is therefore of considerable interest with Wilk and Taupp[71] discovered that 16,17-dihydro-17-*iso*pentyl-15*H*-cyclopenta[a]phenanthrene, obtainable from selenium dehydrogenation of cholesterol at 350°C, was also formed by simply exposing this sterol, adsorbed on silica gel, to iodine vapor at ambient temperature. These authors considered that the oxidation potential in the adsorbed state was substantially lower than that normally required for dehydrogenation in solution, and that this might provide a model for biological dehydrogenation of steroids by quinones naturally present in living cells.

Recently, the application of sophisticated modern analytical methods, particularly high-pressure liquid chromatography and gas chromatography — mass spectrometry with computer-assisted peak sorting, has been used to demonstrate the presence of several cyclopenta[a]phenanthrenes in a remarkably wide range of natural environmental sources. Thus, the nine cyclopenta[a]phenanthrenes derived from sterols[72] shown in Figure 23, as well as the parent hydrocarbon (3)[73,74] and its 17-ethyl-17-methyl derivative (111),[73] have been identified in oil shales in a number of different locations. Another hydrocarbon, namely 16,17-dihydro-15-*iso*-propyl-4-methyl-15*H*-cyclopenta[a]phenanthrene (112), has been identified in river and lake sediments, again from locations as widely separated as Switzerland,[75] the Adirondack lakes, U.S.A.,[76] and the Amazon river.[77] It is thought that this hydrocarbon might arise from triterpenes such as lupeol by microbiological degradation and dehydrogenation. All these discoveries have been made within the last few years, demonstrating that cyclopenta[a]phenanthrenes are much more widely distributed than was hitherto appreciated.

It is now well established that a number of chemical carcinogens occur in various plants and microorganisms, and there seems no *a priori* reason why animals should not also be capable of producing them. Indeed, it is now known that nitrosamines can be formed in vivo by the reaction of dietary nitrite with amines in the gastrointestinal tract, although whether this represents a significant human health hazard is apparently still in question. It still seems entirely possible that small amounts of a carcinogen might be formed as a result of incorrect steroid (especially estrogen) metabolism, but that this has so far remained undetected. We now have, as a result of 50 years of animal research, extensive knowledge of the very long latent periods encountered for tumor induction with a weak carcinogen, or small doses of a potent carcinogen, and these results can be extrapolated to humans with some confidence. Thus, many human tumors which occur predominantly in late middle to old age would be compatible with the lifelong action of trace amounts of an endogenously formed carcinogen. This idea, so popular 40 years ago, has become neglected, although there were persistent reports at that time that various tissue extracts were carcinogenic in laboratory rodents (reviewed by Steiner).[78] The time now seems ripe to apply modern analytical methods to search for endogenously produced carcinogens of low molecular weight, especially those that might have risen from steroids. These ideas have recently been elaborated in a book.[79]

ACKNOWLEDGMENTS

Research on cyclopenta[a]phenanthrenes at the Imperial Cancer Research Fund Laboratories has been in progress now for nearly 20 years, following Marrian's original suggestion. Numerous people whose names appear in the papers quoted have been involved, and have contributed many of the ideas and most of the work. The present author wishes to express his thanks to all of them for their diligence and skill, and in particular to Dr. T. S. Bhatt, who is responsible for most of the animal results. We would also like to remember our colleague Dr. D. C. Livingston, whose premature death in 1983 sadly deprived us of his companionship and sustained collaboration.

REFERENCES

1. **Clar, E.,** Zur kennis merkerniger aromatischer Kohlenwasserstoffe und ihr abkommlinger. 1 Mitt.: Dibenzanthracene und ihr chinone, *Ber. Chem. Ges.*, 62, 350, 1929.
2. **Kennaway, E. L. and Heiger, I.,** Carcinogenic substances and their fluorescence spectra, *Br. Med. J.*, (1), 1044, 1930.
3. **Cook, J. W.,** Discussion on experimental production of malignant tumours, *Proc. R. Soc. London Ser. B*, 113, 273, 1933.
4. **Cook, J. W. and Hazelwood, G. A. D.,** The conversion of a bile acid into a hydrocarbon derived from 1:2-benzanthracene, *Chem. Ind. (London)*, 11, 758, 1933.
5. **Wieland, H. and Dane, E.,** Untersuchungen uber die Konstitution der Gallensauren, *Z. Physiol. Chem.*, 219, 240, 1933.
6. **Fieser, L. F.,** *Chemistry of Natural Products Related to Phenanthrene,* Reinhold, New York, 1936, 86.
7. **Inhoffen, H. H.,** The relationship of natural steroids to carcinogenic aromatic compounds, in *Progress in Organic Chemistry*, Vol. 2, New York, 1953, 131.
8. **Dunlap, C. E. and Warren, S.,** The carcinogenicity of some new derivatives of aromatic hydrocarbons. I. Compounds related to chrysene, *Cancer Res.*, 3, 606, 1943.
9. **Bachmann, W. E., Kennaway, E. L., and Kennaway, N. M.,** The rapid production of tumours by two new hydrocarbons, *Yale J. Biol. Med.*, 11, 97, 1938.
10. **Butenandt, A. and Suranyi, L. A.,** Uberfuhrung von Steroidhormonen in Methylhomologe des Cyclopentenophenanthrens, *Ber. Chem. Ges.*, 75, 597, 1942.

11. **Robinson, R.**, Experiments on the synthesis of substances related to the sterols. XXI. A new synthesis of derivatives of ketocyclopentenophenanthrene, *J. Chem. Soc.*, 1390, 1938.
12. **Butenandt, A., Dannenberg, H., and von Dresler, D.**, Methylhomologe des 1,2-Cyclopentenophenanthrens. II. Mitteilung: Synthese des 3-methyl, 4-methyl, und 3,4-dimethyl-1,2-cyclopentenophenanthrens, *Z. Naturforsch.*, 1, 151, 1946.
13. **Butenandt, A., Dannenberg, H., and von Dresler, D.**, Methylhomologe des 1,2-cyclopentenophenanthrens. V. Mitteilung: Synthese des 10-methyl-1,2-cyclopentenophenanthren, *Z. Naturforsch.*, 4b, 69, 1949.
14. **Hawthorne, J. R. and Robinson, R.**, Experiments on the synthesis of substances related to the sterols. XIII. Hydrocyclopentenophenanthrene derivatives, *J. Chem. Soc.*, 763, 1936.
15. **Butenandt, A., Dannenberg, H., and von Dresler, D.**, Methylhomologe des 1,2-Cyclopentenophenanthrens. IV. Mitteilung: Synthese des 6-methyl und 3,6-dimethyl-1,2-cyclopentenophenanthrens, *Z. Naturforsch.*, 1, 227, 1946.
16. **Butenandt, A., Dannenberg, H., and von Dresler, D.**, Methylhomologe des 1,2-Cyclopentenophenanthrens. VI. Mitteilung: Synthese des 8-methyl und 3,8-dimethyl-1:2-cyclopentenophenanthrens, *Z. Naturforsch.*, 4b, 77, 1949.
17. **Butenandt, A., Dannenberg, H., Bieneck, E., and Steidle, W.**, Methylhomologe des 1,2-Cyclopentenophenanthrens. VII. Mitteilung: Synthese des 5-methyl-1,2-cyclopentenophenanthrens, *Z. Naturforsch.*, 5b, 405, 1950.
18. **Butenandt, A. and Dannenberg, H.**, Untersuchungen uber die Krebszengende Wirksamkeit der Methylhomologe des 1,2-Cyclopentenophenanthrens, *Arch. Geschwulstforsch.*, 6, 1, 1953.
19. **Dannenberg, H.**, Uber Beziehungen zwishen Steroiden und Krebserzeugenden Kohlenwasserstoffen, *Z. Krebsforsch.*, 63, 523, 1960.
20. **Dorfman, R. I. and Unger, F.**, *Metabolism of Steroid Hormones*, Academic Press, New York, 1965, 132.
21. **Loke, K. H., Watson, E. J. D., and Marrian, G. F.**, The isolation of a sixth Kober chromogen from urine of pregnant women and its identification as 18-hydroxyoestrone, *Biochem. J.*, 71, 43, 1959.
22. **Birch, A. J., Jaeger, R., and Robinson, R.**, The synthesis of substances related to the sterols. XLIV. dl-cis-Equilenin, *J. Chem. Soc.*, 585, 1945.
23. **Coombs, M. M. and Jaitly, S. B.**, Potentially carcinogenic cyclopenta[a]phenanthrenes. V. Synthesis of 15,16-dihydro-7-methylcyclopenta[a]phenanthren-17-one, *J. Chem. Soc. C*, 230, 1971.
24. **Coombs, M. M.**, Potentially carcinogenic cyclopenta[a]phenanthrenes. I. A new synthesis of 15,16-dihydro-17-oxocyclopenta[a]phenanthrene and the phenanthrene analogue of 18-noroestrone methyl ether, *J. Chem. Soc. C*, 955, 1966.
25. **Coombs, M. M. and Bhatt, T. S.**, Potentially carcinogenic cyclopenta[a]phenanthrenes. VI. 1,2,3,4-Tetrahydro-17-ketones, *J. Chem. Soc. Perkin Trans. 1*, 1251, 1973.
26. **Coombs, M. M., Hall, M., Siddle, V. A., and Vose, C. W.**, Potentially carcinogenic cyclopenta[a]phenanthrenes. X. Oxygenated derivatives of the carcinogen 15,16-dihydro-11-methylcyclopenta[a]phenanthren-17-one of metabolic interest, *J. Chem. Soc. Perkin Trans. 1*, 265, 1975.
27. **Johnson, W. S. and Peterson, J. W.**, The Stobbe condensation with 1-keto-1,2,3,4,-tetrahydrophenanthrene. A synthesis of 3′-keto-3,4-dihydro-1,2-cyclopentenophenanthrene, *J. Am. Chem. Soc.*, 67, 1366, 1945.
28. **Bachmann, W. E. and Kloetzel, M. C.**, Phenanthrene derivatives. VII. The cyclisation of β-phenanthrylpropionic acids., *J. Am. Chem. Soc.*, 59, 2207, 1937.
29. **Riegel, B., Siegel, S., and Kritchevsky, D.**, The synthesis of 3-alkyl-1,2-cyclopentenophenanthrenes, *J. Am. Chem. Soc.*, 70, 2950, 1948.
30. **Coombs, M. M., Jaitly, S. B., and Crawley, F. E. H.**, Potentially carcinogenic cyclopenta[a]phenanthrenes. IV. Synthesis of 17-ketones by the Stobbe condensation, *J. Chem. Soc. C*, 1266, 1970.
31. **Ribeiro, O., Hadfield, S. T., Clayton, A. F., Vose, C. W., and Coombs, M. M.**, Potentially carcinogenic cyclopenta[a]phenanthrenes. II. Syntheses of the 1-methyl, 1,11-methano, and 7,11-dimethyl derivatives of 15,16-dihydrocyclopenta[a]phenanthren-17-one, *J. Chem. Soc. Perkin Trans. 1*, 87, 1983.
32. **Bhatt, T. S., Hadfield, S. T., and Coombs, M. M.**, Carinogenicity and mutagenicity of some alkoxy cyclopenta[a]phenanthren-17-ones: effect of obstructing the bay region, *Carcinogenesis*, 3, 677, 1982.
33. **Coombs, M. M. and Croft, C. J.**, Carcinogenic cyclopenta[a]phenanthrenes, *Prog. Tumor Res.*, 11, 69, 1969.
34. **Coombs, M. M.**, Potentially carcinogenic cyclopenta[a]phenanthrenes. II. Derivatives containing further unsaturation in ring-D, *J. Chem. Soc. C*, 965, 1966.
35. **Coombs, M. M., Bhatt, T. S., and Croft, C. J.**, Correlation between carcinogenicity and chemical structure in cyclopenta[a]phenanthrenes, *Cancer Res.*, 33, 832, 1973.
36. **Coombs, M. M., Bhatt, T. S., Hall, M., and Croft, C. J.**, The relative carcinogenic activities of a series of 5-methylchrysene derivatives, *Cancer Res.*, 34, 1315, 1974.

37. **Coombs, M. M. and Bhatt, T. S.,** Lack of initiating agents in mutagens which are not carcinogenic, *Br. J. Cancer,* 38, 148, 1978.
38. **Kashino, S., Zacharias, D. E., Peck, R. M., Glusker, J. P., Bhatt, T. S., and Coombs, M. M.,** Bay-region distortions in cyclopenta[a]phenanthrenes, *Cancer Res.,* 46, 1817, 1986.
39. **Hadfield, S. T., Bhatt, T. S., and Coombs, M. M.,** The biological activity and activation of 15,16-dihydro-1,11-methanocyclopenta[a]phenanthren-17-one, a carcinogen with an obstructed bay region, *Carcinogenesis,* 5, 1485, 1984.
39a. **Cox, J.,** Private communication.
40. **Abbott, P. J.,** Strain-specific tumorigenesis in mouse skin induced by the carcinogen, 15,16-dihydro-11-methylcyclopenta[a]phenanthren-17-one, and its relation to DNA adduct formation and persistence, *Cancer Res.,* 43, 2261, 1983.
41. **Coombs, M. M., Bhatt, T. S., and Young, S.,** The carcinogenicity of 15,16-dihydro-11-methylcyclopenta[a]phenanthren-17-one, *Br. J. Cancer,* 40, 914, 1979.
42. **Bhatt, T. S., Coombs, M., and O'Neill, C.,** Biogenic silica fibre promotes carcinogenesis in mouse skin, *Int. J. Cancer,* 34, 519, 1984.
43. **Coombs, M. M., Dixon, C., and Kissonerghis, A.-M.,** Evaluation of the mutagenicity of compounds of known carcinogenicity belonging to the benz[a]anthracene, chrysene, and cyclopenta[a]phenanthrene series using Ames' test, *Cancer Res.,* 36, 4525, 1976.
44. **Coombs, M. M., Hall, M., Siddle, V. A., and Vose, D. W.,** Identification of monohydroxy metabolites of 15,16-dihydrocyclopenta[a]phenanthren-17-one and its carcinogenic 11-methyl homolog produced by rat liver preparations *in vitro, Arch. Biochem. Biophys.,* 172, 434, 1976.
45. **Coombs, M. M., Kissonerghis, A.-M., Allen, J. A., and Vose, C. W.,** Identification of the proximate and ultimate forms of the carcinogen 15,16-dihydro-11-methylcyclopenta[a]phenanthren-17-one, *Cancer Res.,* 39, 4160, 1979.
46. **Coombs, M. M., Bhatt, T. S., Kissonerghis, A.-M., and Vose, C. W.,** Mutagenic and carcinogenic metabolites of the carcinogen 15,16-dihydro-11-methylcyclopenta[a]phenanthren-17-one, *Cancer Res.,* 40, 882, 1980.
47. **Hadfield, S. T., Abbott, P. J., Coombs, M. M., and Drake, A. F.,** The effect of methyl substituents on the *in vitro* metabolism of cyclopenta[a]phenanthren-17-ones: implication for biological activity, *Carcinogenesis,* 5, 1395, 1984.
48. **Coombs, M. M., Russell, J. C., Jones, J. R., and Ribeiro, O.,** A comparative examination of the *in vitro* metabolism of five cyclopenta[a]phenanthrenes of varying carcinogenic potential, *Carcinogenesis,* 6, 1217, 1985.
49. **Coombs, M. M. and Bhatt, T. S.,** High skin tumour initiating activity of the metabolically derived *trans*-3,4-dihydro-3,4-diol of the carcinogen 15,16-dihydro-11-methylcyclopenta[a]phenanthren-17-one, *Carcinogenesis,* 3, 449, 1982.
50. **Abbott, P. J. and Coombs, M. M.,** DNA adducts of the carcinogen, 15,16-dihydro-11-methylcyclopenta[a]phenanthren-17-one, *in vivo* and *in vitro*: high pressure liquid chromatographic separation and partial characterisation, *Carcinogenesis,* 2, 629, 1981.
51. **Weibers, J. L., Abbott, P. J., Coombs, M. M., and Livingston, D. C.,** Mass spectral characterisation of the major DNA-carcinogen adduct formed from the metabolically activated carcinogen 15,16-dihydro-11-methylcyclopenta[a]phenanthren-17-one, *Carcinogenesis,* 2, 637, 1981.
52. **Yang, S. K., McCourt, D. W., Roller, P. P., and Gelboin, H. V.,** Enzymatic conversion of benzo[a]pyrene leading predominantly to the diol-epoxide r-7,t-8-dihydroxy-t-9,10-oxy-7,8,9,10-tetrahydrobenz-[a]pyrene through a single enantiomer of r-7,t-8-dihydroxy-7,8-dihydroxy-7,8-dihydrobenzo[a]pyrene, *Proc. Natl. Acad. Sci. U.S.A.,* 73, 2594, 1976.
53. **Coombs, M. M., Bhatt, T. S., Livingston, D. C., Fisher, S. W., and Abbott, P. J.,** Chemical structure, metabolism, and carcinogenicity in the cyclopenta[a]phenanthrene series, in *Polynuclear Aromatic Hydrocarbons: Chemical Analysis and Biological Fate,* Cook, M. and Dennis, A. J., Eds., Battelle Press, Columbus, 1981, 63.
54. **Coombs, M. M., Hadfield, S. T., and Bhatt, T. S.,** 15,16-Dihydro-1,11-methanocyclopenta[a]phenanthren-17-one: a carcinogen with a bridged bay-region, in *Polynuclear Aromatic Hydrocarbons: Formation, Metabolism and Measurement,* Cook, M. W. and Dennis, A. J., Eds., Battelle Press, Columbus, 1982, 351.
55. **Russell, J. C., Bhatt, T. S., Jones, J. R., and Coombs, M. M.,** Comparison of the binding of some carcinogenic and non-carcinogenic cyclopenta[a]phenanthrenes to DNA *in vitro* and *in vivo, Carcinogenesis,* 6, 1223, 1985.
56. **Buening, M. K., Wislocki, P. G., Levin, W., Yagi, H., Thakker, D. R., Akagi, H., Korreeda, M., Jerina, D. M., and Conney, A. H.,** Tumorigenicity of the optical enantiomers of the diastereomeric benzo[a]pyrene 7,8-diol-9,10-epoxides in new born mice: exceptional activity of (+)-7β,8α-dihydroxy-9α,10α-epoxy-7,8,9,10-tetrahydrobenzo[a]pyrene, *Proc. Natl. Acad. Sci. U.S.A.,* 75, 5358, 1979.

57. **Slaga, T. L., Braken, W. M., Viaje, A., Levin, W., Yagi, H., Jerina, D. M., and Conney, A. H.,** Comparison of the tumour-initiating activities of benzo[a]pyrene arene oxides and diol epoxides, *Cancer Res.,* 37, 4130, 1977.
58. **Bhatt, T. S., Coombs, M. M., DiGiovanni, J., and Diamond, L.,** Mutagenesis in Chinese hamster cells by cyclopenta[a]phenanthrenes activated by a human hepatoma cell line, *Cancer Res.,* 43, 984, 1983.
59. **Lindahl-Kiessling, K., Bhatt, T. S., Karlberg, I., and Coombs, M. M.,** Frequency of sister chromatid exchanges in human lymphocytes cultivated with a human hepatoma cell line as an indicator of the carcinogenic potency of two cyclopenta[a]phenanthrenes, *Carcinogenesis,* 5, 11, 1984.
60. **Abbott, P. J. and Crew, F.,** Repair of DNA adducts of the carcinogen 15,16-dihydro-11-methylcyclopenta[a]phenanthren-17-one in mouse tissues and its relation to tumour induction, *Cancer Res.,* 41, 4115, 1981.
61. **Bhatt, T. S.,** Private communication, 1985.
62. **Coombs, M. M.,** Potentially carcinogenic cyclopenta[a]phenanthrenes. III. Oxidation studies, *J. Chem. Soc. C,* 2484, 1969.
63. **Coombs, M. M. and Hall, M.,** Potentially carcinogenic cyclopenta[a]phenanthrenes. VIII. Bromination of 17-ketones, *J. Chem. Soc. Perkin Trans. 1,* 2236, 1973.
64. **Coombs, M. M.,** Unpublished data, 1985.
65. **Elvidge, J. A., Jones, J. R., Russell, J. C., Wiseman, A., and Coombs, M. M.,** A kinetic investigation of the hydroxide-catalysed detritiation of various [16-^{3}H]-15,16-dihydrocyclopenta[a]phenanthren-17-ones and related compounds, *J. Chem. Soc. Perkin Trans. 2,* 563, 1985.
66. **Clayton, A. F. D., Coombs, M. M., Hewick, K., McPartlin, M., and Trotter, J.,** X-ray structural studies and molecular orbital calculations (CNDO/2) in a series of cyclopenta[a]phenanthrenes: attempts at correlation with carcinogenicity, *Carcinogenesis,* 4, 1659, 1983.
67. **Glusker, J. P.,** X-ray crystallographic studies on carcinogenic polycyclic aromatic hydrocarbons and their derivatives, in *Polycyclic Hydrocarbons and Cancer,* Vol. 3, Ts'o, P. O. P. and Gelboin, H. V., Eds., Academic Press, New York, 1981, 61.
68. **Newman, M. S.,** Synthesis of 7,11,12-trimethylbenz[a]anthracene, *J. Org. Chem.,* 48, 3249, 1983.
69. **Falk, H. L., Goldfein, S., and Steiner, P. E.,** The products of pyrolysis of cholesterol at 360° and their relation to carcinogens, *Cancer Res.,* 9, 438, 1949.
70. **Dannenberg, H.,** Vollstandige Dehydrierung von Sterinen und Steroiden mit Chinonen, *Synthesis,* 2, 74, 1970.
71. **Wilk, M. and Taupp, W.,** Dehydrierung des cholestins in activert adsorbierten Zustand unter normalbedingungen, ein Beitrag zur Frage Endogenese carcinogener, polycyclischer Kohenwasserstoffe, *Z. Naturforsch.,* 24B, 16, 1969.
72. **Ludwig, B., Hussler, G., Wehung, P., and Albrecht, P.,** C_{26}-C_{29} triaromatic steroid derivatives in sediments and petroleum, *Tetrahedron Lett.,* 22, 3313, 1981.
73. **Mackenzie, A. S., Hoffmann, C. F., and Maxwell, J. R.,** Molecular parameters of maturation in Toarcian shales, Paris Basin, France. III. Changes in aromatic steroid hydrocarbons, *Geochim. Cosmochim. Acta,* 45, 1345, 1981.
74. **Chaffee, A. L. and Johns, R. B.,** Polycyclic aromatic hydrocarbons in Australian coals. I. Angularly fused pentacyclic tri- and tetra-aromatic components of Victorian brown coal, *Geochim. Cosmochim. Acta,* 47, 2142, 1983.
75. **Wakeham, S. G., Schaffer, C., and Giger, W.,** Polycyclic aromatic hydrocarbons in recent sediment. II. Compounds derived from biogenic precursors during early diagenesis, *Geochim. Cosmochim. Acta,* 44, 415, 1980.
76. **Tan, Y. L. and Heit, M.,** Biogenic and abiogenic polynuclear aromatic hydrocarbons in sediments from two remote Adirondack lakes, *Geochim. Cosmochim. Acta,* 45, 2267, 1981.
77. **Laflamme, R. E. and Hites, R. A.,** Tetra- and pentacyclic, naturally occurring, aromatic hydrocarbons in recent sediments, *Geochim. Cosmochim. Acta,* 43, 1687, 1979.
78. **Steiner, P. E.,** Cancer-producing agents from human sources, *Int. Abstr. of Surg.,* 76, 105, 1943.
79. **Coombs, M. M. and Bhatt, T. S.,** *Cyclopenta[a]phenanthrenes,* Cambridge University Press, Cambridge, 1987.

Chapter 4

EFFECTS OF METHYL SUBSTITUTION ON THE TUMORIGENICITY AND METABOLIC ACTIVATION OF POLYCYCLIC AROMATIC HYDROCARBONS

Stephen S. Hecht, Assieh A. Melikian, and Shantu Amin

TABLE OF CONTENTS

I. INTRODUCTION

Polynuclear aromatic hydrocarbons (PAH) are formed during the incomplete combustion of organic matter via radical recombination reactions.[1] These complex reactions lead to unsubstituted as well as methylated PAH, and as a result, mixtures of the parent and methylated PAH are detected in the environment. Human exposure to mixtures of these compounds can occur through inhalation of tobacco smoke and polluted air, through ingestion of certain foods or of polluted water, through contact with soots, oils, or tars, and in certain occupational settings.[2] In assessing the potential contribution of such exposures to the risk for tumor development, it is necessary to know the tumorigenic activities of the individual components of the mixtures. This is particularly true for the PAH and methylated PAH, because small changes in structure lead to major changes in tumorigenic activity. The purpose of this review is to summarize our present knowledge of the effects of methyl substitution on the tumorigenicity of PAH and their enzymatic conversion to tumorigenic or inactive metabolites. The review will consist of three sections. First, we will consider the structure-activity relationships for tumorigenicity of methylated PAH, in all systems having four or five condensed rings (the three-ring systems are described in Chapter 6). We will present a generalization on the structural aspects of methylated PAH favoring tumorigenicity and will note exceptions to this. Secondly, we will consider the metabolic activation and detoxification of 5-methylchrysene, as a typical example of a tumorigenic methylated PAH. We will then attempt to demonstrate the relationship between metabolic activation pathways of methylated PAH and the structural requirements favoring their tumorigenicity.

II. TUMORIGENICITY OF METHYLATED PAH

The structural requirements favoring tumorigenicity of methylated PAH can be generalized, for alternant PAH, as follows: tumorigenic activity is favored when the molecule has a bay-region methyl group and a free *peri* position, both adjacent to an unsubstituted angular ring.[3] The bay regions, *peri* positions, and angular rings of chrysene, as an example, are illustrated in Figure 1. Examples of highly tumorigenic alternant PAH having a bay-region methyl group adjacent to an unsubstituted angular ring are illustrated in Figure 2. All of these compounds are more tumorigenic than the corresponding unsubstituted PAH. However, maximum activity is observed only when the compound has a free *peri* position adjacent to the unsubstituted angular ring. Figure 3 illustrates examples of inhibition of tumorigenicity by substitution of a methyl group at the *peri* position adjacent to the angular rings of some tumorigenic PAH.

We will consider the scope and limitations of these structural requirements in more detail in the discussion which follows on the tumorigenic activities of methyl isomers of each ring system. In this discussion, we will focus on the results of mouse skin bioassays because mouse skin has been the most widely used assay system for PAH tumorigenicity. In particular, the data presented in Tables 1 through 10 have been selected because: (1) they are representative of other published data and (2) many of the studies of comparative activities of methyl isomers have been carried out in a single laboratory under identical conditions. For more detailed data on tumorigenic activities of methylated PAH, including bioassays in other experimental animal systems, the reader is referred to the comprehensive series, "Survey of Compounds Which Have Been Tested for Carcinogenic Activity".[4-12]

Mouse-skin bioassays are generally carried out by either an initiation-promotion protocol (denoted I in the tables) or by a complete carcinogenicity protocol (denoted C). In the initiation-promotion protocol, the shaved back of the mouse is treated with a single dose, or in some cases a series of doses over 1 to 3 weeks, of a solution of the PAH. This is the initiation step. Promotion, which leads to tumor development, is carried out by repetitive

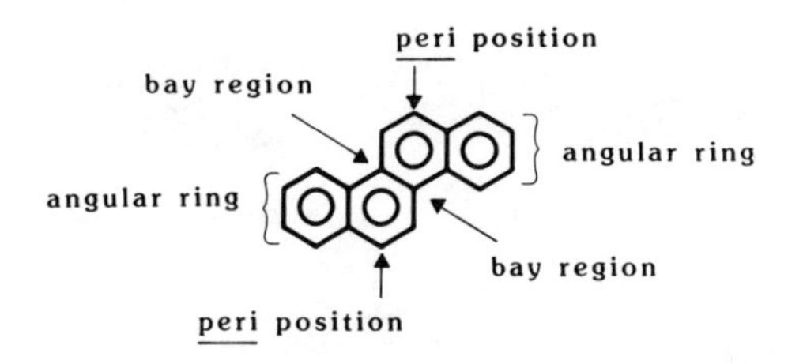

FIGURE 1. Bay regions, angular rings, and *peri* positions of chrysene. (Reprinted with permission from Hecht, S. S., Amin, S., Melikian, A. A., LaVoie, E. J., and Hoffmann, D., *ACS Symp. Ser.*, 283, 85, 1985. Copyright 1985, American Chemical Society.)

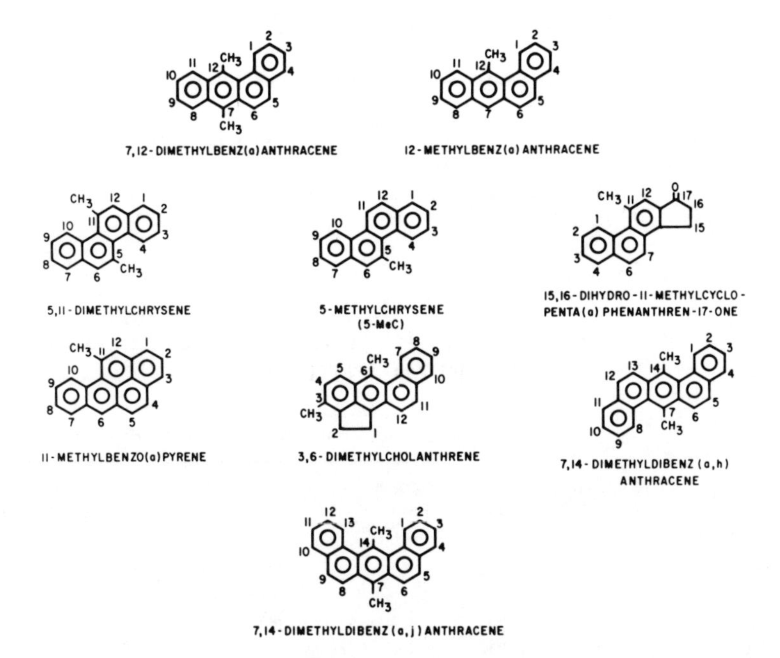

FIGURE 2. Highly tumorigenic methylated PAH having a bay-region methyl group adjacent to an unsubstituted angular ring. All compounds are more tumorigenic than their parent PAH. With the exception of 12-methylbenz[a]anthracene, all compounds are also more tumorigenic than any of the other mono- or dimethyl isomers tested in each ring system. (Reprinted with permission from Hecht, S. S., Amin, S., Melikian, A. A., LaVoie, E. J., and Hoffmann, D., *ACS Symp. Ser.*, 283, 85, 1985. Copyright 1985, American Chemical Society.)

applications of 12-O-tetradecanoylphorbol-13-acetate for 15 to 30 weeks. In this protocol, benign skin tumors are generally observed. In the complete carcinogenicity protocol, the PAH solution is repeatedly applied to the shaved mouse back for 30 to 70 weeks, depending on the compound. Benign and malignant tumors are produced. In many cases, the relative activities of methylated PAH are similar in the two protocol types.

A. Methylbenz[a]anthracenes (MeB[a]A)

Representative mouse-skin tumorigenicity assays of monoMeB[a]A isomers are summarized in Table 1. The strongest tumor initiator is 7-MeB[a]A. The next most active initiators are 12-MeB[a]A and 8-MeB[a]A, followed by 6-MeB[a]A and 9-MeB[a]A. The other isomers and B[a]A itself are either weak tumor initiators or are inactive. The results of the

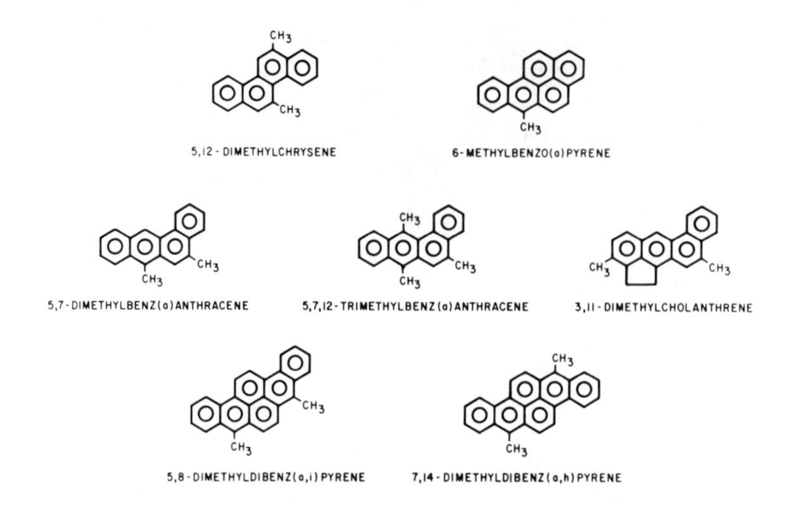

FIGURE 3. Inhibition of tumorigenicity by *peri* methyl substitution. All compounds shown are less tumorigenic than other methyl isomers in the same series or than their parent PAH. (Reprinted with permission from Hecht, S. S., Amin, S., Melikian, A. A., LaVoie, E. J., and Hoffmann, D., *ACS Symp. Ser.*, 283, 85, 1985. Copyright 1985, American Chemical Society.)

Table 1
REPRESENTATIVE MOUSE-SKIN TUMORIGENICITY ASSAYS OF METHYLBENZ[a]ANTHRACENES

			Tumorigenicity on mouse skin				
Parent ring system	**Methyl isomer**	**Strain**	**Protocol type**	**Dose**	**% TBA**	**T/A**	**Ref.**
Benz[a]anthracene	—	CD-1	I	400 nmol	23	0.3	13
	1	CD-1	I	400 nmol	27	0.3	13
	2	CD-1	I	400 nmol	7	0.07	13
	3	CD-1	I	400 nmol	13	0.2	13
	4	CD-1	I	400 nmol	10	0.1	13
	5	CD-1	I	400 nmol	17	0.23	13
	6	CD-1	I	30 nmol	27	0.3	13
				100 nmol	10	0.27	
				400 nmol	33	0.6	
	7	CD-1	I	30 nmol	28	0.55	13
				100 nmol	70	1.6	
				400 nmol	77	4.9	
	8	CD-1	I	30 nmol	17	0.23	13
				100 nmol	23	0.73	
				400 nmol	40	1.03	
	9	CD-1	I	400 nmol	28	0.55	13
	10	CD-1	I	400 nmol	13	0.23	13
	11	CD-1	I	400 nmol	10	0.1	13
	12	CD-1	I	30 nmol	23	0.3	13
				100 nmol	17	0.27	
				400 nmol	52	1.03	
	1	C3H	C	2000 nmol 2 × weekly 91 weeks	0	NG	14
	2	C3H	C	2000 nmol 2 × weekly 91 weeks	0	0	14
	3	C3H	C	2000 nmol 2 × weekly 91 weeks	0	0	14

Table 1 (continued)
REPRESENTATIVE MOUSE-SKIN TUMORIGENICITY ASSAYS OF METHYLBENZ[a]ANTHRACENES

				Tumorigenicity on mouse skin			
Parent ring system	Methyl isomer	Strain	Protocol type	Dose	% TBA	T/A	Ref.
	4	C3H	C	2000 nmol 2 × weekly 91 weeks	0	0	14
	5	C3H	C	2000 nmol 2 × weekly 91 weeks	0	0	14
	6	C3H	C	2000 nmol 2 × weekly 91 weeks	55	NG	14
	7	C3H	C	2000 nmol 2 × weekly 91 weeks	75	NG	14
		Taconic	C	290 nmol 2 × weekly 20 weeks	55	1.3	15
	8	C3H	C	2000 nmol 2 × weekly 91 weeks	60	NG	14
	9	C3H	C	2000 nmol 2 × weekly 91 weeks	5	NG	14
	10	C3H	C	2000 nmol 2 × weekly 91 weeks	0	0	14
	11	C3H	C	2000 nmol 2 × weekly 91 weeks	10	NG	14
	12	C3H	C	2000 nmol 2 × weekly 91 weeks	60	NG	14
	7, 12	CD-1	I	9 nmol	97	8.1	13
		CD-1	I	10 nmol	72	3.0	16
		CD-1	I	1 nmol	10	0.2	16
		CD-1	I	0.1 nmol	0	0	16
		SENCAR	I	10 nmol	100	24	16
		SENCAR	I	1 nmol	95	3.8	16
		SENCAR	I	0.1 nmol	20	0.6	16
	6, 8	CD-1	I	200 nmol	40	0.7	17
	3, 9	CD-1	I	200 nmol	3	0.07	17
	5, 7	CD-1	I	200 nmol	3	0.03	17
	5, 7	SENCAR	I	400 nmol	41	0.6	18
	2, 8	NG	C	0.3% 2 × weekly	0	0	19
	2, 9	Taconic	C	290 nmol 2 × weekly, 48 weeks	13	NG	20
	2, 10	NG	C	0.3% 2 × weekly	0	0	19
	3, 9	NG	C	0.3% 2 × weekly	0	0	19
	3, 10	NG	C	0.3% 2 × weekly	0	0	19
	8, 9	NG	C	0.3% 2 × weekly	80	NG	19
	9, 10	NG	C	0.3% 2 × weekly	40	NG	19
	9, 11	CAF_1	C	0.4% 2 × weekly 37 weeks	3	NG	21
	1, 7, 12	CD-1	I	200 nmol	3	0.03	22
	2, 7, 12	CD-1	I	200 nmol	7	0.10	22
	5, 7, 12	CD-1	I	200 nmol	34	0.40	22
	7, 11, 12	CD-1	I	45 nmol	41	1.1	23
				5 nmol	7	0.07	23
	7, 8, 12	NG	C	0.3% 2 × weekly 17 weeks	15	NG	24
	7, 9, 12	NG	C	0.3% 2 × weekly 29 weeks	55	NG	24

Table 1 (continued)
REPRESENTATIVE MOUSE-SKIN TUMORIGENICITY ASSAYS OF METHYLBENZ[a]ANTHRACENES

			Tumorigenicity on mouse skin				
Parent ring system	Methyl isomer	Strain	Protocol type	Dose	% TBA	T/A	Ref.
	7, 8, 9, 12	NG	C	0.3% 2 × weekly 29 weeks	45	NG	24

Note: I, initiation-promotion. C, complete carcinogenicity. TBA, tumor-bearing animals. T/A, tumors per animal. NG, not given.

complete carcinogenicity assays are generally in agreement with the assays for tumor initiation; 7-MeB[a]A is the most active followed by 12-MeB[a]A, 6-MeB[a]A, and 8-MeB[a]A. Similar results have been obtained upon subcutaneous injection in mice and rats.[4-12]

The high tumorigenicity of 7-MeB[a]A is perhaps the most striking exception to the generalization on the structural requirements favoring tumorigenicity. 7-MeB[a]A has no bay-region methyl group, and is definitely more tumorigenic than 12-MeB[a]A, which has a bay-region methyl group adjacent to an unsubstituted angular ring. Although extensive elegant metabolic studies have been carried out on the MeB[a]A series, as summarized in Chapter 5, the reason for the exceptional activity of 7-MeB[a]A is presently not clearly understood. Although not as striking as 7-MeB[a]A, the tumorigenic activities of 6-MeB[a]A and 8-MeB[a]A are significant, and appear to be comparable to that of 12-MeB[a]A. These two compounds can also be considered as exceptions to the structural generalization cited above.

Among the dimethylB[a]A isomers (diMeB[a]A) tested on mouse skin, 7,12-diMeB[a]A is by far the most tumorigenic. Its tumorigenic activity is remarkable. In CD-1 mice, 9 nmol (2.3 μg) of DMBA induced tumors in 97% of the animals (8.1 tumors/mouse) in an initiation-promotion assay, whereas 400 nmol of B[a]A gave tumors in only 23% (0.3 tumors/mouse). In SENCAR mice, which are more sensitive to skin-tumor induction than are CD-1 mice, a dose of 1 nmol of 7,12-diMeB[a]A (0.25 μg) was still sufficient to induce a tumor incidence of 95% (3.8 tumors/mouse). By comparison, a comparable tumor incidence was obtained with 100 nmol of benzo[a]pyrene.[16] The potent activity of 7,12-diMeB[a]A has been observed in many systems.[4-12] It has been used extensively for the induction of breast tumors in female Sprague-Dawley rats because only a single dose is required.[25] 7,12-DiMeB[a]A fulfills the structural generalization favoring tumorigenicity because it has a bay-region methyl group and free *peri* position, both adjacent to an unsubstituted angular ring. As in 7-MeB[a]A, however, the methyl group at the 7 position definitely has an enhancing effect on its tumorigenicity.

Among the other diMeB[a]A isomers tested on mouse skin, the 6,8-, and 8,9-isomers appear to have relatively high activity, which is perhaps expected based on the initiating activities of 6-, 8-, and 9-MeB[a]A.

B. Methylcholanthrenes

Cholanthrene can be considered as a derivative of B[a]A. 3-Methylcholanthrene has been extensively used as a model compound for carcinogenesis studies and as an inducer of P-450 enzymes. Relatively little data are available on the tumorigenic activities on mouse skin of other methyl- or dimethylcholanthrenes. As indicated in Table 2, the available data are in accord with the structural requirements favoring tumorigenicity. 3,6-Dimethylcholan-

Table 2
REPRESENTATIVE MOUSE-SKIN TUMORIGENICITY ASSAYS OF METHYLCHOLANTHRENES

				Tumorigenicity on mouse skin			
Parent ring system	Methyl isomer	Strain	Protocol type	Dose	% TBA	T/A	Ref.
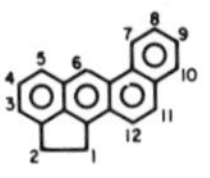		SENCAR	I	100 nmol	97	7.30	26
	3	CD-1	I	15 nmol	50	1.03	23
		CD-1	I	5 nmol	10	0.10	23
		SENCAR	I	100 nmol	97	17.57	26
	3, 6	CD-1	I	5 nmol	57	1.27	23
		CD-1	I	15 nmol	82	3.57	23
	3, 11	SENCAR	I	100 nmol	10	0.2	27

Note: I, initiation-promotion. C, complete carcinogenicity. TBA, tumor-bearing animals. T/A, tumors per animal. NG, not given.

threne, which has a bay-region methyl group adjacent to the 7-10 angular ring, is a more potent tumor initiator than is 3-methylcholanthrene. In contrast, 3,11-dimethylcholanthrene, in which the *peri* position adjacent to the unsubstituted angular ring is occupied, is only weakly tumorigenic or inactive.

C. Cyclopenta[a]phenanthrenes

These compounds could technically be considered as phenanthrene derivatives because the fourth ring is not aromatic. However, they are reviewed here because of their similarities in activity to the methlchrysenes. Among the monomethyl derivatives of 16,17-dihydrocyclopenta[a]phenanthrene (Table 3), only the 7- and 11-isomers showed weak tumorigenicity on mouse skin. Weak activity was also noted among some of the dimethyl derivatives. In the 15,16-dihydrocyclopenta[a]phenanthren-17-one series, the tumorigenic activities are more impressive. In particular, the 11-methyl isomer is by far the most active, with tumorigenicity similar to that of benzo[a]pyrene (B[a]P). Among the other monomethyl compounds, only the 7-methyl isomer showed slight activity. These results are in agreement with the structural requirements favoring tumorigenicity. Among the monomethyl isomers of 15,16-dihydrocyclopenta[a]phenanthren-17-one, only the 11-methyl isomer has a bay-region methyl group and a free *peri* position, both adjacent to the unsubstituted angular (1-4) ring.

D. Methylchrysenes (MeC)

Comparative assays of chrysene and its six monomethyl isomers (Table 4) have shown that 5-MeC is the strongest tumor initiator and complete carcinogen on mouse skin. At the relatively high initiating dose of 4130 nmol, all the isomers showed some initiating activity, with 5-MeC and 3-MeC being the most active. At lower initiating doses, only 5-MeC showed significant activity. Comparative studies have shown that its tumor initiating activity is about the same as that of B[a]P.[36] It is a stronger tumor initiator than 7-MeB[a]A (Table 1). In complete carcinogenicity assays, 5 MeC is clearly the most active isomer, inducing tumors in 90% of the mice after 20 weeks; at this point, none of the other MeC isomers had caused a significant number of tumors. Among the other isomers, 2-MeC showed weak carcinogenic activity; chrysene, 3-MeC, 4-MeC, and 6-MeC induced only a few tumors after 72 weeks, and 1-MeC was inactive. These results are in agreement with the structural requirements favoring tumorigenicity. Only 5-MeC has a bay-region methyl group and a free *peri* position, both adjacent to the unsubstituted (1-4) angular ring. 4-MeC has a bay-region methyl group,

Table 3
REPRESENTATIVE MOUSE-SKIN TUMORIGENICITY ASSAYS OF METHYLCYCLOPENTA[a]PHENANTHRENES

			Tumorigenicity on mouse skin				
Parent ring system	**Methyl isomer**	**Strain**	**Protocol type**	**Dose**	**% TBA**	**T/A**	**Ref.**
16,17-Dihydrocyclopenta[a]phenanthrene	—	Bl.H.	C	0.4%, 2 × weekly 78 weeks	5	NG	28
	1	NG	C	0.4%, 2 × weekly 70 weeks	0	0	28
	2	Bl.H.	C	0.4%, 2 × weekly 85 weeks	0	0	28
	4	Bl.H.	C	0.4%, 2 × weekly 102 weeks	0	0	28
	7	NG	C	0.4%, 2 × weekly 66 weeks	14	NG	28
	11	Bl.H.	C	0.4%, 2 × weekly 62 weeks	18	NG	28
	12	Bl.H.	C	0.4%, 2 × weekly 39 weeks	5	NG	28
	17	Bl.H.	C	0.4%, 2 × weekly 80 weeks	4	NG	28
	2, 12	Bl.H.	C	0.4%, 2 × weekly 76 weeks	0	0	28
	4, 12	NG	C	0.4%, 2 × weekly 79 weeks	0	0	28
	4, 17	NG	C	0.4%, 2 × weekly 39 weeks	0	0	28
	6, 7	Bl.H.	C	0.4%, 2 × weekly 93 weeks	18	NG	28
	6, 17	Bl.H.	C	0.4%, 2 × weekly 48 weeks	0	0	28
	11, 12	Bl.H.	C	0.4%, 2 × weekly 39 weeks	10	NG	28
	11, 17	TO	C	0.5%, 2 × weekly 52 weeks	10	NG	29
	6, 17, 17	Bl.H.	C	0.4%, 2 × weekly 69 weeks	9	NG	28
	11, 12, 17	TO	C	0.5%, 2 × weekly 52 weeks	30	NG	29
15,16-Dihydrocyclopenta[a]phenanthren-17-one	—	TO	I	1740 nmol	0	0	30
	3	TO	I	1640 nmol	0	0	30
	11	TO	I	1640 nmol	90	NG	30
	—	TO	C	120 nmol, 2 × weekly 52 weeks	0	0	31
	2	TO	C	120 nmol, 2 × weekly 50 weeks	0	0	32
	3	TO	C	120 nmol, 2 × weekly 50 weeks	0	0	32
	4	TO	C	120 nmol, 2 × weekly 50 weeks	0	0	32
	6	TO	C	120 nmol, 2 × weekly 50 weeks	0	0	32
	7	TO	C	120 nmol, 2 × weekly 50 weeks	20	NG	32

Table 3 (continued)
REPRESENTATIVE MOUSE-SKIN TUMORIGENICITY ASSAYS OF METHYLCYCLOPENTA[a]PHENANTHRENES

Parent ring system	Methyl isomer	Strain	Protocol type	Tumorigenicity on mouse skin: Dose	% TBA	T/A	Ref.
	11	TO	C	120 nmol, 2 × weekly 50 weeks	70	NG	32
		TO	C	102 nmol, 2 × weekly 50 weeks	90	NG	33
		TO	C	41 nmol, 2 × weekly 50 weeks	80	NG	33
		TO	C	20 nmol, 2 × weekly 50 weeks	90	NG	33
	12	TO	C	120 nmol, 2 × weekly 52 weeks	0	0	31

Note: I, initiation-promotion. C, complete carcinogenicity. TBA, tumor-bearing animals. T/A, tumors per animal. NG, not given.

but it is not adjacent to an unsubstituted angular ring; none of the other isomers have a bay-region methyl group.

Relatively little information is available on the tumorigenic activity of dimethylchrysenes on mouse skin. However, comparisons of 5-MeC, 5,11-diMeC, and 5,12-diMeC have been informative. A comparative complete carcinogenicity assay clearly showed that 5,11-diMeC was more active than was 5-MeC, and that 5,12-diMeC was inactive. In 5,11-diMeC, there are two bay-region methyl groups both adjacent to unsubstituted angular rings, and both *peri* positions are free. However, in 5,12-diMec, the *peri* position adjacent to the 1-4 angular ring is occupied, and the compound is inactive both as a complete carcinogen and tumor initiator.

E. Methylbenzo[c]phenanthrenes (MeB[c]P)

B[c]P does not have a three-sided bay region analogous to that of B[a]A and chrysene, but rather a four-sided 1-12 region, called a fjord. Thus, the structural requirements discussed above may not be expected to apply to MeB[c]P, and the results summarized in Table 5 indicate that they do not. The most active compounds appear to be 3-, 4-, 5-, and 6-MeB[c]P. The isomer containing a fjord region methyl group, 1-MeB[c]P, has minimal activity. In attempting to rationalize these results, it may be better to view B[c]P as a phenanthrene derivative. As described in Chapter 6, methylated phenanthrenes are readily detoxified by metabolism at the K-region, and methyl substitution in that region appears to enhance activity by blocking detoxification. This may be occurring in 4-, 5-, and 6-MeB[c]P. Thus, 1,4-dimethylphenanthrene but not 4-methylphenanthrene is tumorigenic, and by analogy one might expect 1,6-diMeB[c]P to be exceptionally active.

F. Methylpyrenes

Table 6 summarizes data on the mouse-skin tumorigenicity of pyrene and its three possible methyl isomers, as well as two dimethylpyrenes. None of these compounds are significantly tumorigenic. Since pyrene has no bay region, the results might have been expected based on the structural generalization favoring tumorigenicity.

Table 4
REPRESENTATIVE MOUSE-SKIN TUMORIGENICITY ASSAYS OF METHYLCHRYSENES

Parent ring system	Methyl isomer	Strain	Tumorigenicity on mouse skin: Protocol type	Dose	% TBA	T/A	Ref.
Chrysene	—	CD-1	I	4390 nmol	55	0.55	34
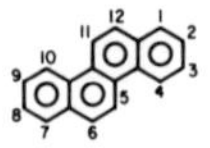	1	CD-1	I	4130 nmol	30	0.3	34
	2	CD-1	I	4130 nmol	42	0.68	34
	3	CD-1	I	4130 nmol	70	1.30	34
				1240 nmol	20	0.40	34
				410 nmol	15	0.15	34
	4	CD-1	I	4130 nmol	35	0.45	34
	5	CD-1	I	4130 nmol	85	4.80	34
				1240 nmol	100	8.00	34
				410 nmol	100	5.50	34
				100 nmol	90	5.20	35
				33 nmol	80	3.90	35
	6	CD-1	I	4130 nmol	35	0.55	34
	—	CD-1	C	440 nmol, 3 × weekly 72 weeks	20	0.20	34
	1	CD-1	C	410 nmol, 3 × weekly 72 weeks	0	0	34
	2	CD-1	C	410 nmol, 3 × weekly 72 weeks	55	1.05	34
	3	CD-1	C	410 nmol, 3 × weekly 72 weeks	25	0.30	34
	4	CD-1	C	410 nmol, 3 × weekly 72 weeks	15	0.25	34
	5	CD-1	C	410 nmol, 3 × weekly 35 weeks	100	4.95	34
				41 nmol, 3 × weekly 55 weeks	75	1.90	36
				21 nmol, 3 × weekly 62 weeks	45	1.10	36
	6	CD-1	C	410 nmol, 3 × weekly 72 weeks	15	0.15	34
	5, 11	CD-1	I	41 nmol	70	2.70	3
	5, 12	CD-1	I	124 nmol	5	0.1	37
	5, 11	CD-1	C	21 nmol, 3 × weekly 50 weeks	87	4.9	38
	5, 12	CD-1	C	21 nmol, 3 × weekly 50 weeks	3	0.03	38
	5, 6	NG	C	0.3%, 2 × weekly 55 weeks	35	NG	24
	5, 7	TO	C	117 nmol, 2 × weekly 50 weeks	55	NG	39

Note: I, initiation-promotion. C, complete carcinogenicity. TBA, tumor-bearing animals. T/A, tumors per animal. NG, not given.

G. Methylfluoranthenes (MeF)

A comparative assay of all the MeF isomers (Table 7) showed that 2-MeF and 3-MeF are the most active as tumor initiators. They are, however, weak tumor initiators, giving only about the same response as 1-MeC, 4-MeC, and 6-MeC. Fluoranthene is a nonalternant

Table 5
REPRESENTATIVE MOUSE-SKIN TUMORIGENICITY ASSAYS OF METHYLBENZO[c]PHENANTHRENES

Parent ring system	Methyl isomer	Strain	Protocol type	Dose	% TBA	T/A	Ref.
			Tumorigenicity on mouse skin				
Benzo[c]phenanthrene	—	C3H	C	2200 nmol, 2 × weekly 91 weeks	25	NG	14
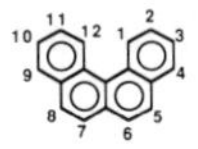	1	C3H	C	2000 nmol, 2 × weekly 91 weeks	5	NG	14
	2	C3H	C	2000 nmol, 2 × weekly 91 weeks	10	NG	14
	3	C3H	C	2000 nmol, 2 × weekly 91 weeks	60	NG	14
	4	C3H	C	2000 nmol, 2 × weekly 91 weeks	80	NG	14
	5	C3H	C	2000 nmol, 2 × weekly 91 weeks	90	NG	14
	6	C3H	C	2000 nmol, 2 × weekly	70	NG	14

Note: C, complete carcinogenicity. TBA, tumor-bearing animals. T/A, tumors per animal. NG, not given.

Table 6
REPRESENTATIVE MOUSE-SKIN TUMORIGENICITY ASSAYS OF METHYLPYRENES

Parent ring system	Methyl isomer	Strain	Protocol type	Dose	% TBA	T/A	Ref.
			Tumorigenicity on mouse skin				
Pyrene	—	CD-1	I	4950 nmol	5	0.1	40
	1	CD-1	I	4630 nmol	20	NG	41
	1	CD-1	I	4630 nmol	13	0.17	42
	2	CD-1	I	4630 nmol	3	0.03	42
	4	CD-1	I	4630 nmol	3	0.07	42
	1, 3	CD-1	I	4350 nmol	4	0.04	42
	1, 6	CD-1	I	4350 nmol	17	0.21	42

Note: I, initiation-promotion. TBA, tumor-bearing animals. T/A, tumors per animal. NG, not given.

PAH without a bay region and, as discussed further in Section II.J, the structural requirements for tumorigenicity appear to be different from those of alternant PAH.

Among the diMeF isomers tested, 2,3-diMeF and 7,8-diMeF were the most active, with tumorigenicity comparable to that of 2-MeC. The tumorigenic activity of 2,3-diMeF could be expected, since both 2- and 3-MeF were active, but the activity of 7,8-diMeF is not readily explained. Both 7,10-diMeF and 8,9-diMeF were inactive.

H. Methylbenzo[a]pyrenes (MeB[a]P)

Table 8 summarizes representative data on MeB[a]P isomers, which have been tested under identical conditions, as tumor initiators in SENCAR mice. The least active isomers were 7-, 8-, 9-, and 10-MeB[a]P, which were essentially devoid of tumor initiating activity; 6-MeB[a]P, 5-MeB[a]P, and 2-MeB[a]P were also less tumorigenic than was B[a]P. The

Table 7
REPRESENTATIVE MOUSE-SKIN TUMORIGENICITY ASSAYS OF METHYLFLUORANTHENES

			Tumorigenicity on mouse skin				
Parent ring system	**Methyl isomer**	**Strain**	**Protocol type**	**Dose**	**% TBA**	**T/A**	**Ref.**
Fluoranthene	—	CD-1	I	4950 nmol	3	0	43
	1	CD-1	I	4630 nmol	0	0	43
	2	CD-1	I	4630 nmol	30	0.5	43
	3	CD-1	I	4630 nmol	30	0.5	43
	7	CD-1	I	4630 nmol	9	0.1	43
	8	CD-1	I	4630 nmol	3	0.1	43
	2, 3	CD-1	I	4350 nmol	47	0.8	43
	7, 8	CD-1	I	4350 nmol	43	0.8	43
	7, 10	CD-1	I	4350 nmol	0	0	43
	8, 9	CD-1	I	4350 nmol	13	0.1	43

Note: I, initiation-promotion. TBA, tumor-bearing animals. T/A, tumors per animal.

most tumorigenic isomer was 11-MeB[a]P; the other compounds had activity comparable to that of B[a]P. These data agree well with the structural generalization. Only 11-MeB[a]P has a bay-region methyl group and a free *peri* position, both adjacent to the unsubstituted angular (7-10) ring. Compared to B[a]P, isomers substituted in the angular ring or at the *peri* (6) position, had decreased activity. These results are also in accord with the structural requirements favoring tumorigenicity, and can be explained at least partially based on known metabolic activation pathways of B[a]P (see Section IV of this chapter and Chapter 5). The relatively low activities of 2-MeB[a]P and 5-MeB[a]P cannot be readily explained based upon currently available data.

Among the dimethyl isomers, diminished activity compared to B[a]P was noted for 1,2-, 1,6-, 3,6-, and 7,10-diMeB[a]P. Except for 1,2-diMeB[a]P, the results are in agreement with the structural generalization because diminished activity is observed upon substitution in the angular ring or at the adjacent *peri* position.

I. Methyldibenzanthracenes (MediBA)

Table 9 summarizes data on the limited comparative mouse-skin assays on MediBA. In the diB[a,h]A system, the 7- and 7,14-isomers are more active as tumor initiators than the parent compound; 7,14-diB[a,h]A is also a more potent complete carcinogen than is diB[a,h]A. These results provide another example of the enhancing effect on tumorigenicity of a bay region methyl group adjacent to an unsubstituted angular ring. Similar results have been obtained in comparative assays of 7-MediB[a,j]A, 7,14-diMediB[a,j]A, and the parent hydrocarbon. In contrast, 9,14-diMediB[a,c]A which has two bay-region methyl groups adjacent to the unsubstituted (1-4 and 5-8) angular rings but does not have adjacent free *peri* positions, is essentially inactive or only weakly active as a tumor initiator.

J. Methylbenzofluoranthenes (MeBF)

Benzo[b]fluoranthene (B[b]F) and benzo[k]fluoranthene (B[k]F) are nonalternant hydrocarbons; tumorigenicity data are summarized in Table 10. B[b]F has a bay region between the 1 and 12 positions and, if the structural generalization applied to such nonalternant hydrocarbons, 1-MeB[b]F should be more active than B[b]F because it has a bay-region methyl group adjacent to the unsubstituted (9-12) angular ring. However, the data in Table 10 demonstrate that this is not the case; 1-MeB[b]F is less tumorigenic than is B[b]F.

Table 8
REPRESENTATIVE MOUSE-SKIN TUMORIGENICITY ASSAYS OF METHYLBENZO[a]PYRENES

Parent ring system	Methyl isomer	Strain	Tumorigenicity on Mouse Skin: Protocol type	Dose	% TBA	T/A	Ref.
Benzo[a]pyrene	—	SENCAR	I	200 nmol	67	2.2	44
	1	SENCAR	I	200 nmol	80	3.62	44
	2	SENCAR	I	200 nmol	38	0.55	44
	3	SENCAR	I	200 nmol	76	2.62	44
	4	SENCAR	I	200 nmol	67	2.03	44
	5	SENCAR	I	200 nmol	27	0.37	44
	6	SENCAR	I	200 nmol	24	0.35	44
	7	SENCAR	I	200 nmol	0	0	44
	8	SENCAR	I	200 nmol	3	0.3	44
	9	SENCAR	I	200 nmol	0	0	44
	10	SENCAR	I	200 nmol	0	0	44
	11	SENCAR	I	200 nmol	90	5.6	44
	12	SENCAR	I	200 nmol	69	1.9	44
	1, 2	SENCAR	I	200 nmol	39	0.5	44
	1, 6	SENCAR	I	200 nmol	23	0.3	44
	3, 6	SENCAR	I	200 nmol	40	0.57	44
	4, 5	SENCAR	I	200 nmol	60	0.97	44
	7, 10	CD-1	I	360 nmol	0	0	45
	—	Eppley Swiss	C	100 nmol, 2 × weekly 20 weeks	79	1.1	46
	6	Eppley Swiss	C	100 nmol, 2 × weekly 20 weeks	30	0.3	46

Note: I, initiation-promotion. C, complete carcinogenicity. TBA, tumor-bearing animals. T/A, tumors per animal.

Table 9
REPRESENTATIVE MOUSE-SKIN TUMORIGENICITY ASSAYS OF METHYLDIBENZANTHRACENES

Parent ring system	Methyl isomer	Strain	Tumorigenicity on mouse skin: Protocol type	Dose	% TBA	T/A	Ref.
Dibenz[a,h]anthracene	—	SENCAR	I	100 nmol	97	6.67	26
	7	SENCAR	I	100 nmol	100	9.10	26
	7, 14	SENCAR	I	100 nmol	100	15.02	26
				10 nmol	93	6.20	26
	—	Taconic	C	0.5%, 2 × weekly 60 weeks	57	NG	20
	7, 14	Taconic	C	0.5%, 2 × weekly 40 weeks	77	NG	20
Dibenz[a,j]anthracene	—	SENCAR	I	400 nmol	50	0.93	26
	7	SENCAR	I	400 nmol	79	2.79	26
	7, 14	SENCAR	I	400 nmol	100	14.39	26
Dibenz[a,c]anthracene	—	CD-1	I	180 nmol	20	0.23	47
	9, 14	SENCAR	I	2000 nmol	25	0.39	26

Note: I, initiation-promotion. C, complete carcinogenicity. TBA, tumor-bearing animals. T/A, tumors per animal.

Table 10
REPRESENTATIVE MOUSE-SKIN TUMORIGENICITY ASSAYS OF METHYLBENZOFLUORANTHENES

			Tumorigenicity on mouse skin				
Parent ring system	**Methyl isomer**	**Strain**	**Protocol type**	**Dose**	**% TBA**	**T/A**	**Ref.**
Benzo[b]fluoranthene		CD-1	I	100 nmol	95	3.3	
	—	CD-1	I	40 nmol	45	0.9	48
	1	CD-1	I	100 nmol	75	3.0	48
				40 nmol	15	0.3	
	3	CD-1	I	40 nmol	75	2.5	48
	7	CD-1	I	40 nmol	15	0.2	48
	8	CD-1	I	40 nmol	25	0.3	48
	9	CD-1	I	100 nmol	25	0.4	48
			I	40 nmol	33	0.4	
	12	CD-1	I	40 nmol	5	0.2	48
	1, 3	CD-1	I	40 nmol	80	3.6	48
	5, 6	CD-1	I	40 nmol	21	0.2	48
Benzo[k]fluoranthene	—	CD-1	I	4000 nmol	37	0.7	49
	2	CD-1	I	4000 nmol	32	0.6	49
	8	CD-1	I	4000 nmol	16	0.2	49
	9	CD-1	I	4000 nmol	37	0.6	49
	7, 12	CD-1	I	4000 nmol	37	0.8	49

Note: I, initiation-promotion. TBA, tumor-bearing animals. T/A, tumors per animal.

Substitution in the 9-12 angular ring decreases tumorigenicity, as does substitution at the 8 *peri* position. In this respect, methyl B[b]F isomers are similar to methylated alternant PAH. A remarkable finding is the relatively high activity of 3-MeB[b]F, which is the most tumorigenic methylB[b]F tested. There is presently no explanation for this enhanced activity, which is also seen in 1,3-diMeB[b]F. The latter may be as strong as 5-MeC as a tumor initiator. Three monomethyl B[k]F isomers have been tested and all appear to be weak tumor initiators, as is the parent hydrocarbon.

III. METABOLIC ACTIVATION OF 5-MeC

Our studies have focused on 5-MeC because of its unique tumorigenicity among the MeC isomers. 5-MeC is also structurally unique because it has two bay regions, one of which contains a methyl group. By understanding the metabolic activation pathways of 5-MeC, which fulfills the structural requirements favoring tumorigenicity of methylated PAH, it should be possible to better understand the factors controlling the effects of methyl substitution on tumorigenicity in other PAH systems. In this section, we describe a variety of experimental approaches which have indicated that a major pathway of metabolic activation of 5-MeC is formation of *anti*-1,2-dihydroxy-3,4-epoxy-1,2,3,4-tetrahydro-5-MeC (*anti*-DE-I).

A. Structure-Activity Studies

Our initial approach focused on the synthesis and bioassay of a series of fluorinated 5-MeC derivatives. Since fluorine is a relatively small, highly electronegative atom, it was assumed that fluorine substitution would inhibit enzymatic oxidation at the position to which it was attached, or perhaps at neighboring positions. This approach is well-known in medicinal chemistry and has been used in studies of the metabolic activation of other PAH.[50]

Table 11
TUMOR-INITIATING ACTIVITY OF MONOFLUORO-5-MeC

Compound	Initiating dose (nmol)	% TBA (compound/5-MeC)	T/A (compound/5-MeC)
1-F-5-MeC	115	10/60[a]	0.1/1.5[a]
	385	40/80[a]	0.6/1.9[a]
3-F-5-MeC	115	0/60[a]	0/1.5[a]
	385	5/80[a]	0/1.9[a]
6-F-5-MeC	115	80/95	4.1/3.0
	385	85/80	2.8/5.2
7-F-5-MeC	115	60/50	0.9/1.4
	385	90/80	4.8/4.7
9-F-5-MeC	115	45/50	0.8/1.4
	385	85/80	3.2/4.7
11-F-5-MeC	115	45/50	0.6/1.4
	385	65/80	2.5/4.7
12-F-5-MeC	115	18/67[a,b]	0.2/1.8[a,b]
	385	77/82[b]	2.1/4.9[b]

Note: TBA, tumor-bearing animals. T/A, tumors per animal.

[a] Significant, $p < 0.01$.
[b] Average of 3 experiments.

Thus, if a fluorinated 5-MeC were less active than 5-MeC, the results would suggest that the position where fluorine was substituted might be involved in metabolic activation. In contrast, if fluorine substitution blocked metabolic detoxification, an increase in tumorigenicity should be expected. The monofluoro-5-MeC were synthesized by unambiguous routes[51] and were tested for tumor initiating activity and complete carcinogenicity on mouse skin. The results are summarized in Table 11 and in Figure 4.[37,52] The results of both assay types clearly showed that 12-F-5-MeC and, in particular, 1-F-5-MeC and 3-F-5-MeC, were significantly less tumorigenic than was 5-MeC. In addition, 9-F-5-MeC was somewhat less tumorigenic than was 5-MeC in both assays, and 6-F-5-MeC was more tumorigenic than was 5-MeC in the complete carcinogenicity assay. The results of these studies strongly indicated that the 1-4 angular ring of 5-MeC was involved in its metabolic activation.

The tumor initiating activities of some other 5-MeC analogues are summarized in Table 12. Substitution of a methoxy group at the 12 position decreased activity, as in 5,12-diMeC (Table 4) and 12-F-5-MeC (Table 11). Substitution of a methoxy group at the 6 position also decreased activity. There is presently no explanation for this observation. Saturation of the 5,6 positions eliminated activity and bridging the 4,5 positions decreased activity. The latter observation is in accord with the hypothesis that the 1-4 angular ring is involved in metabolic activation. 5-Ethylchrysene had similar tumorigenic activity as 5-MeC, whereas 5-methoxychrysene was somewhat less active.

B. Identification of 5-MeC Metabolites

Identified metabolites of 5-MeC are illustrated in Figure 5. With the exception of the dihydrodiol epoxides, all metabolites have been identified by incubating 5-MeC or its metabolites with cofactors and liver 9000 $\times$ g supernatant from rats or mice. The metabolites were isolated by HPLC, and characterized by their NMR, UV, and mass spectra and by comparison to synthetic standards.[53-62] The formation of the dihydrodiol epoxides in mouse skin has been detected by isolation of DNA adducts and tetraols, as discussed further below.

The general pattern of 5-MeC metabolism is similar to that of other PAH.[63] The initial

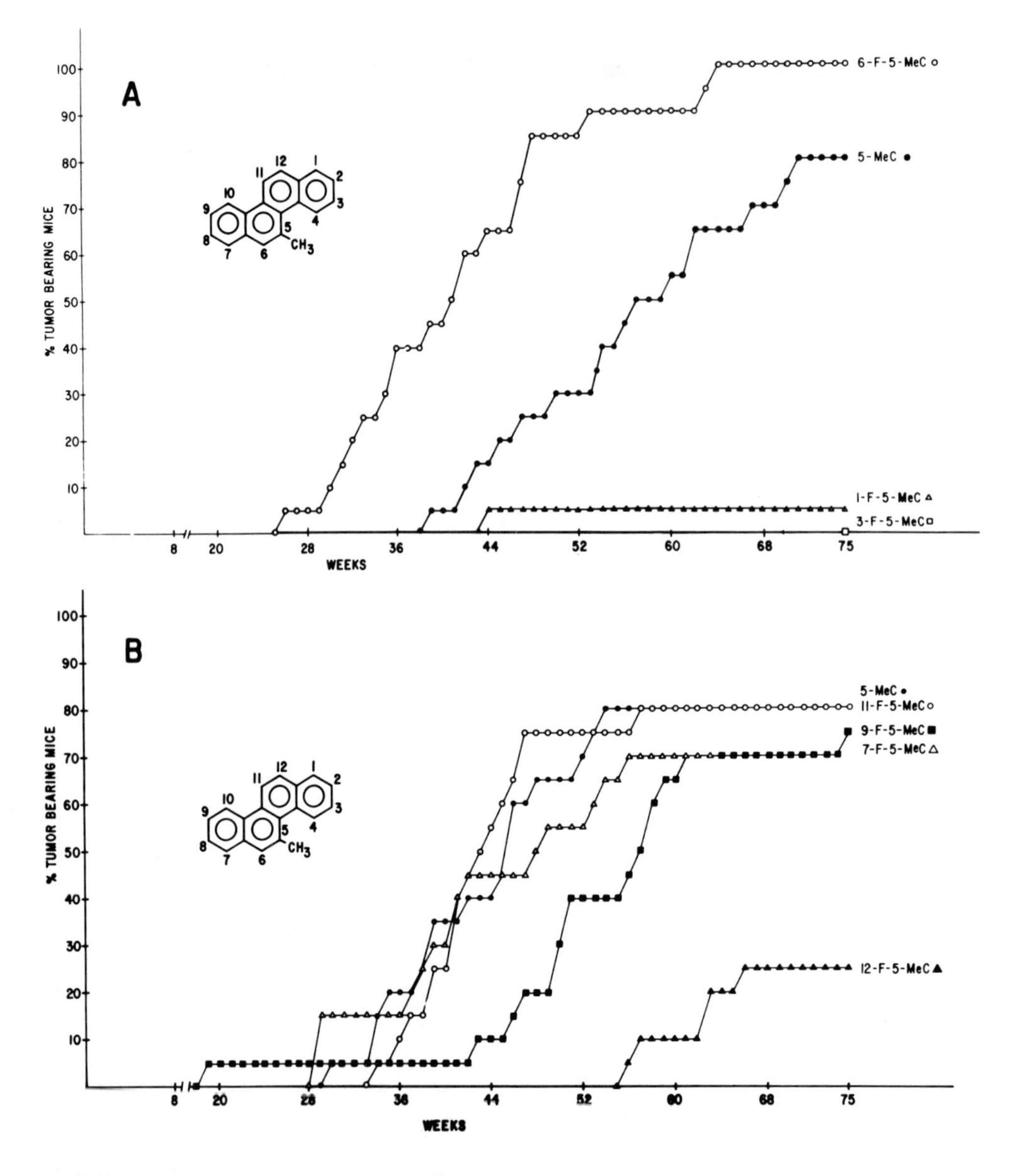

FIGURE 4. Complete carcinogenicity of fluorinated 5-MeC on CD-1 mouse skin. Each compound (19 nmol) was applied 3 times weekly for the duration of the assay. (A) Comparative assays of 1-F-5-MeC, 6-F-5-MeC, and 5-MeC; (B) Comparative assays of 7-F-5-MeC, 9-F-5-MeC, 11-F-5-MeC, 12-F-5-MeC, and 5-MeC. (From Hecht, S. S., LaVoie, E. J., Mazzarese, R., Hirota, N., Ohmori, T., and Hoffmann, D., *J. Natl. Cancer Inst.*, 63, 855, 1979.)

oxidation products are likely to be arene oxides, which can be hydrated to dihydrodiols or can rearrange nonenzymatically to phenols. As in the metabolism of chrysene, dihydrodiol formation in the angular rings is preferred;[64] K-region metabolites of 5-MeC have not been detected. Initial oxidation at the 3,4 position is also not favored, presumably due to the methyl group. 5-Hydroxymethylchrysene (5-HOMeC) is further oxidized to dihydrodiols and phenols.[56] 5-MeC-1,2-diol and 5-MeC-7,8-diol are of special interest because they are the precursors to anti-DE-I and anti-DE-II, which lead to DNA adduct formation in mouse skin.[57,58] The syn dihydrodiol epoxides are apparently also formed in mouse skin.[60] Glucuronide and sulfate conjugates of the dihydrodiols and phenols have also been observed in mouse skin in vivo.[58]

Table 12
TUMOR-INITIATING ACTIVITY OF SOME 5-MeC ANALOGUES

	Initiating dose (nmol)	% TBA compound/5-MeC	T/A compound/5-MeC	Ref.
12-MeO-5-MeC	110	5/95	0.1/3.0	37
	368	10/80	0.1/5.2	
6-MeO-5-MeC	110	10/70	0.1/1.6	36
	368	45/90	0.75/2.9	(and unpubl. data)
5,6-Dihydro-5-MeC	123	0/70	0/1.6	(unpubl. data)
	410	0/90	0/2.9	
4,5-Ethylenechrysene	118	15/70	0.15/1.6	(unpubl. data)
	394	25/90	0.25/2.9	
5-Ethylchrysene	100	90/90	4.3/5.1	(unpubl. data)
5-Methoxychrysene	116	40/55	0.5/1.3	36
	388	55/80	1.8/3.1	

Note: Assays were carried out with CD-1 mice. Promotion by 3 times weekly application of 12-O-tetradecanoylphorbol-13-acetate for 20 weeks. TBA, Tumor-bearing animals. T/A, tumors per animal.

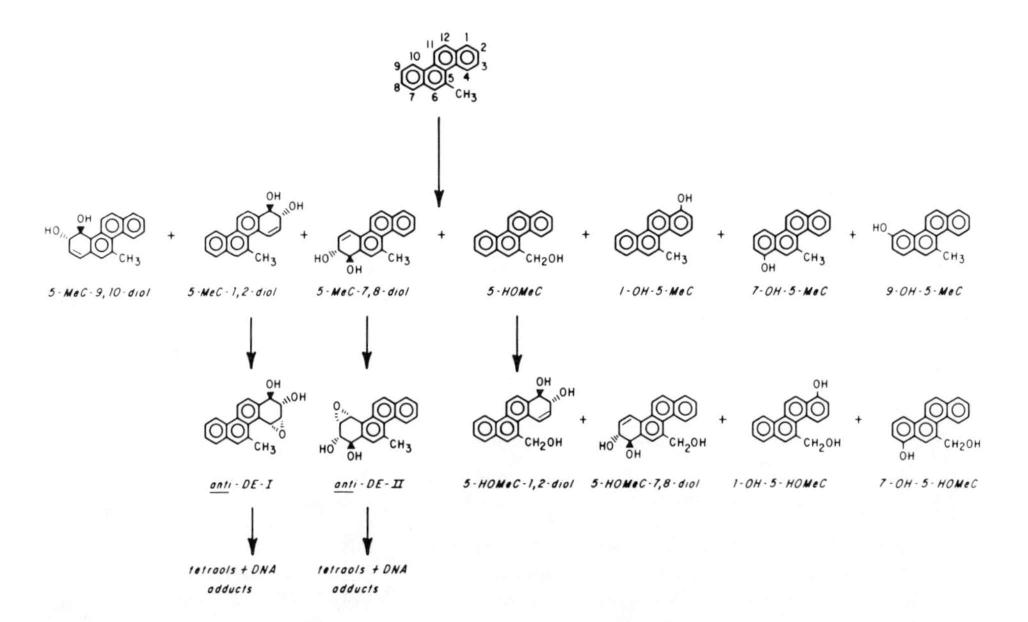

FIGURE 5. Identified metabolites of 5-MeC. *Syn*-isomers of the dihydrodiol epoxides are also formed.[60] Glucuronide, sulfate, and glutathione conjugates of 5-MeC metabolites have been detected (Reference 58 and unpublished data). (Reprinted with permission from Hecht, S. S., Melikian, A. A., and Amin, S., *Chem. Res.*, 19, 174, 1986. Copyright 1986, American Chemical Society.)

The relative quantities of metabolites depend on the tissue studied and on the inducer pretreatment. Major quantitative differences have been observed in comparative experiments with mouse or rat liver 9000 $\times$ *g* supernatant from untreated animals or animals pretreated with 3-methylcholanthrene, Aroclor® 1254, or phenobarbital.[53,56,58,65] These differences are undoubtedly due to differing levels of various cytochrome P-450 isozymes in the various systems. Therefore, we prefer to carry out our metabolism studies in mouse epidermis in vivo, when the appropriate labeled substrate is available. Results obtained under these in vivo conditions should be more readily interpreted with respect to potential metabolic activation and detoxification pathways than are results obtained in vitro using liver homogenates. A comparison of [^{3}H]5-MeC metabolism in vivo in mouse epidermis and in vitro in

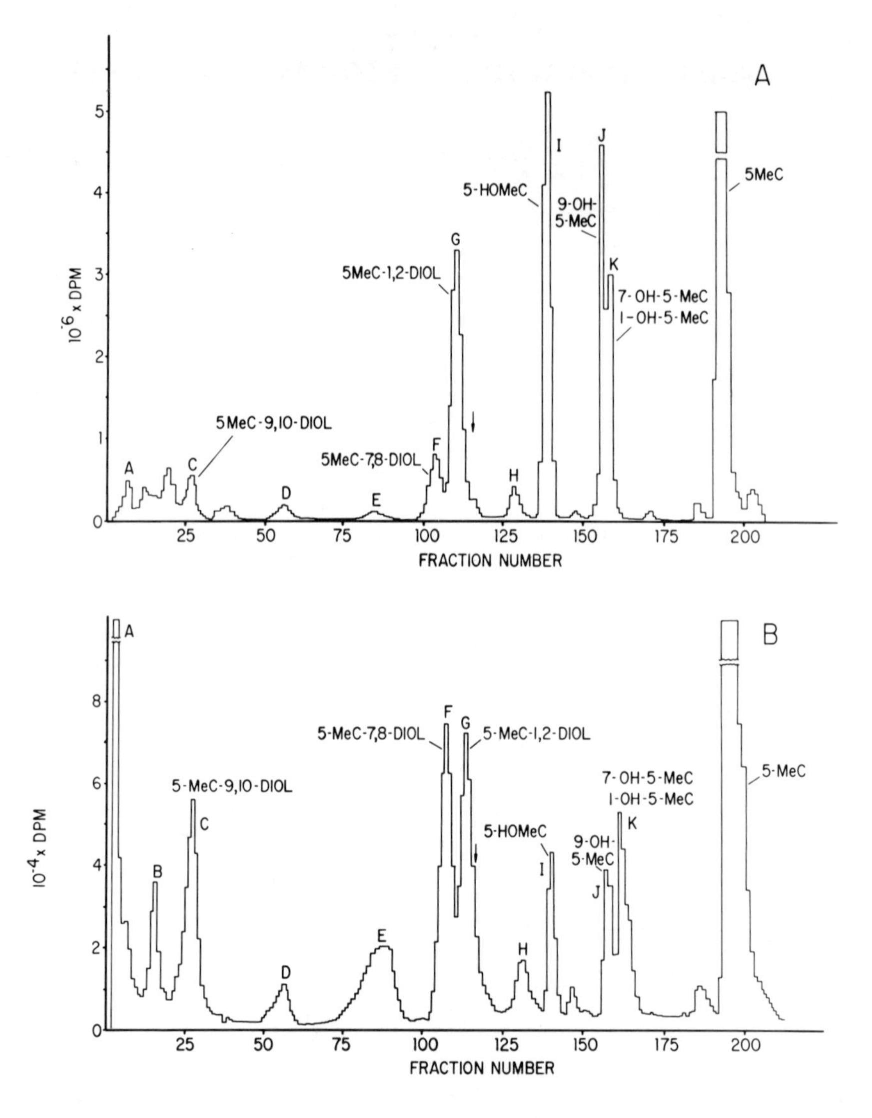

FIGURE 6. High pressure liquid chromatograms obtained upon analysis of ethyl acetate extracts of (A) mouse liver 9000 *g* supernatant incubated with [^{3}H]5-MeC and (B) epidermis isolated from mice sacrificed 2 hr after topical administration of [^{3}H]5-MeC. (From Melikian, A. A., LaVoie, E. J., Hecht, S. S., and Hoffmann, D., *Carcinogenesis,* 4, 843, 1983. With permission.)

mouse liver is illustrated in Figure 6. Whereas levels of 5-MeC-1,2-diol exceeded those of 5-MeC-7,8-diol in mouse liver, the two dihydrodiols were present in approximately equal quantities 2 hr after application of [^{3}H]5-MeC to mouse skin. Figure 7 illustrates the time course of [^{3}H]5-MeC metabolite formation in mouse epidermis. An important point is that the levels of 5-MeC-1,2-diol and 5-MeC-7,8-diol were similar over the entire period. Thus, it is clear that quantitation of metabolite formation in liver preparations may not provide a reliable guide to the events occurring in the target tissue, mouse epidermis.

There are many unknown aspects of 5-MeC metabolism. The stereochemistry of dihydrodiol formation and dihydrodiol epoxide formation has not been elucidated. The further metabolites of the 5-MeC-dihydrodiols have not been thoroughly characterized. A detailed study of the phenolic metabolites has not yet been carried out. The details of glucuronide, sulfate, and glutathione conjugation in vivo have not been established, although it is known that these phase II reactions do occur.

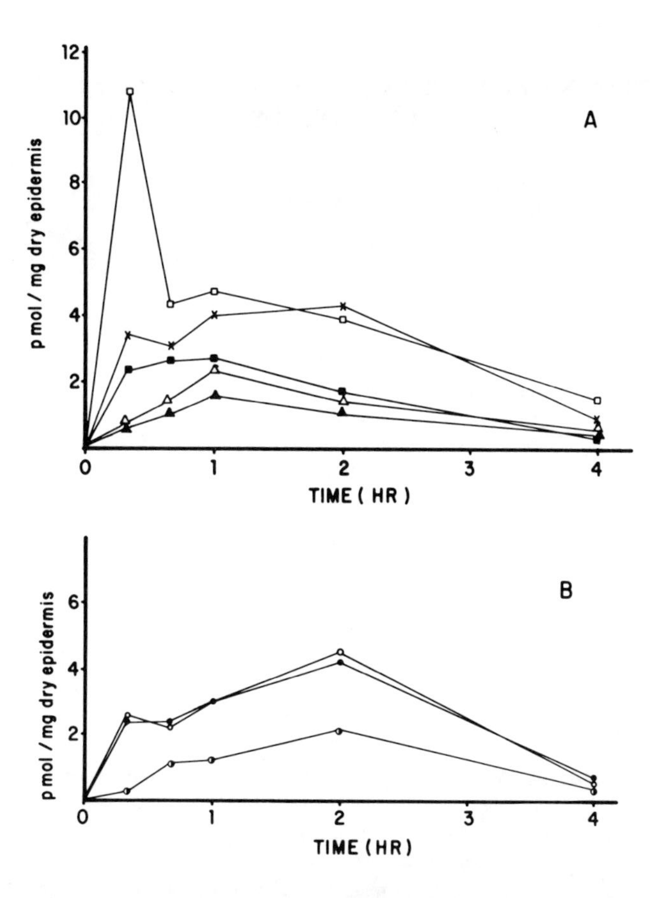

FIGURE 7. Levels of metabolites of [^{3}H]5-MeC in mouse epidermis at various time intervals up to 4 hr. after topical treatment. (A) □—□, metabolite A (primarily sulfate conjugates) of Figure 6B; ■—■ , 5-HOMeC; x—x, total of 1-OH-5-MeC, 7-OH-5-MeC, and 9-OH-5-MeC; △—△, ethyl acetate extractable metabolites obtained upon treatment of water soluble metabolites with β-Glucuronidase; ▲—▲ , ethyl acetate extractable metabolites obtained upon treatment of water soluble metabolites with arylsulfatase. (B) ●—●, 5-MeC-1,2-diol; ○—○, 5-MeC-7, 8-diol; ◐—◐, 5-MeC-9,10-diol. The level of 5-MeC-1,2-diol present 2 hr after treatment corresponds to 1.33% of the dose of [^{3}H]5-MeC.

C. Tumorigenicity of Metabolites; Identification of Anti-DE-I as an Ultimate Carcinogen

With the basic metabolic pattern having been established, it became important to determine which metabolites were involved in metabolic activation and which could be considered as detoxification products. This was examined by carrying out comparative assays for tumor-initiating activity of 5-MeC and its known and potential metabolites. The data are summarized in Table 13. Since the response to 5-MeC between assays can vary, it is only useful to compare the activity of the metabolite to that of 5-MeC within a given assay. The results clearly showed that 5-MeC-1,2-diol was a stronger tumor initiator than was 5-MeC, but that 5-MeC-7,8-diol was a weaker tumor initiator than was 5-MeC. 5-MeC-9,10-diol was inactive. 5-HOMeC had comparable tumorigenic activity to that of 5-MeC. The synthetic analogue, 5-AcOMeC, also had activity similar to that of 5-MeC. All of the phenolic compounds were tumorigenic, but none was as active as 5-MeC. Among the known phenolic metabolites, 9-OH-5-MeC had appreciable activity. These results indicated that 5-MeC-1,2-diol was a likely major proximate carcinogen of 5-MeC, and that 5-HOMeC could also be involved in 5-MeC activation.

Table 13
TUMOR-INITIATING ACTIVITY OF 5-MeC METABOLITES AND RELATED COMPOUNDS

Compound	Dose (nmol)	% TBA (compound/5-MeC)	T/A (compound/5-MeC)	Ref.
5-MeC-1,2-diol	109	100/75	7.3/3.0	66
5-MeC-7,8-diol	109	50/75	1.1/3.0	66
5-MeC-9,10-diol	109	0/75	0/3.0	66
5-HOMeC	116	95/90	8.8/5.9	56
	39	45/55	2.6/5.6	56
5-AcOMeC[a]	100	85/100	4.3/5.9	(unpubl. data)
1-OH-5-MeC	116	40/100	1.1/8.9	(unpubl. data)
2-OH-5-MeC	116	85/100	3.7/8.9	(unpubl. data)
3-OH-5-MeC	116	75/100	4.4/8.9	(unpubl. data)
7-OH-5-MeC	116	50/100	1.0/8.9	(unpubl. data)
8-OH-5-MeC	116	60/100	2.5/8.9	(unpubl. data)
9-OH-5-MeC	116	95/100	3.7/8.9	(unpubl. data)

Note: Assays were carried out with CD-1 mice. Promotion by 3 times weekly application of 12-O-tetradecanoylphorbol-13-acetate for 20 weeks. TBA, tumor-bearing animals. T/A, tumors per animal.

[a] 5-acetoxymethylchrysene.

Table 14
TUMOR-INITIATING ACTIVITY OF 5-MeC METABOLITES

	Dose (nmol)	% TBA		T/A	
		15 weeks[a]	25 weeks	15 weeks	25 weeks
5-MeC	100	50	90	1.2	5.2
	33	45	80	1.1	3.9
5-MeC-1,2-diol	100	85	100	4.3	12.7
	33	70	85	2.4	9.9
5-MeC-7,8-diol	100	10	75	0.1	1.3
	33	5	30	0.1	0.3
Anti-DE-I	100	60	80	1.8	4.4
	33	35	65	0.4	1.3
Anti-DE-II	100	0	0	0	0
	33	0	5	0	0.1
Acetone	—	0	10	0	0.1

Note: Groups of 20 female CD-1 mice (age 50 — 55 days) were shaved and treated with a single dose of each compound in 0.1 mℓ acetone. Ten days later, each group was treated 3 times weekly with 2.5 μg of 12-O-tetradecanoylphorbol-13-acetate in 0.1 mℓ acetone, for 25 weeks. See Reference 35. TBA, tumor-bearing animals. T/A, tumors per animal.

[a] Preliminary reading at 15 weeks of treatment with 12-O-tetradecanoylphorbol-13-acetate.

The high activity of 5-MeC-1,2-diol was not unexpected, since studies on a number of unsubstituted PAH including chrysene had shown that such angular-ring dihydrodiols were major proximate carcinogens, and that the corresponding dihydrodiol epoxides were ultimate carcinogens.[67,68] However, the higher tumorigenicity of 5-MeC-1,2-diol than of 5-MeC-7,8-diol, which was confirmed in a subsequent bioassay (see Table 14), was intriguing, since both dihydrodiols could form bay-region dihydrodiol epoxides.

Table 15
TUMORIGENICITY OF 5-MeC METABOLITES IN NEWBORN MICE

Compound	No. of mice injected	Effective no. of mice		Pulmonary tumors		Hepatic tumors	
				% TBA	T/A	% TBA	T/A
5-MeC	100	Female	48	21	0.25	12	0.29
		Male	35	20	0.26	23	0.43
		Total	83	21	0.25	17	0.34
5-MeC-1,2-diol	100	Female	43	12	0.14	7	0.23
		Male	44	11	0.18	25	0.52
		Total	87	12	0.16	16	0.38
5-MeC-7,8-diol	100	Female	45	18	0.24	11	0.49
		Male	46	13	0.13	2	0.02
		Total	91	15	0.19	7	0.25
Anti-DE-I	100	Female	48	81	5.6	4	0.13
		Male	38	82	3.3	34	2.6
		Total	86	81	4.6	19	1.2
Syn-DE-I	100	Female	41	29	0.34	7	0.46
		Male	49	6	0.06	14	0.20
		Total	90	17	0.17	11	0.37
Anti-DE-II	100	Female	50	6	0.06	4	0.04
		Male	49	18	0.18	2	0.02
		Total	99	12	0.12	3	0.03
DMSO	100	Female	41	7	0.07	2	0.02
		Male	48	4	0.04	2	0.04
		Total	89	6	0.06	2	0.03

Note: Ha/ICR mice were given i.p. injections of each compound (total dose, 56 nmol) in DMSO on the 1st, 8th, and 15th days of life. Mice were weaned at age 21 days, separated by sex, and sacrificed at age 35 weeks. TBA, tumor-bearing animals. T/A, tumors per animal. Tests were done on racemic compounds.

In the next step, the tumorigenic activities of the corresponding dihydrodiol epoxides were tested. These assays were carried out in mouse skin and in newborn mice. The newborn mouse assay was used because previous studies had shown that newborn mice were more sensitive to dihydrodiol epoxides than was mouse skin.[67] In many PAH systems, dihydrodiol epoxide metabolites are less tumorigenic on mouse skin than the parent hydrocarbon or proximate carcinogen dihydrodiols, even though persuasive evidence, partly from newborn mouse assays, exists that they are major ultimate carcinogens.[67] The reason for this discrepancy is not known. The results of the assays in mouse skin and in newborn mice are summarized in Tables 14 and 15.[69] The mouse-skin assays demonstrated that *anti*-DE-I was more tumorigenic than was *anti*-DE-II, although it was less active than 5-MeC and 5-MeC-1,2-diol. The newborn mouse assays clearly showed that, among the compounds tested, *anti*-DE-I was exceptionally tumorigenic. Taken together, these data provide strong evidence that a major activation pathway for 5-MeC is 5-MeC → 5-MeC-1,2-diol → *anti*-DE-I. The results show that a bay-region dihydrodiol epoxide isomer, *anti*-DE-I, with the methyl group and epoxide ring in the same bay region, is a potent tumorigen. In contrast, no significant tumorigenicity was observed for *anti*-DE-II, which is also a bay-region dihydrodiol epoxide, but does not have the epoxide ring and the methyl group in the same bay region. The lack of tumorigenicity of *syn*-DE-I indicates that there are subtle steric requirements for activity.

The results of these bioassays agreed remarkably well with the results of the structure-activity studies of fluorinated 5-MeC. Substitution of fluorine at the 1 and 3 positions should block formation of *anti*-DE-I; preliminary metabolic studies supported this hypothesis.[52] Therefore, the inactivity of 1-F-5-MeC and 3-F-5-MeC can be explained on this basis. The

Table 16
TUMORIGENICITY ASSAYS OF STRUCTURAL ANALOGUES OF 5-HOMeC

Compound	Protocol type	Dose	% TBA	T/A	Ref.
5-HOMeC	I	116 nmol	90	9.5	56
6-HOMeC	I	116 nmol	0	0	56
5-HOMeC	C	19 nmol, 3 × weekly 61 weeks	55	2.0	(unpubl. data)
6-HOMeC	C	19 nmol, 3 × weekly 61 weeks	0	0	(unpubl. data)
5-MeC	C	21 nmol, 3 × weekly 61 weeks	77	4.4	(unpubl. data)
7-F-5-HOMeC	I	109 nmol	95	7.9	56
3-F-5-HOMeC	I	109 nmol	5	0.1	56

Note: Assays carried out with CD-1 mice. I, initiation-promotion. C, complete carcinogenicity. TBA, tumor-bearing animals. T/A, tumors per animal.

low activity of 12-F-5-MeC can also be explained by inhibition of 5-MeC-1,2-diol formation, as discussed below in more detail.[65] In contrast, 7-F-5-MeC and 9-F-5-MeC had tumorigenic activities similar to that of 5-MeC, because 5-MeC-7,8-diol does not contribute significantly to 5-MeC tumorigenesis. Thus, the fluorine probe studies provide strong supportive evidence for the hypothesis that *anti*-DE-I is a major ultimate carcinogen of 5-MeC.

D. The Role of 5-HOMeC

As described above, the tumor-initiating activity on mouse skin of 5-HOMeC was comparable to that of 5-MeC, and it therefore seemed likely that 5-HOMeC was involved in 5-MeC activation. It has been suggested that a possible mechanism of methylated PAH activation is one electron oxidation or hydroxymethylation followed by conjugation of the -CH_2OH group with sulfate or a related acid. This could result in formation of a stabilized benzylic radical or carbonium ion which could interact with DNA.[70-72] Some evidence for this putative activation pathway has been presented, although DNA binding studies with 7,12-diMeB[a]A do not seem to support the concept.[73] We carried out structure-activity studies to examine the mechanism of activation of 5-HOMeC. The results are summarized in Table 16. 5-HOMeC was active as a tumor initiator and complete carcinogen. Its activity as a complete carcinogen was somewhat less than that of 5-MeC. In contrast, 6-HOMeC was inactive in both assays. This suggests that activation does not proceed via esterification of the hydroxymethyl group followed by reaction with DNA, as such a process should be expected to occur with both compounds. The assays of the fluorinated 5-HOMeC derivatives were striking, in that 3-F-5-HOMeC was inactive but 7-F-5-HOMeC had activity similar to that of 5-HOMeC. These results are similar to those obtained with the fluorinated analogues of 5-MeC, and strongly suggest that 5-HOMeC is activated via its 1,2-dihydrodiol-3,4-epoxide, like 5-MeC. Metabolic studies supported this hypothesis.[56]

E. DNA Adduct Formation from 5-MeC and Its Dihydrodiol Epoxides

Since anti-DE-I appeared to be a major ultimate carcinogen of 5-MeC, its reactions with DNA were investigated, and compared to those of *anti*-DE-II. Reaction of each racemic dihydrodiol epoxide with DNA gave at least 7 adducts, but in each case, one adduct predominated. These adducts were identified by their spectral properties as having arisen by attack of the N^2 of deoxyguanosine at the benzylic-carbon terminus of the epoxide ring.[74] The structures are shown in Figure 8. The structural features are similar to those observed

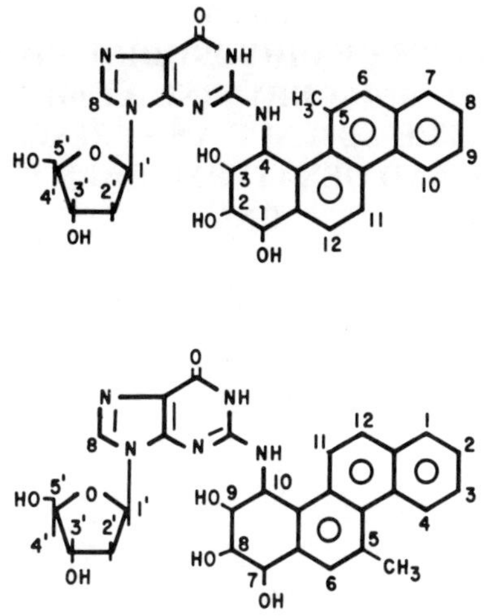

FIGURE 8. Structures of the major adducts formed upon reaction of (top) *anti*-DE-I or (bottom) *anti*-DE-II with DNA in vitro. (Reprinted with permission from Hecht, S. S., Amin, S., Melikian, A. A., LaVoie, E. J., and Hoffmann, D., *ACS Symp. Ser.*, 283, 85, 1985. Copyright 1985, American Chemical Society.)

for other PAH dihydrodiol epoxide-DNA adducts,[75] and suggest that the lower tumorigenicity of *anti*-DE-II compared to *anti*-DE-I is not due to a major difference in adduct structure, although the stereochemical aspects of these reactions with DNA require further investigation.

To investigate the formation of these adducts in vivo, mice were treated topically with [^{3}H]5-MeC and the DNA was isolated from epidermis, hydrolyzed to deoxyribonucleosides, and the HPLC and Sephadex® LH-20 chromatographic properties of the labeled adducts were compared with those of the modified deoxyribonucleosides illustrated in Figure 8. Major adducts, chromatographically indistinguishable from these compounds, were observed. The ratio of the *anti*-DE-I adduct to the *anti*-DE-II adduct was 3 to 1, 4 to 48 hr after application of [^{3}H]5-MeC to mouse skin.[57,58] The predominance of the *anti*-DE-I adducts over the *anti*-DE-II adducts parallels the results described above which indicate that *anti*-DE-I, but not *anti*-DE-II, is a major ultimate carcinogen of 5-MeC. Some possible reasons for the preferential formation of *anti*-DE-I adducts were investigated.[58] As described above, the extents of formation of 5-MeC-1,2-diol and 5-MeC-7,8-diol in mouse epidermis were the same. Differences in metabolism of 5-MeC to these two potential proximate carcinogens would therefore not account for the observed differences in DNA adduct formation. The comparative metabolism of the dihydrodiols to *anti*-DE-I and *anti*-DE-II in mouse skin has not been examined. On steric grounds, preferential formation of *anti*-DE-II would be expected. Studies of the reactivity of *anti*-DE-I and *anti*-DE-II with DNA in vitro indicated that *anti*-DE-I reacted more extensively. This could be partially responsible for the higher in vivo levels of anti-DE-I DNA adducts, and was investigated in more detail as described below. Since the ratio of the major *anti*-DE-I-DNA adduct to the *anti*-DE-II-DNA adduct was constant over the 4 to 48 hr period investigated, differential repair of the two adducts would not seem to be involved in their differing levels.

These results indicated that dihydrodiol epoxide reactivity with DNA could play a role in the preferential formation of *anti*-DE-I DNA adducts in vivo. To examine this in more detail, the rates of hydrolysis and extents of DNA binding of *syn*- and *anti*-DE-I, *syn*- and *anti*-DE-II, and *anti*-1,2-dihydroxy-3,4-epoxy-1,2,3,4-tetrahydrochrysene (*anti*-chrysene-DE) were investigated.[76,77] The half-lives of the dihydrodiol epoxides at pH 7.0 and 37°C are summarized in Table 17 and their relative extents of binding to DNA in Table 18. It is clear

Table 17
HALF LIVES OF DIHYDRODIOL EPOXIDES OF 5-MeC AND CHRYSENE AT pH 7.0 AND 37°C IN THE ABSENCE AND PRESENCE OF NATIVE AND DENATURED CALF THYMUS DNA

	$t_{1/2}$(min)		
Compound	Buffer solution only	Denatured DNA	Native DNA
Anti-DE-I	59	24	3.5
Syn-DE-I	62	48	22
Anti-DE-II	17.5	9	2
Syn-DE-II	5.4	4.9	2.8
Anti-chrysene-DE	104	77	21

Table 18
RELATIVE EXTENTS OF BINDING OF DIHYDRODIOL EPOXIDES OF 5-MeC AND CHRYSENE TO NATIVE CALF-THYMUS DNA AT pH 7.0 AND 37°C

Compound	Relative extents of binding
Anti-DE-I	4.9
Syn-DE-I	2.0
Anti-DE-II	2.6
Syn-DE-II	1
Anti-chrysene-DE	2.4

that the rates of hydrolysis of the dihydrodiol epoxides do not correlate with their DNA binding properties. When the hydrolyses were carried out in the presence of denatured DNA, a rate enhancement of 1.1- to 2.5-fold was observed, while in the presence of native DNA, the enhancement was 2 to 17 fold (see Table 17). The ratios of the rates of hydrolysis in the presence of native DNA to the rates of hydrolysis in the presence of denatured DNA did correlate with the extents of binding to DNA as illustrated in Figure 9. These results suggest that intercalation of the dihydrodiol epoxide in DNA precedes reaction, as has been observed with B[a]P-7,8-dihydrodiol-9,10-epoxides.[78-80] This may be a key factor in determining extents of binding of dihydrodiol epoxides to DNA in vitro. It is of interest that the greatest rate enhancement and highest extent of binding were observed for *anti*-DE-I. This parallels the greater in vivo DNA binding and tumorigenicity of *anti*-DE-I compared to *anti*-DE-II and the other dihydrodiol epoxides. While these results suggest that extents of DNA binding of the dihydrodiol epoxides are a determinant of tumorigenic activity, the exceptional tumorigenicity of *anti*-DE-I can probably not be explained on this basis alone.

F. Summary of the Metabolic Activation of 5-MeC

The experiments described above provide strong evidence that a major pathway of 5-MeC activation involves formation of 5-MeC-1,2-diol as a proximate carcinogen and *anti*-DE-I as an ultimate carcinogen. A similar activation pathway apparently occurs upon further

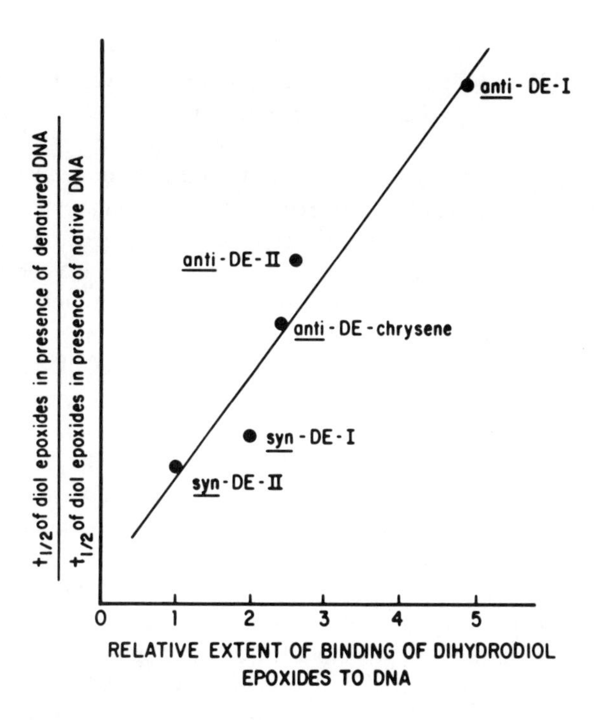

FIGURE 9. Plot of the ratio of the half-lives of dihydrodiol epoxides in the presence of denatured DNA to their half-lives in the presence of native DNA vs. their extents of covalent binding to calf-thymus DNA. (From Melikian, A. A., Leszynska, J. M., Amin, S., Hecht, S. S., Hoffmann, D., Pataki, J., and Harvey, R. G., *Cancer Res.*, 45, 1990, 1985. With permission.)

metabolism of 5-HOMeC, although this sequence of reactions appears to be a quantitatively less-important aspect of 5-MeC metabolism. Metabolic activation via *anti*-DE-I is in accord with the bay-region theory of PAH activation. However, 5-MeC can form two bay-region dihydrodiol epoxides, *anti*-DE-I and *anti*-DE-II, and it is apparent that the latter is not important in the overall activation of 5-MeC. The difference in the two pathways appears to be expressed at the level of the dihydrodiol epoxide and seems to result in part from the greater reactivity with DNA of *anti*-DE-I than of *anti*-DE-II. An important structural feature of *anti*-DE-I is the presence of a methyl group in the same bay region as the epoxide ring. Among the six MeC isomers, only 5-MeC can form such a metabolite, and it is uniquely tumorigenic.

Many aspects of 5-MeC metabolism and activation require further study. The absolute configurations of the important dihydrodiol and dihydrodiol epoxide metabolites remain to be determined. Eight bay-region dihydrodiol epoxide enantiomers can be formed from 5-MeC, and differences in tumorigenicity and DNA interactions among these potential metabolites can be expected. In particular, the relative chemical and biological properties of the four DE-I isomers require evaluation. Relatively little is known about the conjugation of 5-MeC dihydrodiol epoxides with glutathione; this process is likely to be important in controlling the balance of activation and detoxification. Considerable amounts of apparently polar DNA adducts are formed from 5-MeC in vivo; these have not been identified nor are their biological consequences known. The basis of the high sensitivity of newborn mouse lung, but not mouse skin, to *anti*-DE-I needs to be investigated. The pharmacokinetic factors which control the potential tumorigenicity of dihydrodiol epoxide metabolites in mouse skin require further study.

IV. RELATIONSHIP OF METABOLIC ACTIVATION PATHWAYS OF METHYLATED PAH TO THE STRUCTURAL REQUIREMENTS FAVORING TUMORIGENICITY

Among the alternant PAH, the structural features favoring tumorigenicity are the presence of a bay-region methyl group and a free *peri* position, both adjacent to an unsubstituted angular ring. Although there are important exceptions to this generalization, as noted in Section II, there are many methylated PAH which do conform and the generalization is consequently a useful framework for considering structure-activity relationships for methylated PAH tumorigenicity. The requirement for an unsubstituted angular ring is essentially a corollary of the bay-region hypothesis. In fact, consideration of the relative tumorigenic activities of methylated PAH led partially to its formulation.[81] Angular-ring bay-region dihydrodiol epoxides have been shown to be or have been implicated as ultimate carcinogens of many unsubstituted alternant PAH, including B[a]A, chrysene, B[a]P, dibenzanthracenes, and dibenzopyrenes.[67,68] They are also known to be ultimate carcinogens of methylated PAH in the chrysene, benz[a]anthracene (see Chapter 1), and cyclopenta[a]phenanthrene (see Chapter 3) systems. Whereas it had previously been assumed that substitution of a methyl group in the required angular ring would block dihydrodiol and dihydrodiol epoxide formation, more recent studies have shown that this is not the case. Dihydrodiols can form at substituted double bonds in the benz[a]anthracene and cyclopenta[a]phenanthrene systems (see Chapter 3), and a 7,8-dihydrodiol is a metabolite of 7-MeB[a]P.[82,83] However, the formation of dihydrodiols at methyl-substituted double bonds seems to occur to a lesser extent than at unsubstituted positions[82,83] In addition, methyl substitution in an angular ring may direct metabolism toward other regions of the PAH molecule. Although the requirement for an unsubstituted angular ring is apparently evident from these considerations, the mechanisms by which an adjacent bay-region methyl group enhances tumorigenicity or by which a *peri* substituent inhibits tumorigenicity are less clear.

A. Enhancement of Tumorigenicity by a Bay-Region Methyl Group

5-MeC is a useful model for studying this effect because it has two bay regions, but only one of them has a methyl group. Thus, comparisons of the chemical and biological properties of the dihydrodiols and dihydrodiol epoxides that can form in the two angular rings of 5-MeC should provide insight into the mechanism of the enhancing effect on tumorigenicity of a bay-region methyl group. These studies are described in some detail in Section III. They have shown that the bay-region methyl group has no apparent effect on the extents of formation of 5-MeC-1,2-diol and 5-MeC-7,8-diol in mouse skin, but that DNA adduct formation by the pathway 5-MeC-1,2-diol → *anti*-DE-I exceeds DNA adduct formation via 5-MeC-7,8-diol → *anti*-DE-II. These results are in accord with tumorigenicity studies which demonstrate that 5-MeC-1,2-diol and *anti*-DE-I are more tumorigenic than are 5-MeC-7,8-diol and *anti*-DE-II. The enhanced reactivity of *anti*-DE-I with DNA seems to account partially for these observations. This reactivity in turn appears to be controlled by a preliminary physical interaction of *anti*-DE-I with DNA. The position of the methyl group adjacent to the bay-region epoxide may be important in the proper alignment of the dihydrodiol epoxide in the physical complex which precedes covalent binding to DNA.

While the comparative studies of 5-MeC activation via the DE-I and DE-II pathways provide an intramolecular probe for the enhancing effect of a bay-region methyl group, comparison of the metabolic activation pathways of 5-MeC and 6-MeC provides an intermolecular probe.[61] The structural difference between 5-MeC and 6-MeC is that the former contains a bay-region methyl group and the latter does not, and of course 5-MeC is a strong tumorigen whereas 6-MeC is not. Comparison of the metabolism of $[^3H]$6-MeC and $[^3H]$5-MeC in mouse skin showed that $[^3H]$6-MeC-1,2-diol was the major metabolite of the former and that its concentration was greater than that of [^{3}H]5-MeC-1,2-diol, produced from [^{3}H]5-MeC. In fact, 1,2-dihydrodiol formation was the major pathway of [^{3}H]6-MeC metabolism in mouse skin. Thus, the bay-region methyl group of 5-MeC did not enhance 1,2-dihydrodiol formation relative to 6-MeC. The tumor-initiating activities of 5-MeC-1,2-diol and 6-MeC-1,2-diol were compared. 5-MeC-1,2-diol was significantly more tumorigenic than was 6-MeC-1,2-diol. Metabolic studies showed that the conversions in mouse skin of the two dihydrodiols to tetraols, as a monitor for dihydrodiol epoxide formation, were similar (although major differences were observed in rat liver). Comparison of the extents of DNA adduct formation from [^{3}H]5-MeC and [^{3}H]6-MeC in mouse skin showed that dihydrodiol epoxide-DNA adducts from [^{3}H]5-MeC exceeded those from [^{3}H]6-MeC by about 20-fold. Taken together, these results present a picture similar to that obtained by the comparative studies of the 5-MeC-1,2-diol vs. 5-MeC-7,8-diol pathways. They indicate that the strong tumorigenicity of 5-MeC is due to the unique reactivity and tumorigenicity of *anti*-DE-I, a dihydrodiol epoxide with the methyl group and epoxide ring in the same bay region. It can be concluded from these studies that the enhancing effect of a bay-region methyl group on PAH tumorigenicity results from the unique properties of such bay-region dihydrodiol epoxide metabolites.

These bay-region dihydrodiol epoxides having a methyl group and epoxide ring in the same bay region are important in the metabolic activation of other methylated PAH. Studies on 11-methyl cyclopenta[a]phenanthren-17-one have shown that its 3,4-dihydrodiol-1,2-epoxide is an ultimate carcinogen.[84] Comparative studies of the metabolic activation of the various methylcyclopenta[a]phenanthren-17-one isomers have led to conclusions similar to those described above, although differences in repair of DNA adducts from the 11- and 12-methyl isomers also seem to be involved.[85,86] These experiments are described in more detail in Chapter 3. Investigations of the metabolic activation of 7,12-diMeB[a]A have clearly shown that the highly tumorigenic 3,4-dihydrodiol-1,2-epoxide metabolite, which has a methyl group and epoxide ring in the same bay region, is involved in the formation of major DNA adducts in mouse skin. Adducts are formed from both the *anti*- and *syn*-dihydrodiol epoxides, and it has been suggested that the *syn*-adducts, which result mainly from interaction with deoxyadenosine, may be partially responsible for the unique tumorigenicity of 7,12-diMeB[a]A.[87-90] Comparative studies of DNA adduct formation and tumorigenicity of the 3,4-dihydrodiol-1,2-epoxides of B[a]A, 7-MeB[a]A, 12-MeB[a]A, and 7,12-diMeB[a]A would be extremely interesting and informative. Such studies have been hindered, to some extent, by the relative instability of some of the appropriate dihydrodiol epoxides. Comparative investigations of MeB[a]A metabolic activation are described in Chapter 5.

X-ray crystallographic studies of methylated PAH having a bay-region methyl group have shown that distortions from normal PAH geometry occur.[91] In 5,12-diMeC and 5-MeC, in-plane and out-of-plane distortions occur to accomodate the methyl group in the bay region. The in-plane distortions result from widening of the bay region and the out-of-plane distortions from torsion about the bay-region bonds. Out-of-plane distortions are greater in 7,12-diMeB[a]A than in 5-MeC. The extent to which these distortions may play a role in the enhancing effect on tumorigenicity of a bay-region methyl group is not known, but based on the discussion above, it would be expected that distortions in the dihydrodiol epoxide metabolite would be of greatest importance. Although the x-ray crystallographic structures of these dihydrodiol epoxides have not been determined, the structures of *anti*-DE-I and *anti*-DE-II have been approximated by computer modeling. The results showed that very little additional distortion would be caused by introduction of the epoxide ring.[91] The geometry of the dihydrodiol epoxides having a methyl group and epoxide ring in the same bay region may be critical in controlling their DNA interactions. Calculations have suggested that destabilization of the *syn*-diaxial diastereomer by the DE-I methyl group might be important.[92]

B. Inhibition of Tumorigenicity by Substitution *peri* to an Angular Ring

The conformation of angular-ring dihydrodiols is important in determining their tumorigenic properties. Angular-ring dihydrodiols with the hydroxy groups in the diequatorial conformation are generally converted readily to the corresponding bay-region dihydrodiol epoxides which may be ultimate carcinogens. However, dihydrodiols with their hydroxyl groups in the diaxial conformation are frequently not converted enzymatically to the corresponding dihydrodiol epoxides and/or the dihydrodiol epoxides are not exceptionally tumorigenic.[68,93] When a methyl group occupies the *peri* position adjacent to an angular ring, as in 6-MeB[a]P or 5,12-diMeC, the conformation of the angular-ring dihydrodiol will be diaxial because of the steric effects of the methyl group. A fluorine atom has the same effect. This phenomenon has been reviewed[93] and is also discussed in Chapter 5. Thus, it has been suggested that a *peri*-methyl group inhibits tumorigenicity by altering the conformation of the dihydrodiol percursor to a bay-region dihydrodiol epoxide from diequatorial to diaxial. Extensive investigations of the metabolic activation of 6-MeB[a]P and 6-fluoroB[a]P have provided convincing evidence that a conformational change of the 7,8-dihydrodiol from diequatorial to diaxial is one reason for their lower activities compared to B[a]P.[94,95,95a]

A second possible explanation for the inhibitory effect of a *peri*-methyl group would be that it inhibits formation of the dihydrodiol in the adjacent angular ring. Comparative metabolic studies of the strong tumorigen 5,11-diMeC and the inactive *peri*-substituted 5,12-diMeC provide support for this hypothesis.[96] These experiments, which were carried out using rat and mouse liver preparations, indicated that 5,12-diMeC was preferentially metabolized at its 7,8 positions compared to its 1,2 positions. The ratio of 7-hydroxy-5,12-diMeC to 1-hydroxy-5,12-diMeC was as high as 100 to 1, and formation of 5,12-diMeC-7,8-diol greatly exceeded formation of 5,12-diMeC-1,2-diol which could not be detected as a metabolite. In 5,11-diMeC, the 1,2-dihydrodiol was formed. Although comparative studies of 5,11-diMeC and 5,12-diMeC metabolism have not been carried out in mouse skin, the results of the in vitro studies strongly suggest that a shift in metabolism from the 1,2 to the 7,8 bond is the basis for the lower tumorigenicity of 5,12-diMeC compared to 5-MeC and 5,11-diMeC. Similar results were obtained in a comparative study of 6-MeB[a]P and B[a]P.[95a] Comparative studies of [^{3}H]5-MeC and [^{3}H]12-F-5-MeC metabolism have been carried out in mouse skin.[65] Whereas the ratio of [^{3}H]5-MeC-1,2-diol to [^{3}H]5-MeC-7,8-diol was 1:1, 2 hr after topical application of [^{3}H]5-MeC, the ratio of [^{3}H]12-F-5-MeC-1,2-diol to [^{3}H]12-F-5-MeC-7,8-diol was 1:68. These results provide an explanation for the low tumorigenicity of 12-F-5-MeC noted in Section III.A. In addition, comparative metabolic studies in mouse skin of [^{3}H]6-MeC and [^{3}H]5-MeC have clearly demonstrated that, compared to [^{3}H]5-MeC, 1,2-dihydrodiol formation from [^{3}H]6-MeC is greatly favored over 7,8-dihydrodiol formation.[61] The relative symmetry of the chrysene ring system, compared to B[a]A or B[a]P, may influence the course of the enzymatic oxidations. In the absence of substitution, the 1,2 and 7,8 positions of chrysene are equivalent; a shift in metabolism upon substitution adjacent to the 1,2 or 7,8 bonds would therefore seem reasonable. This may not occur in less symmetric PAH systems, where the inhibition of tumorigenicity by *peri*-methyl substitution may result primarily from a change in dihydrodiol conformation.

V. PROSPECTS FOR FURTHER RESEARCH

With some exceptions, such as 7-MeB[a]A, the basic relationships between structure and tumorigenicity among methylated PAH and the general metabolic bases for these relationships are now reasonably well understood. This is not the case for the methylated nonalternant PAH. Nevertheless, further research is required before the detailed steps of methylated PAH activation leading to tumor initiation can be fully understood. For example, methylated PAH such as 7,12-diMeB[a]A form multiple DNA adducts in target tissues. Which adducts are responsible for tumor initiation? Do certain adducts activate proto-oncogenes, as has been suggested for the *syn*-deoxyadenosine adducts of 7,12-diMeB[a]A?[89] What are the structural features, if any, which may favor oncogene activation by methylated PAH? The stereochemical details of PAH dihydrodiol epoxide-DNA interactions are essential in determining biological activity, but these details are not well understood for methylated PAH. In addition, very little is known about methylated PAH distribution, pharmacokinetics, and excretion particularly as conjugated metabolites. Most metabolic studies have been carried out in vitro or in particular tissues such as breast or skin in vivo, but the picture in the whole animal is not clear.

Almost all bioassays of methylated PAH have been carried out by topical application in mice or by subcutaneous or intravenous injection in rats. Very little, if anything, is known about the consequences of inhalation exposure to methylated PAH, yet this is probably the major route of human exposure to these compounds. Although such assays are expensive, they could be informative if a few important methylated PAH were judiciously chosen for study. An added complication is the effects of other components of the mixtures in which the methylated PAH are always found. Some studies have been carried out on the effects of other PAH on the tumorigenicity and metabolism of methylated PAH;[97,98] more work in this area is needed. Cocarcinogens, promoters, and inhibitors of tumorigenesis can occur in the diet and in the general environment. More research is needed on their effects on methylated PAH tumorigenicity. In particular, the general conclusion that dihydrodiol epoxides are important ultimate carcinogens of methylated PAH suggests new approaches for the identification of naturally occurring or synthetic dihydrodiol epoxide scavengers, which may be effective in vivo. Extensive studies of this type are already underway for B[a]P.[99,100]

Approaches for the estimation of individual human susceptibility to methylated PAH need to be developed. With the basic metabolic patterns of the more tumorigenic members of the class having been established, it should be possible to develop sensitive assays to detect key metabolites or adducts in blood, urine, or exfoliated cells. Such assays are already available for B[a]P-DNA adducts.[101] The biological significance of particular measurable adducts or metabolites, with respect to tumor formation, needs to be established in animal studies. These results can possibly be used to provide an index of human susceptibility to methylated PAH tumorigenesis.

ACKNOWLEDGMENTS

Our studies on methylated PAH are supported by Grant No. CA-44377 from the National Cancer Institute. We thank Dr. Dietrich Hoffmann for his strong intellectual support throughout these studies. We thank Ms. Gail Thiede for typing and editing the manuscript and Mr. Keith Huie and Ms. Noreen Sweeney for helping with literature searches.

NOTE ADDED IN PROOF

In two recent studies (*Cancer Res.*, 47, 3613, 1987 and *Cancer Res.*, 47, 5310, 1987), the stereochemistry of 5-MeC metabolic activation via the highly tumorigenic 5-MeC-IR, 2S, 4R-epoxide has been established. These studies have shown that a methyl group in the same bay region as the epoxide ring is the key structural feature associated with the high tumorigenicity of bay region dihydrodiol epoxides of methylated PAH.

REFERENCES

1. **Badger, G. M., Buttery, R. G., Kimber, R. W. L., Lewis, G. E., Moritz, A. G., and Napier, I. M.,** The formation of aromatic hydrocarbons at high temperatures. Part I. Introduction, *J. Chem. Soc.*, 2449, 1952.
2. International Agency for Research on Cancer, IARC Monographs on the Evaluation of the Carcinogenic Risk of Chemicals to Humans, Polynuclear Aromatic Compounds. Part I. Chemical, Environmental and Experimental Data, Vol. 32, International Agency for Research on Cancer, Lyon, 1983, 33.
3. **Hecht, S. S., Amin, S., Rivenson, A., and Hoffmann, D.,** Tumor initiating activity of 5,11-dimethylchrysene and the structural requirements favoring carcinogenicity of methylated polynuclear aromatic hydrocarbons, *Cancer Lett.*, 65, 1979.
4. **Hartwell, J. L.,** Survey of Compounds Which Have Been Tested for Carcinogenic Activity, Publ. No. 149, U.S. Public Health Service, Washington, D.C., 1951.
5. **Shubik, P. and Hartwell, J. L.,** Survey of Compounds Which Have Been Tested for Carcinogenic Activity, Publ. No. 149, Suppl. 1, U.S. Public Health Service, Washington, D.C., 1957.
6. **Shubik, P. and Hartwell, J. L.,** Survey of Compounds Which Have Been Tested for Carcinogenic Activity, Publ. No. 149, Suppl. 2, U.S. Public Health Service, Washington, D.C., 1969.
7. **Thompson, J. I.,** Survey of Compounds Which Have Been Tested for Carcinogenic Activity, Publ. No. 149, 1968—1969 Vol., U.S. Public Health Service, Washington, D.C., 1972.
8. **Thompson, J. I.,** Survey of Compounds Which Have Been Tested for Carcinogenic Activity, Publ. No. 149, 1970—1971 Vol., U.S. Public Health Service, Washington, D.C., 1974.
9. National Cancer Institute, Survey of Compounds Which Have Been Tested for Carcinogenic Activity, Publ. No. 149, 1972—1973 Vol., U.S. Public Health Service, Washington, D.C., 1975.
10. National Cancer Institute, Survey of Compounds Which Have Been Tested for Carcinogenic Activity, Publ. No. 149, 1974—1975 Vol., U.S. Public Health Service, Washington, D.C., 1983.
11. National Cancer Institute, Survey of Compounds Which Have Been Tested for Carcinogenic Activity, Publ. No. 149, 1976—1977 Vol., U.S. Public Health Service, Washington, D.C. 1984.
12. National Cancer Institute, Survey of Compounds Which Have Been Tested for Carcinogenic Activity, Publ. No. 149, 1978 Vol., U.S. Public Health Service, Washington, D.C., 1980.
13. **Wislocki, P. G., Fiorentini, K. M., Fu, P. P., Yang, S. K., and Lu, A. Y. H.,** Tumor-initiating ability of the twelve monomethylbenz[a]anthracenes, *Carcinogenesis*, 3, 215, 1982.
14. **Stevenson, J. L. and Von Haam, E.,** Carcinogenicity of benz[a]anthracene and benzo[c]phenanthrene derivatives, *Am. Ind. Hyg. Assoc. J.*, 475, 1965.
15. **Miller, E. C. and Miller, J. A.,** The carcinogenicity of fluoro derivatives of 10-methyl-1,2-benzanthracene. I. 3- and 4′-monofluoro derivatives. *Cancer Res.*, 20, 133, 1960.
16. **DiGiovanni, J., Slaga, T. J., and Boutwell, R. K.,** Comparison of the tumor-initiating activity of 7,12-dimethylbenz[a]anthracene and benzo[a]pyrene in female SENCAR and CD-1 mice, *Carcinogenesis*, 1, 381, 1980.
17. **Slaga, T. J., Gleason, G. L., DiGiovanni, J., Berry, D. L., Juchau, M. R., Fu, P. P., Sukumaron, K. B., and Harvey, R. G.,** Tumor-initiating activities of various derivatives of benz[a]anthracene and 7,12-dimethylbenz[a]anthracene in mouse skin, in *Polynuclear Aromatic Hydrocarbons*, Jones, P. W. and Leber, P., Eds., Ann Arbor Science, Ann Arbor, Mich., 1979, 753.
18. **Wood, A. W., Levin, W., Chang, R. L., Conney, A. H., Slaga, T. J., O'Malley, R. F., Newman, M. S., Buhler, D. R., and Jerina, D. M.,** Mouse skin tumor-initiating activity of 5-, 7-, and 12-methyl- and fluorine-substituted benz[a]anthracenes, *J. Natl. Cancer Inst.*, 69, 725, 1982.
19. **Barry, G., Cook, J. W., Haslewood, G. A. D., Hewett, C. L., Hieger, I., and Kenneway, E. L.,** The production of cancer by pure hydrocarbons. III, *Proc. R. Soc. London, Ser. B*, 117, 318, 1935.
20. **Heidelberger, C., Baumann, M. E., Griesbach, L., Ghobar, A., and Vaughan, T. M.,** The carcinogenic activities of various derivatives of dibenzanthracene, *Cancer Res.*, 22, 78, 1962.
21. **Hill, W. T., Stanger, D. W., Pizzo, A., Riegel, B., Shubik, P., and Wartman, W. B.,** Inhibition of 9,10-dimethyl-1,2-benzanthracene skin carcinogenesis in mice by polycyclic hydrocarbons, *Cancer Res.*, 11, 892, 1951.
22. **Slaga, T. J., Huberman, E., DiGiovanni, J., Gleason, G., and Harvey, R. G.,** The importance of the "bay region" diol-epoxide in 7,12-dimethylbenz[a]anthracene skin tumor initiation and mutagenesis, *Cancer Lett.*, 6, 213, 1979.
23. **Levin, W., Wood, A. W., Chang, R. L., Newman, M. S., Thakker, D. R., Conney, A. H., and Jerina, D. M.,** The effect of steric strain in the bay region of polycyclic aromatic hydrocarbons: tumorigenicity of alkyl-substituted benz[a]anthracenes, *Cancer Lett.*, 20, 139, 1983.
24. **Badger, G. M., Cook, J. W., Hewett, C. L., Kennaway, E. L., Kennaway, N. M., Martin, R. H., and Robinson, A. M.,** The production of cancer by pure hydrocarbons, *Proc. R. Soc. London*, 129, 439, 1940.

25. **Welsch, C. W.,** Host factors affecting the growth of carcinogen-induced rat mammary carcinomas: a review and tribute to Charles Brenton Huggins, *Cancer Res.,* 45, 3415, 1985.
26. **DiGiovanni, J., Diamond, L., Harvey, R. G., and Slaga, T. J.,** Enhancement of the skin tumor-initiating activity of polycyclic aromatic hydrocarbons by methyl-substitution at non-benzo 'bay-region' positions, *Carcinogenesis,* 4, 403, 1983.
27. **Slaga, T. J., Gleason, G. L., and Hardin, L.,** Comparison of the skin tumor initiating activity of 3-methylcholanthrene and 3,11-dimethylcholanthrene in mice, *Cancer Lett.,* 7, 97, 1979.
28. **Butenandt, A. and Dannenberg, H.,** Untersuchungen über die krebserzeugende Wirksamkeit der Methylhomologen des 1,2-Cyclopentenophenanthrens, *Arch. Geschwulstforsch.,* 6, 1, 1953.
29. **Coombs, M. M. and Croft, C. J.,** Carcinogenic cyclopenta[a]phenanthrenes, *Progr. Exp. Tumor Res.,* 11, 69, 1969.
30. **Coombs, M. M. and Bhatt, T. S.,** Lack of initiating activity in mutagens which are not carcinogenic, *Br. J. Cancer,* 38, 148, 1978.
31. **Coombs, M. M. and Croft, C. J.,** Carcinogenic derivatives of cyclopenta[a]phenanthrene, *Nature,* 210, 1281, 1966.
32. **Coombs, M. M., Bhatt, T. S., and Croft, C. J.,** Correlation between carcinogenicity and chemical structure in cyclopenta[a]phenanthrenes, *Cancer Res.,* 33, 832, 1973.
33. **Coombs, M. M., Bhatt, T. S., and Young, S.,** The carcinogenicity of 15,16-dihydro-11-methylcyclopenta[a]phenanthren-17-one, *Br. J. Cancer,* 40, 914, 1979.
34. **Hecht, S. S., Bondinell, W. E., and Hoffmann, D.,** Chrysene and methylchrysenes: presence in tobacco smoke and carcinogenicity. *J. Natl. Cancer Inst.,* 53, 1121, 1974.
35. **Hecht, S. S., Radok, L., Amin, S., Huie, K., Melikian, A. A., Hoffmann, D., Pataki, J., and Harvey, R. G.,** Tumorigenicity of 5-methylchrysene dihydrodiols and dihydrodiol epoxides in newborn mice and on mouse skin, *Cancer Res.,* 45, 1449, 1985.
36. **Hecht, S. S., Loy, M., Maronpot, R. R., and Hoffmann, D.,** Comparative carcinogenicity of 5-methylchrysene, benzo[a]pyrene, and modified chrysenes, *Cancer Lett.,* 1, 147, 1976.
37. **Hecht, S. S., Hirota, N., Loy, M., and Hoffmann, D.,** Tumor initiating activity of fluorinated 5-methylchrysenes, *Cancer Res.,* 38, 1694, 1978.
38. **Hecht, S. S. and Amin, S.,** unpublished data.
39. **Coombs, M. M., Bhatt, T. S., Hall, M., and Croft, C. J.,** The relative carcinogenic activities of a series of 5-methylchrysene derivatives, *Cancer Res.* 34, 1315, 1974.
40. **El-Bayumy, K., Hecht, S. S., and Hoffmann, D.,** Comparative tumor-initiating activity on mouse skin of 6-nitrobenzo[a]pyrene, 6-nitrochrysene, 3-nitroperylene, 1-nitropyrene, and their parent hydrocarbons, *Cancer Lett.,* 16, 333, 1982.
41. **Van Duuren, B. L., Sivak, A., Segal, A., Orris, L., and Langseth, L.,** The tumor-promoting agents of tobacco leaf and tobacco smoke condensate, *J. Natl. Cancer Inst.,* 37, 519, 1966.
42. **Rice, J. and LaVoie, E. J.,** unpublished results.
43. **Hoffman, D., Rathkamp, G., Nesnow, S., and Wynder, E. L.,** Fluoranthenes: quantitative determination in cigarette smoke, formation by pyrolysis, and tumor-initiating activity, *J. Natl. Cancer Inst.,* 49, 1165, 1972.
44. **Iyer, R. P., Lyga, J. W., Secrist, J. A., III, Daub, G. H., and Slaga, T. J.,** Comparative tumor-initiating activity of methylated benzo[a]pyrene derivatives on mouse skin, *Cancer Res.,* 40, 1073, 1980.
45. **Hecht, S. S., Hirota, N., and Hoffmann, D.,** Comparative tumor initiating activity of 10-methylbenzo[a]pyrene, 7,10-dimethylbenzo[a]pyrene, and benzo[a]pyrene, *Cancer Lett.,* 5, 179, 1978.
46. **Cavalieri, E., Roth, R., Grandjean, C., Althoff, J., Patil, K., Liakus, S., and Marsh, S.,** Carcinogenicity and metabolic profiles of 6-substituted benzo[a]pyrene derivatives on mouse skin, *Chem. Biol. Interact.,* 22, 53, 1978.
47. **Chouroulinkov, I., Coulomb, H., MacNicoll, A. D., Grover, P. L., and Sims, P.,** Tumour-initiating activities of dihydrodiols of dibenz[a,c]anthracene, *Cancer Lett.,* 19, 21, 1983.
48. **Amin, S., Hussain, N., Balanikas, G., Huie, K., and Hecht, S. S.,** Mutagenicity and tumor initiating activity of methylated benzo[b]fluoranthenes, *Carcinogenesis,* 1023, 1985.
49. **Amin, S., Hussain, N., Balanikas, G., Huie, K., and Hecht, S. S.,** Mutagenicity and tumor initiating activity of methylated benzo[k]fluoranthenes, *Cancer Lett.,* 26, 343, 1985.
50. **Hecht, S. S., Amin, S., Melikian, A. A., LaVoie, E. J., and Hoffmann, D.,** Effects of methyl and fluorine substitution on metabolic activation and tumorigenicity of polycyclic aromatic hydrocarbons, in *Polycyclic Hydrocarbons and Carcinogenesis, ACS Symposium Series 283,* Harvey, R. G., Ed., American Chemical Society, Washington, D.C., 1985, 85.
51. **Hecht, S. S., Loy, M., Mazzarese, R., and Hoffmann, D.,** Synthesis and mutagenicity of modified chrysenes related to the carcinogen, 5-methylchrysene, *J. Med. Chem.,* 21, 38, 1978.
52. **Hecht, S. S., LaVoie, E. J., Mazzarese, R., Hirota, N., Ohmori, T., and Hoffmann, D.,** Comparative mutagenicity, tumor initiating activity, carcinogenicity, and *in vitro* metabolism of fluorinated 5-methylchrysenes, *J. Natl. Cancer Inst.,* 63, 855, 1979.

53. **Hecht, S. S., LaVoie, E. J., Mazzarese, R., Amin, S., Bedenko, V., and Hoffmann, D.,** 1,2-Dihydro-1,2-dihydroxy-5-methylchrysene, a major activated metabolite of the environmental carcinogen, 5-methylchrysene, *Cancer Res.*, 38, 2191, 1978.
54. **Amin, S., Hecht, S. S., LaVoie, E., and Hoffmann, D.,** Synthesis and mutagenicity of 5,11-dimethylchrysene and some methyl oxidized derivatives of 5-methylchrysene, *J. Med. Chem.*, 22, 1336, 1979.
55. **Amin, S., Hecht, S. S., and Hoffmann, D.,** Synthesis of angular ring methoxy-5-methylchrysenes, and 5-methylchrysenols, *J. Org. Chem.*, 46, 2394, 1981.
56. **Amin, S., Juchatz, A., Furuya, K., and Hecht, S. S.,** Effects of fluorine substitution on the tumor initiating activity and metbolism of 5-hydroxymethylchrysene, a tumorigenic metabolite of 5-methylchrysene, *Carcinogenesis,* 2, 1027, 1981.
57. **Melikian, A., LaVoie, E. J., Hecht, S. S., and Hoffmann, D.,** Influence of a bay region methyl group on formation of 5-methylchrysene dihydrodiol epoxide: DNA adducts in mouse skin, *Cancer Res.*, 42, 1239, 1982.
58. **Melikian, A. A., LaVoie, E. J., Hecht, S. S., and Hoffmann, D.,** 5-Methylchrysene metabolism in mouse epidermis *in vivo,* diol epoxide-DNA adduct persistence, and diol epoxide reactivity with DNA as potential factors influencing the predominance of 5-methylchrysene-1,2-diol-3,4-epoxide-DNA adducts in mouse epidermis, *Carcinogenesis,* 4, 843, 1983.
59. **Amin, S., Camanzo, J., Huie, K., and Hecht, S. S.,** Improved photochemical synthesis of 5-methylchrysene and its application to the preparation of 7,8-dihydro-7,8-diydroxy-5-methylchrysene, *J. Org. Chem.*, 49, 831, 1984.
60. **Melikian, A. A., Hecht, S. S., Hoffmann, D., Pataki, J., and Harvey, R. G.,** Analysis of *syn-* and *anti-*1,2-dihydroxy-3,4-epoxy-1,2,3,4-tetrahydro-5-methylchrysene-deoxyribonucleoside adducts by boronate chromatography, *Cancer Lett.*, 27, 91, 1985.
61. **Amin, S., Huie, K., Melikian, A. A., Leszczynska, J. M., and Hecht, S. S.,** Comparative metabolic activation of the weak carcinogen 6-methylchrysene and the strong carcinogen 5-methylchrysene, *Cancer Res.*, 45, 6406, 1985.
62. **Pataki, J., Lee, H., and Harvey, R. G.,** Carcinogenic metabolites of 5-methylchrysene, *Carcinogenesis,* 4, 399, 1983.
63. International Agency for Research on Cancer, IARC Monographs on the Evaluation of the Carcinogenic Risk of Chemicals to Humans, Polynuclear Aromatic Compounds. I. Chemical, Environmental and Experimental Data, Vol. 32, International Agency for Research on Cancer, Lyon, 1983, 57.
64. **Hodgson, R. M., Pal, K., Grover, P. L., and Sims, P.,** The metabolic activation of chrysene by hamster embryo cells, *Carcinogenesis,* 3, 1051, 1982.
65. **Amin, S., Camanzo, J., and Hecht, S. S.,** Inhibition by a perifluorine atom of 1,2-dihydrodiol formation as a basis for the lower tumorigenicity of 12-fluoro-5-methylchrysene than of 5-methylchrysene, *Cancer Res.*, 44, 3772, 1984.
66. **Hecht, S. S., Rivenson, A., and Hoffmann, D.,** Tumor-initiating activity of dihydrodiols formed metabolically from 5-methylchrysene, *Cancer Res.*, 40, 1396, 1980.
67. **Conney, A. H.,** Induction of microsomal enzymes by foreign chemicals and carcinogenesis by polycyclic aromatic hydrocarbons: G. H. A. Clowes Memorial Lecture, *Cancer Res.*, 42, 4875, 1982.
68. **Lehr, R. E., Kumar, S., Levin, W., Wood, A. W., Chang, R. L., Conney, A. H., Yagi, H., Sayer, J. M., and Jerina, D. M.,** The bay region theory of polycyclic aromatic hydrocarbon carcinogenesis, in *Polycyclic Hydrocarbons and Carcinogenesis,* ACR Symp. Ser. 283, Harvey, R. G., Ed., American Chemical Society, Washington, D.C., 1985, 63.
69. **Hecht, S. S., Radok, L., Amin, S., Huie, K., Melikian, A. A., Hoffmann, D., Pataki, J., and Harvey, R. G.,** Tumorigenicity of 5-methylchrysene dihydrodiols and dihydrodiol epoxides in newborn mice and on mouse skin, *Cancer Res.*, 45, 1449, 1985.
70. **Cavalieri, E. and Rogan, E.,** One-electron oxidation in aromatic hydrocarbon carcinogenesis, in *Polycyclic Hydrocarbons and Carcinogenesis,* ACS Symposium Ser. 283, Harvey, R. G., Ed., American Chemical Society, Washington, D.C., 1985, 289.
71. **Flesher, J. W. and Sydnor, K. L.,** Carcinogenicity of derivatives of 7,12-dimethylbenz[a]anthracene, *Cancer Res.*, 31, 1951, 1971.
72. **Watabe, T., Ishizuka, T., Isobe, M., and Ozawa, N.,** A 7-hydroxymethyl sulfate ester as an active metabolite of 7,12-dimethylbenz[a]anthracene, *Science,* 215, 403, 1982.
73. **Dipple, A., Tomaszewski, J. E., Moschel, R. C., Bigger, C. A. H., Nebzydoski, J. A., and Egan, M.,** Comparison of metabolism-mediated binding to DNA of 7-hydroxymethyl-12-methylbenz[a]anthracene and 7,12-dimethylbenz[a]anthracene, *Cancer Res.*, 39, 1154, 1979.
74. **Melikian, A. A., Amin, S., Hecht, S. S., Hoffmann, D., Pataki, J., and Harvey, R. G.,** Identification of the major adducts formed by reaction of 5-methylchrysene anti-dihydrodiol epoxides with DNA *in vitro, Cancer Res.*, 44, 2524, 1984.
75. **Dipple, A., Moschel, R. C., and Bigger, C. A. H.,** Polynuclear aromatic hydrocarbons, in *Chemical Carcinogens,* 2nd ed. ACS Monograph 182, Searle, C. E., Ed., American Chemical Society, Washington, D.C., 1984, 41.

76. **Melikian, A. A., Leszczynska, J. M., Amin, S., Hecht, S. S., Hoffmann, D., Pataki, J., and Harvey, R. G.,** Rates of hydrolysis and extents of DNA binding of 5-methylchrysene dihydrodiol epoxides, *Cancer Res.,* 45, 1990, 1985.
77. **Kim, M-H., Geacintov, N. E., Pope, M., Pataki, J., and Harvey, R. G.,** Reaction mechanisms of *trans*-1,2-dihydroxy-*anti*-3,4-epoxy-1,2,3,4-tetrahydro-5-methylchrysene with DNA in aqueous solutions, *Carcinogenesis,* 6, 121, 1985.
78. **Geacintov, N. E., Yoshida, H., Ibanez, V., and Harvey, R. G.,** Noncovalent binding of 7β,8α-dihydroxy-9α,10α-epoxy tetrahydrobenzo(a)pyrene to deoxyribonucleic acid and its catalytic effect on hydrolysis of the diol epoxide to tetrol, *Biochemistry,* 21, 1864, 1982.
79. **MacLeod, M. C. and Selkirk, J. K.,** Physical interactions of isomeric benzo[a]pyrene diol-epoxides with DNA, *Carcinogenesis,* 3, 287, 1982.
80. **Meehan, T., Gamper, H., and Becker, J. H.,** Characterization of reversible physical binding of benzo[a]pyrene to DNA. *J. Biol. Chem.,* 257, 10479, 1982.
81. **Jerina, D. M. and Daly, J. W.,** Oxidation at carbon, in *Drug Metabolism — from Microbe to Man,* Parke, D. V. and Smith, R. L., Eds., Taylor and Francis, London, 1977, 13.
82. **Wong, T. K., Chiu, P-L., Fu, P. P., and Yang, S. K.,** Metabolic study of 7-methylbenzo[a]pyrene with rat liver microsomes: separation by reverse-phase and normal-phase high performance liquid chromatography and characterization of metabolites, *Chem. Biol. Interact.,* 36, 153, 1981.
83. **Kinoshita, T., Konieczny, M., Santella, R., and Jeffrey, A. M.,** Metabolism and covalent binding to DNA of 7-methylbenzo[a]pyrene, *Cancer Res.,* 42, 4032, 1982.
84. **Coombs, M. M. and Bhatt, T. S.,** High skin tumour initiating activity of the metabolically derived *trans*-3,4-dihydrodiol of the carcinogen 15,16-dihydro-11-methylcyclopenta[c]phenanthren-17-one, *Carcinogenesis,* 3, 449, 1982.
85. **Coombs, M. M., Russell, J. C., Jones, J. R., and Ribeiro, O.,** A comparative examination of the *in vitro* metabolism of five cyclopenta[a]phenanthrenes of varying carcinogenic potential, *Carcinogenesis,* 6, 1217, 1985.
86. **Russell, J. C., Bhatt, T. S., Jones, J. R., and Coombs, M. M.,** Comparison of the binding of some carcinogenic and non-carcinogenic cyclopenta[a]phenanthrenes to DNA *in vitro* and *in vivo, Carcinogenesis,* 6, 1223, 1985.
87. **Slaga, T. J., Gleason, G. L., DiGiovanni, J., Sukumaran, K. B., and Harvey, R. G.,** Potent tumor-initiating activity of the 3,4-dihydrodiol of 7,12-dimethylbenz[a]anthracene in mouse skin, *Cancer Res.,* 39, 1934, 1979.
88. **Sawicki, J. T., Moschel, R. C., Dipple, A.,** Involvement of both *syn*- and *anti*-dihydrodiol-epoxides in the binding of 7,12-dimethylbenz[a]anthracene to DNA in mouse embryo cell cultures, *Cancer Res.,* 43, 3212, 1983.
89. **Dipple, A., Pigott, M., Moschel, R. C., Costantino, N.,** Evidence that binding of 7,12-dimethylbenz[a]anthracene to DNA in mouse embryo cell cultures results in extensive substitution of both adenine and guanine residues, *Cancer Res.,* 43, 4132, 1983.
90. **Moschel, R. C., Pigott, M. A., Costantino, N., Dipple, A.,** Chromatographic and fluorescence spectroscopic studies of individual 7,12-dimethylbenz[a]anthracene-deoxyribonucleoside adducts, *Carcinogenesis,* 4, 1201, 1983.
91. **Zacharias, D. E., Kashino, S., Glusker, J. P., Harvey, R. G., Amin, S., and Hecht, S. S.,** The bay region geometry of some 5-methylchrysenes: steric effects in 5,6- and 5,12-dimethylchrysenes, *Carcinogenesis,* 5, 1421, 1984.
92. **Silverman, B. D.,** Effect of methylation on the conformer stability and reactivity of the bay-region diol-epoxides of chrysene, *Chem. Biol. Interact.,* 53, 313, 1985.
93. **Yang, S. K., Chou, M. W., and Fu, P. P.,** Metabolic and structural requirements for the carcinogenic properties of unsubstituted and methyl-substituted polycyclic aromatic hydrocarbons, in: *Carcinogenesis: Fundamental Mechanisms and Environmental Effects,* Pullmann, B., Ts'o, P. O. P., and Gelboin, H. V., Eds., D. Reidel, London, 1980, 143.
94. **Buhler, D. R., Ünlü, F., Thakker, D. R., Slaga, T. J., Conney, A. H., Wood, A. W., Chang, R. L., and Jerina, D. M.,** Effect of a 6-fluoro substituent on the metabolism and biological activity of benzo[a]pyrene, *Cancer Res.,* 43, 1541, 1983.
95. **Chiu, P-L., Fu, P. P., and Yang, S. K.,** Effect of a peri fluoro substituent on the conformation of dihydrodiol derivatives of polycyclic aromatic hydrocarbons, *Biochem. Biophys. Res. Commun.,* 106, 1405, 1982.

95a. **Hamernik, K. L., Chin, P.-L., Chou, M. N., Fu, P. P., and Yang, S. K.,** Metabolic activation of 6-methylbenzo[a]pyrene, in *Polynuclear Aromatic Hydrocarbons: Formation, Metabolism, and Measurement,* Cooke, M. and Dennis, A. J., Eds., Battelle Press, Columbus, Ohio, 1983, 583.

96. **Amin, S., Camanzo, J., and Hecht, S. S.,** Identification of metabolites of 5,11-dimethylchrysene and 5,12-dimethylchrysene and the influence of a peri-methyl group on their formation, *Carcinogenesis,* 3, 1159, 1982.

97. **Baird, W. M., Salmon, C. P., and Diamond, L.,** Benzo(e)pyreneinduced alterations in the metabolic activation of benzo[a]pyrene and 7,12-dimethylbenz[a]anthracene by hamster embryo cells. *Cancer Res.,* 44, 1445, 1984.
98. **Slaga, T. J., Jecker, L., Bracken, W. M., and Weeks, C. E.,** The effects of weak or non-carcinogenic polycyclic hydrocarbons on 7,12-dimethylbenz[a]anthracene and benzo[a]pyrene skin tumor-initiation. *Cancer Lett.* 7, 51, 1979.
99. **Chang, R. L., Huang, M.-T., Wood, A. W., Wong, C.-Q., Newmark, H. L., Yagi, H., Sayer, J. M., Jerina, D. M., and Conney, A. H.,** Effect of ellagic acid and hydroxylated flavonoids on the tumorigenicity of benzo[a]pyrene and (±)-7β,8α-dihydroxy-9α, 10α-epoxy-7,8,9,10-tetrahydrobenzo[a]pyrene on mouse skin and in the newborn mouse. *Carcinogenesis,* 6, 1127, 1985.
100. **Huang, M-T., Chang, R. L., Wood, A. W., Newmark, H. L., Sayer, J. M., Yagi, H., Jerina, D. M., and Conney, A. H.,** Inhibition of the mutagenicity of bay-region diol-epoxides of polycyclic aromatic hydrocarbons by tannic acid, hydroxylated anthraguinones and hydroxylated cinnamic acid derivatives, *Carcinogenesis,* 6, 237, 1985.
101. **Garner, R. C.,** Assessment of carcinogen exposure in man, *Carcinogenesis,* 6, 1071, 1985.

Chapter 5

METABOLISM AND ACTIVATION OF BENZ[a]ANTHRACENE AND METHYLBENZ[a]ANTHRACENES

Shen K. Yang

TABLE OF CONTENTS

I. INTRODUCTION

The most exciting period of time in the history of polycyclic aromatic hydrocarbon (PAH) carcinogenesis research, has been 1973—1977, ever since PAHs were discovered as the components in soot which may be responsible for the induction of cancer in man. The landmark findings during this period of time are

1. K-region epoxides are not responsible for the covalent binding to DNA in cells in culture which had been treated with 7-methylbenz[a]anthracene[1] (7-MBA; other methylbenz[a]anthracenes are similarly abbreviated) or benzo[a]pyrene[2] (BaP)
2. BaP 7,8-dihydrodiol (a *trans* isomer), a metabolite of BaP, is further metabolized to products that bind more strongly to DNA in vitro than BaP[3]
3. A BaP 7,8-dihydrodiol-9,10-epoxide was found to be responsible for covalent binding to DNA in cultured cells pretreated with BaP[4,5]
4. An *anti*-7,8-dihydrodiol-9,10-epoxide (*anti* and *syn* indicate that the benzylic hydroxyl group are *trans* and *cis* to the epoxide oxygen, respectively) was found to be the most potent mutagenic derivative of BaP to mammalian cells in culture[6-9]
5. A bay-region theory was proposed[10,11] and was supported by subsequent carcinogenicity tests of many PAHs and their derivatives[9]
6. Chemical synthesis of *anti* and *syn* stereoisomers of BaP 7,8-dihydrodiol-9,10-epoxides[12,13]
7. 7*R*,8*R*-dihydrodiol is the metabolic precursor of the ultimate carcinogenic metabolite of BaP[14-17]
8. A specific stereoisomer of BaP 7,8-dihydrodiol-9,10-epoxide, (+)7*R*,8*S*-dihydroxy-9*S*,10*R*-epoxy-7,8,9,10-tetrahydrobenzo[a]pyrene (*anti*-(+)-7,8-dihydrodiol-9,10-epoxide), was found to be the predominant metabolite[14,17,18] that binds to the exocyclic 2-amino group of guanine in DNA both in vitro and in vivo[17,19-23]
9. The 7*R*,8*R*-dihydrodiol and *anti*-(+)-7*R*,8*S*-dihydrodiol-9*S*,10*R*-epoxide are the most carcinogenic dihydrodiol and dihydrodiol-epoxide derivatives of BaP, respectively[24-29]

During this time, K-region (or K-L region) theory[30,31] quickly fell into disfavor. In retrospect, it is interesting to note that the K-L region theory played such a dominant role in influencing PAH carcinogenesis research for almost 30 years.

Since 1978, a large number of review articles[32-40] that dealt with various aspects of PAH carcinogenesis have appeared in the literature. In addition to the simple molecular orbital calculation which forms the basis of the bay-region theory,[10,11] there have also been many publications that dealt with simple as well as sophisticated approaches in calculations, attempting to provide a theoretical basis for the structure-activity relationship of methylbenz[a]anthracenes.[41-43]

In this chapter, current advances on the metabolic studies will be presented in an attempt to understand the molecular basis for the observed differences in carcinogenic potencies of methylbenz[a]anthracenes. Since metabolism is required for carcinogenic PAHs to express their mutagenic and carcinogenic activities, we will examine what metabolic and structural factors affect the activation and detoxification pathways. The bay-region theory of hydrocarbon-induced carcinogenesis and other hypotheses of PAH carcinogenesis have been discussed elsewhere,[44-46] and details of these discussions will not be included in this chapter.

II. NOMENCLATURE

Various regions of BA are defined in Figure 1. These are K-region, bay region, M-region, and L-region. The K-region is the 5,6-double bond and is the most electron-rich region of

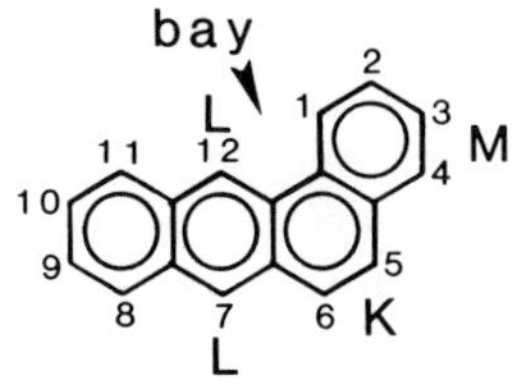

FIGURE 1. Numbering system and designations of various regions of benz[a]anthracene (BA).

the molecule. The L-region includes both C_7 and C_{12} and is the most reactive (with respect to substitution and addition reactions) region of the molecule. The bay region is the angular region between and including the C_1 and C_{12} positions. The M-region is the site where two important metabolic reactions (epoxidation and hydration) occur, producing a *trans*-dihydrodiol (all *trans*-dihydrodiols will be abbreviated in this chapter as dihydrodiols), which is the metabolic precursor of the bay region dihydrodiolepoxides. Thus, BaP 7,8-dihydrodiol, chrysene 1,2-dihydrodiol, BA 3,4-dihydrodiol, and benzo[c]phenanthrene (BcPh) 3,4-dihydrodiol are all M-region dihydrodiols. Unfortunately, many investigators have mistakenly called these M-region dihydrodiols "bay-region dihydrodiols". Dihydrodiols such as BaP 9,10-dihydrodiol, chrysene 3,4-dihydrodiol, BA 1,2-dihydrodiol, and BcPh 1,2-dihydrodiol are true bay-region dihydrodiols, since for these molecules one of the two hydroxyl groups is actually in the bay region. The unique structural feature of bay-region dihydrodiols is that both of hydroxyl groups adopt quasiaxial conformations. It should be emphasized, however, that although the letter "M" is also used here, the M-region is not necessarily the "metabolical most-reactive part of the molecule" as suggested by Pullman.[31] Although the M-region of a PAH may be the major site of metabolism, it is not the most chemically reactive site. Another important term in this discussion is *"peri"*; C_4 and C_5, C_6 and C_7, and C_7 and C_8 in BA are each a pair of carbon atoms that are *peri* to each other.

The rings including the 1,2,3,4 positions and the 8,9,10,11 positions are called angular benzo-ring and nonangular benzo-ring, respectively. "Bay-region dihydrodiol-epoxides" should only be reserved to compounds such as BA 3,4-dihydrodiol-1,2-epoxide and BaP 7,8-dihydrodiol-9,10-epoxide, in which the epoxide oxygen is situated in the bay region. It is possible to have a "double bay-region" dihydrodiol-epoxides (e.g., benzo[e]pyrene 9,10-dihydrodiol-10,11-epoxide). Other types of dihydrodiol-epoxides are called nonbay-region dihydrodiol-epoxide. In this chapter, dihydrodiol-epoxide always refers to a vicinal(benzo-ring) dihydrodiol-epoxide. It is possible to have a dihydrodiol-epoxide that is not a vicinal dihydrodiol-epoxide, e.g., BA 3,4-dihydrodiol-8,9-epoxide.

III. CARCINOGENIC AND TUMOR-INITIATING ACTIVITIES

The relative carcinogenic and tumor-initiating potencies of methylbenz[a]anthracenes on mouse skin have been known for some years and have often been cited in the context of structure-activity relationship of PAHs.[9,47] The relative activities are listed below.

1. Relative carcinogenic activity:[47] 7,12-DMBA $>$ 7-MBA $>$ 8-MBA $\sim$ 6-MBA $\sim$ 12-MBA $>$ 9-MBA $\sim$ 10-MBA $\sim$ 11-MBA $>$ 5-MBA $\sim$ 1-MBA $\sim$ 2-MBA $\sim$ 3-MBA $\sim$ 4-MBA $\sim$ BA.
2. Relative Tumor-initiating activity:[9] 7,12-DMBA $>$ 7-MBA $>$ 12-MBA $>$ 8-MBA $\sim$ 6-MBA $\sim$ 9-MBA $> \sim$ 10-MBA $\sim$ 11-MBA $>$ 5-MBA $\sim$ 1-MBA $\sim$ 3-MBA $>$ 2-MBA $\sim$ 4-MBA $\sim$ BA.

IV. METABOLIC FATES OF EPOXIDES

Arene epoxides — Addition of an oxygen atom across any aromatic double bond results in an arene epoxide. It is often simply called an epoxide. Arene epoxides are important initial metabolic oxidation products of PAHs. The major metabolic fates of an enzymically formed arene epoxide are (1) nonenzymatic (or spontaneous) rearrangement to one or both of two possible phenolic products, (2) enzymatic hydration to a *trans*-dihydrodiol, and (3) enzymatic conjugation with glutathione.[48,49]

Benzo-ring tetrahydro-epoxides — Benzo-ring tetrahydro-epoxides include vicinal dihydrodiol-epoxides and are a class of PAH derivatives that received more attention since 1974.[4] Bay-region tetrahydro-epoxides (e.g., 1,2-epoxy-1,2,3,4-tetrahydro-BA) and dihydrodiol-epoxides (e.g., BA 3,4-dihydrodiol-1,2-epoxide) are chemically the most reactive derivatives of any particular PAH known. Nonbay region tetrahydro-epoxides (e.g., 3,4-epoxy-1,2,3,4-tetrahydro-BA) are much less reactive chemically. The fates of metabolically formed tetrahydro-epoxides (formed via an epoxidation reaction at the isolated double bond of a dihydro-PAH such as 1,2-epoxy-1,2,3,4-tetrahydro-BA) and dihydrodiol-epoxides (formed via three enzyme-catalyzed reactions at a benzo-ring of a PAH) are (1) non-enzymatic and enzymatic hydrolyses to form tetrahydrodiols or tetrahydrotetrols, respectively.[12,18,50] (2) reaction with proteins and nucleic acids through the benzylic carbocation via S_N1 and/or S_N2 mechanisms, (3) enzymatic conjugation with glutathione.[51] Nonbay region dihydrodiol-epoxides are generally more stable than bay region dihydrodiol-epoxides and may have a long halflife at neutral or slightly basic pH aqueous solutions. Some non-K-region dihydrodiol-epoxides are enzymatically hydrated to tetrols by the action of microsomal epoxide hydrolase.[50,52] It is important to emphasize that arene epoxides and tetrahydro-epoxides are two different kinds of epoxide derivatives and have very different chemical properties. The term "epoxide" should not be used indiscriminately to include both types of epoxides.

V. METABOLIC REACTIONS

Many enzymes are responsible for the activation and detoxification of PAHs. These are cytochrome P-450 isozymes, epoxide hydrolase, NADPH cytochrome P-450 reductase, glutathione transferase, glucuronide transferase, sulfate transferase, and phospholipid. NADPH is required for P-450-catalyzed oxidation reactions. There are multiple forms of cytochrome P-450 isozymes.[53] P-450s are the sites for oxygen and substrate activation, Different forms of cytochrome P-450 isozymes may have different substrate specificity. The regioselectivity and stereoselectivity of cytochrome P-450 isozymes have important consequences in the metabolic activation and detoxification pathways of PAHs (see below). Epoxides hydrolase has dual roles in PAH carcinogenesis; it catalyzes the hydration of M-region epoxides to M-region dihydrodiols, which are the metabolic precursors of the reactive bay-region dihydrodiol-epoxides, and converts other arene epoxides to noncarcinogenic dihydrodiols.[54] NADPH-cytochrome P-450 reductase catalyzes the transfer of electrons from NADPH to cytochrome P-450. Phospholipid facilitates the interaction between reductase and cytochrome P-450s. Transferases generally convert epoxides, phenols, and dihydrodiols to more polar detoxified products. However, some examples are known to indicate that conjugates resulting from transferase action may possess high biological activities.[46,55]

VI. CONFORMATION OF DIHYDRODIOLS AND VICINAL DIHYDRODIOL-EPOXIDES

The conformation of a non-K-region dihydrodiol has important consequences in determining whether a dihydrodiol is further metabolized to a vicinal dihydrodiol-epoxide.[56] The conformation is also an important determinant of the chemical reactivity of vicinal bay-region dihydrodiol-epoxides.[57-59]

It is very simple to predict the conformation of dihydrodiols.[56]

1. All bay-region dihydrodiols (e.g., 1,2-dihydrodiols of BA and MBA) adopt quasidiaxial (or pseudodiaxial) conformations. To date, only one bay-region dihydrodiol, 5-methylchrysene *trans*-5,6-dihydrodiol, is known to adopt a quasidiaxial conformation.[60]
2. All dihydrodiols with a *peri* substituent (methyl, hydroxymethyl, fluoro, chloro, bromo, etc.) adopt quasidiaxial conformations. These dihydrodiols include 4-MBA 5,6-dihydrodiol, 7-MBA 5,6- and 8,9-dihydrodiols, 12-MBA 10,11-dihydrodiol.
3. Dihydrodiols not included in 1. and 2. above adopt quasidiequatorial (or pseudodiequatorial) conformation. These include dihydrodiols formed at methyl-substituted double bonds (e.g., 4-MBA 3,4-dihydrodiol, 8-MBA 8,9-dihydrodiol, 6-MBA 5,6-dihydrodiol, and 11-MBA 10,11-dihydrodiol).

There are four possible conformations of vicinal dihydrodiol-epoxides:[57-59]

1. *Anti*-diequatorial (e.g., BaP *anti*-7,8-dihydrodiol-9,10-epoxide and BA *anti*-3,4-dihydrodiol-1,2-epoxide)
2. *Anti*-diaxial (BcPh *anti*-3,4-dihydrodiol-1,2-epoxide and benzo[e]pyrene *anti*-9,10-dihydrodiol-10,11-epoxide)
3. *Syn*-diequatorial (e.g., BcPh *syn*-3,4-dihydrodiol-1,2-epoxide)
4. *Syn*-diaxial (e.g., BaP *syn*-7,8-dihydrodiol-9,10-epoxide, BA *syn*-3,4-dihydrodiol-1,2-epoxide, and benzo[e]pyrene *syn*-9,10-dihydrodiol-10,11-epoxide).

BaP *syn*-7,8-dihydrodiol-9,10-epoxide (*syn*-diaxial) is chemically more reactive than BaP *anti*-7,8-dihydrodiol-9,10-epoxide (anti-diequatorial). However, BaP *anti*-7,8-dihydrodiol-9,10-epoxide is more mutagenic to mammalian cells and more tumorigenic on mouse skin than BaP *syn*-7,8-dihydrodiol-9,10-epoxide.[9] Both *anti*- and *syn* isomers of double bay-region benzo[e]pyrene 9,10-dihydrodiol-10,11-epoxide have their hydroxyl groups locked in a quasidiaxial conformation, and exhibit little or no tumorigenic activity.[9] In contrast, BcPh *syn*-3,4-dihydrodiol-1,2-epoxide (*syn*-diequatorial) and *anti*-3,4-dihydrodiol-1,2-epoxide (*anti*-diaxial) have equally high tumorigenic activity on mouse skin.[9]

Due to steric crowding of the methyl group in the bay region (at either C_1 or C_{12} of BA), the angle between the plane of the bay-region benzo-ring (1,2,3,4-benzo-ring) and the plane of the anthracenic ring of 1-MBA, and 12-MBA, and 7,12-DMBA is 21°, 21°, and 24°, respectively.[61] Thus, analogous to that of BcPh,[57-59] the bay-region *anti*- and *syn*-3,4-dihydrodiol-1,2-epoxides of 12-MBA and 7,12-DMBA may also adopt *anti*-diaxial and *syn*-diequatorial conformations and have similar tumorigenic activities. The effects of a bay-region methyl group on the carcinogenic activities of various PAHs are discussed in more detail by Hecht et al.[47]

VII. FACTORS THAT AFFECT METABOLIC ACTIVATION PATHWAYS

The factors that influence the metabolic pathway of PAH → arene epoxide → dihydrodiol → dihydrodiol-epoxide will be discussed in this section.

A. Regioselectivity of Cytochrome P-450 Isozymes

1. Regioselective Metabolism of Parent Hydrocarbons

The initial metabolic reactions, the formation of arene epoxides and direct-hydroxylation

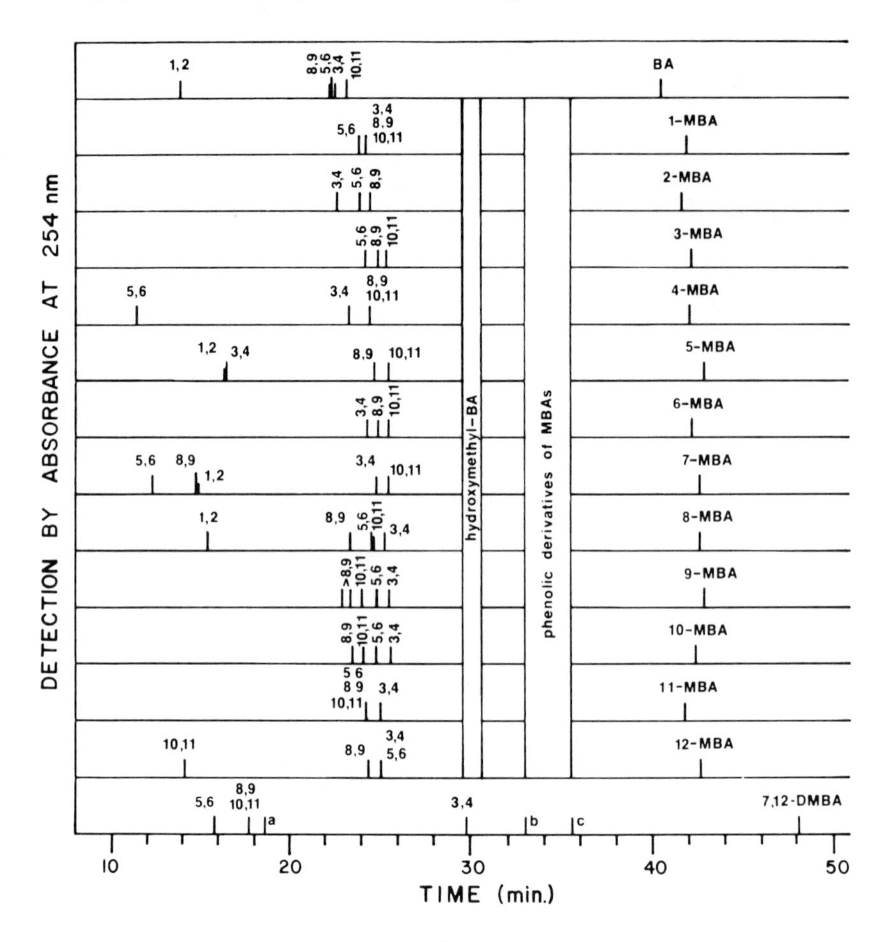

FIGURE 2. Reversed-phase HPLC retention times of *trans*-dihydrodiols of BA, MBAs, and 7,12-DMBA relative to the retention times of other phenolic and hydroxymethyl derivatives. The phenolic and dihydrodiol derivatives of hydroxymethyl-BAs are not indicated. All MBA dihydrodiols indicated in this figure have been found to be metabolites of the respective parent hydrocarbons. a, 7,12-dihydroxymethyl-BA; b, 7-hydroxymethyl-12-methyl-BA; and c, 7-methyl-12-hydroxymethyl-BA. HPLC conditions are described in Reference 62, 6-MBA 5,6-dihydrodiol, with quasidiequatorial conformation, is now known to be a metabolite of 6-MBA. The closely eluted quasidiequatorial dihydrodiols (eluted between 22 and 26 min) can be separated by normal-phase HPLC[87,97] or by reversed-phase HPLC using a Vydac C18 column.[63] (From Yang, S. K., Chou, M. W., and Fu, P. P., in *Polynuclear Aromatic Hydrocarbons: Chemical Analyses and Biological Fate,* Cook, M. and Dennis, A. J., Eds., Battelle Press, Columbus, Ohio, 1981, 253.)

products, are catalyzed by cytochrome P-450 isozymes which are contained in the drug-metabolizing enzyme complex. Various forms of cytochrome P-450 isozymes differ in their catalytic activities as well as in their regioselectivities. The regioselectivity of a cytochrome P-450 isozyme determines the relative amounts of metabolites formed at various regions of a PAH. The amounts of direct-hydroxylation products, such as some phenolic products and hydroxymethyl derivatives, can be determined by high-performance liquid chromatography (HPLC) by using one or a combination of columns (reversed-phase HPLC separations of metabolites of BA and MBA are shown in Figure 2).[62-64] However, most non-K-region arene epoxides are unstable and their metabolic formation cannot be determined under ordinary in vitro and in vivo conditions.[65,66] In the absence of glutathione transferase and glutathione, arene epoxides are either hydrated to dihydrodiols and/or nonenzymically rearranged to phenolic products.[48,49] The ratio of dihydrodiol to phenol found is dependent on the relative amount and catalytic activities of the cytochrome P-450 isozymes and an epoxide hydrolase.

Table 1
REGIOSELECTIVE METABOLISM OF BA BY RAT-LIVER MICROSOMES

	Dihydrodiol (% of all metabolites formed)				
Microsomes[a]	**1,2-**	**3,4-**	**5,6-**	**8,9-**	**10,11-**
Control	<2	4.0	44.4	41.4	4.0
PB	<2	2.5	43.9	34.6	4.6
3-MC	<2	1.6	41.6	49.0	3.2

[a] Liver microsomes prepared from untreated (control), phenobarbital (PB)-treated, and 3-methylcholanthrene (3-MC)-treated immature Long-Evans rats.

Data from Thakker, D. R., Levin, W., Yagi, H., Turujman, S., Kapadia, D., Connery, A. H., and Jerina, D. M., *Chem. Biol. Interact.*, 27, 145, 1979.

In the absence of glutathione transferase, a high (cytochrome P-450) to (epoxide hydrolase) ratio favors the formation of phenolic products. High epoxide hydrolase activity in an enzyme system favors the formation of dihydrodiols. The extent of epoxidation occurring at a double bond can be determined by the sum of the dihydrodiol and phenolic products. For example, the extent of epoxidation that occurred at the 3,4-double bond of BA is the sum of the BA 3,4-dihydrodiol, 3-hydroxy-BA, and 4-hydroxy-BA. Technically, it is difficult to determine the exact regioselectivity of a cytochrome P-450 isozyme, even under in vitro conditions. This is due to the lack of a single analytical system that separates all the metabolites formed.[61,62] However, under conditions in which a small proportion of only a few phenolic products are formed, the regioselectivity of a cytochrome P-450 isozyme can be fairly accurately determined by the amounts of dihydrodiols formed. Table 1 indicates the estimated regioselectivity of the cytochrome P-450 isozymes present in liver microsomes from untreated (control), phenobarbital-treated, and 3-methylcholanthrene-treated rats at the various double bonds of BA.[65] The 8,9- and 5,6-double bonds are the major sites, 10,11- and 3,4-double bonds are the minor sites, and 1,2-double bond is the least favored site of oxidative metabolism by cytochrome P-450 isozymes contained in various liver microsomal preparations (Table 1). Loew et al.[43] attempted to use π-bond reactivities of BA and MBAs to predict major metabolities formed (5,6- > 8,9- > 10,11- > 3,4- > 1,2-). As indicated in Table 1, 1,2-, 3,4-, and 10,11-double bonds are all minor sites of oxidative metabolism. Although the 10,11-double bond of 12-MBA has the highest π-bond reactivity,[43] it is a minor site of oxidative metabolism (Table 2).[67] The 1,2- and 5,6-double bonds of 8-MBA are metabolized to a similar extent, whereas the 3,4-double bond is a major site of metabolism.[69] These metabolism studies of BA and MBAs thus indicate that π-bond reactivities cannot be used to reliably predict major sites of metabolism.

A binding site model was proposed for cytochrome P-448 (P-450_c) to explain why a particular region of a substrate molecule is a minor site of oxidative metabolism.[70,71] This model successfully explained why the 1,2- and 3,4-double bonds of BA are minor sites of metabolism, since according to this model these double bonds do not effectively fit into the binding site. However, it failed to explain why the 10,11-double bond, which fits the binding site as effectively as the 5,6- and 8,9-double bonds of BA, is a minor site of metabolism. The relative extent of metabolism at various double bonds of BA and MBAs by liver microsomes from 3-methylcholanthrene-treated rats were estimated by HPLC analyses (Table 2). Cytochrome P-448 (P-450_c) is the major P-450 isozyme contained in liver microsomes from 3-methylcholanthrene-treated rats. The results of quantitative 6-MBA and 7,12-DMBA

Table 2
REGIOSELECTIVE METABOLISM AT VARIOUS POSITIONS OF BA, MBAs, AND 7,12-DMBA BY LIVER MICROSOMES FROM 3-METHYLCHOLANTHRENE-TREATED MALE SPRAGUE-DAWLEY RATS

	Relative extent of metabolism at various positions[a]					
Hydrocarbon	1,2-	3,4-	5,6-	8,9-	10,11-	-CH_3
BA	±	+	+++++	+++++	++	−
1-MBA	ND	+	+++++	++++	++	±
2-MBA	ND	++	+++	+++++	ND	±
3-MBA	ND	+++++[b]	++	++++	++	+++
4-MBA[c]	ND	++	++++	+++++	++	+++
5-MBA	+	+	ND	++++	+++	+++++
6-MBA[c]	ND	+++++[b]	+	+++	+++	±
7-MBA	+	++	+++	+++++	+	++
8-MBA[c]	+	+++++	+	+++	+	++++
9-MBA[c]	ND	++	+++++	+	++	++
10-MBA[c]	ND	++	+++++	+++++	+	+
11-MBA[c]	ND	++	+++++	+++	+	±
12-MBA	ND	+	++++	+++++	++	±
7,12-DMBA	+	+	++	+++++	+++	++++

[a] Relative occurrence of methyl-hydroxylation and phenolic and dihydrodiol metabolite formations at indicated positions is estimated for each hydrocarbon based on the results of reversed-phase HPLC (e.g., Figure 2) and normal-phase HPLC analyses.[62,87,97] Metabolites derived from hydroxymethyl-BAs are not included. The "±" indicates a very minor amount of metabolite formation. ND, not detected.

[b] 4-Phenol was the most abundant metabolite of both 3-MBA and 6-MBA. No 3,4-dihydrodiol was detected from 3-MBA.

[c] Dihydrodiol is formed at the methyl-substituted double bond.

Modified from Yang, S. K., Chou, M. W., and Fu, P. P., in *Polynuclear Aromatic Hydrocarbons: Chemical Analysis and Biological Fate,* Cook, M. and Dennis, A. J., Eds., Battelle Press, Columbus, Ohio, 1981, 253.

metabolism studies using other liver-microsomal preparations, containing different cytochrome P-450 isozymes, are available in the literature.[72,73] It is clear from the results summarized above and given in Table 2 that, although the 3,4-double bond is the major site of epoxidation in the metabolism of 3-MBA, 6-MBA, and 8-MBA, only for 8-MBA metabolism is the 3,4-dihydrodiol the major metabolite (Table 2). Based on the extensive metabolism studies using rat-liver microsomes as the enzyme source, there does not appear to be a simple and reliable rule to predict the major site(s) in the oxidative metabolism of BA and MBA.

2. Regioselective Metabolism of Dihydrodiols

The vicinal double bond (or "isolated double bond") of a non-K-region dihydrodiol is the most olefinic double bond of a PAH dihydrodiol. However, it may or may not be the major site of oxidative metabolism. Again, the major site(s) of epoxidation reactions is not determined by the electronegativity of the vicinal double bond, but by the regioselectivity of the cytochrome P-450 isozymes. Examples of dihydrodiols whose vicinal double bonds are the major site of metabolism are benzo[a]pyrene 7,8-dihydrodiol,[14,15] BA 1,2-dihydrodiol,[52,74,75] chrysene 3,4-dihydrodiol,[77] dibenz[a,h]anthracene 1,2-dihydrodiol,[76] 7-MBA 10,11-dihydrodiol,[56] 12-MBA 8,9-dihydrodiol, [56] 4-MBA 8,9-dihydrodiol,[56] chrysene 1,2-dihydrodiol,[78] and phenanthrene 1,2-dihydrodiol.[78] Examples of dihydrodiols whose vicinal double bonds are minor sites of epoxidation reactions are benzo[a]pyrene 9,10-dihydrodiol,[79] benzo[e]pyrene 9,10-dihydrodiol,[80] BA 3,4-dihydrodiol,[81] BcPh 3,4-dihydrodiol,[70] 10-MBA 8,9-dihydrodiol,[56] and 7,12-DMBA 8,9-dihydrodiol.[56]

3. Metabolism at the Methyl Substituent

Hydroxylation reaction at the methyl group of methylated PAHs is catalyzed by cytochrome P-450 isozymes. Hydroxymethyl-BA, like MBA itself, may have tumor-initiating activities.[9] The tumorigenic activity of a hydroxymethyl-BA may also be due to its metabolic conversion to a bay-region 3,4-dihydrodiol-1,2-epoxide(s).[9] Another possibility is the formation of sulfate conjugate in vivo.[44-46] Sulfate is a good leaving group. Thus, a benzylic carbonium ion may be formed from a hydroxymethyl-BA/sulfate conjugate. A sulfate conjugate of 7-hydroxymethyl-BA has been demonstrated to be a potent bacterial mutagen, although it has a very short half-life in an aqueous system.[11] The delocalization energies ($\Delta E_{deloc}/\beta$) of benzylic carbonium ions that could be generated from various sulfate/hydroxymethyl-benz[a]anthracene conjugates can be easily calculated in the same manner as those calculated for the carbonium ions derived from bay-region tetrahydroepoxides of various PAHs.[11] The $\Delta E_{deloc}/\beta$ values for the carbonium ions (in decreasing stability) are 7-MBA (0.880) > 12-MBA (0.830) > 8-MBA (0.738) > 6-MBA = 11-MBA = 5-MBA (0.722) > 4-MBA (0.647) > 10-MBA (0.610) > 2-MBA (0.600) > 9-MBA = 1-MBA (0.592) > 3-MBA (0.540). It is not known if the sulfate conjugate of hydroxymethylbenz[a]anthracene isomers has any tumorigenic activities. If sulfate conjugates of hydroxymethylbenz[a]anthracenes are responsible for the tumorigenic activities of the parent MBAs, one would expect to detect covalent binding adducts resulting from the interaction of benzylic carbonium ions with DNA. However, analysis of DNA isolated from cells that had been pretreated with 7-MBA did not reveal any binding adduct that has a UV absorption spectrum indicating an intact BA nucleus.[82]

It is interesting to note that the extent of hydroxylation by cytochrome P-448 at the methyl substituent varies significantly with the location of the methyl group. A high degree of hydroxylation at the methyl group indicates that the methyl group of an MBA is favorably oriented in the active site of the cytochrome P-450. However, the relationship between the hydroxylation at the methyl group and carcinogenic potency of the parent MBA is not apparent.

B. Stereoselectivity of Cytochrome P-450 Isozymes

The stereoselectivity of cytochrome P-450 isozymes are known to play important roles in the metabolic activation of benzo[a]pyrene. The formation of both 7*R*,8*S*-epoxide and *anti*-7*R*,8*S*-dihydrodiol-9*S*,10*R*-epoxide are results of stereoselective epoxidation reactions catalyzed by cytochrome P-450 isozymes. As we shall see in the discussion that follows, the activation pathways in the metabolisms of BA and MBAs may not be the same as those found in BaP metabolism.

1. Stereoselective Metabolism of Parent Hydrocarbons

Each double bond of BA and MBA has two stereoheterotopic faces. Depending on which stereoheterotopic face is favored in its interaction with the catalytic site of a cytochrome P-450 isozyme, an epoxide consisting of unequal amounts of the two enantiomers can be formed. Thus a cytochrome P-450 isozyme can have a stereoselective property toward one of the two stereoheterotopic faces of a double bond. Recently developed analytical methods[65,66,83] allowed us to determine the stereoselectivity of cytochrome P-450 isozymes contained in three rat-liver microsomal preparations in the formation of K-region epoxides (Table 3). Cytochrome P-448 (P-450_c), the major isozyme in liver microsomes from 3-methylcholanthrene-treated rats, has the highest stereoselectivity in catalyzing the formation

Table 3
EFFECT OF METHYL SUBSTITUENT ON THE STEREOSELECTIVE FORMATION OF K-REGION EPOXIDE BY RAT-LIVER MICROSOMES

	5*S*,6*R* (or 4*S*,5*R*) enantiomer (%)						
Microsomes	**BA**[a]	**6-MBA**[b]	**7-MBA**[b]	**12-MBA**[c]	**7,12-DMBA**[b]	**BcPh**[b]	**BaP**[a]
Control	52 (75)[b]	64	93	73	76	60	52
PB	55 (79)	50	72	79	80	71	60
MC	96 (96)	25	91	99	97	81	95

[a] From References 65 and 66. Data in parentheses were from an experiment using a different microsomal preparation.[66]
[b] Experimental details to be published elsewhere.
[c] From Reference 84.

of 5*S*,6*R*-epoxide in BA and all MBA studied except 6-MBA (Table 3). Apparently a methyl substituent at C_6 of BA can alter the stereoheterotopic interaction between cytochrome P-448 and the 5,6-double bond of BA. These results are in general agreement with the steric binding model proposed by Jerina et al.[70,71] who predicted that a 5*S*,6*R*-epoxide is the major enantiomer formed in the metabolism of unsubstituted PAHs such as BA by the cytochrome P-448 (P-450$_c$) isozyme. It is also apparent from the results in Table 3 that a methyl substituent at various positions of BA can alter the stereoselectivity of the cytochrome P-450 isozymes. It should be pointed out that cytochrome P-448 is not the only isozyme relevant in drug metabolism and chemical carcinogenesis. The results shown in Table 3 clearly indicate that the steric model proposed for cytochrome P-448[71] cannot be generalized to include other cytochrome P-450 isozymes.

Non-K-region arene epoxides are much less stable than K-region arene epoxides. At the present time, only the 8,9- and 10,11-epoxides of BA[66] and 12-MBA[84] have been directly isolated as epoxides under certain experimental conditions. Circular dichroism spectral analyses of the metabolically formed arene epoxides and their subsequent hydration products indicate that 8*R*,9*S*-epoxide and 10*S*,11*R*-epoxide are stereoselectively formed by the cytochrome P-450 contained in several rat-liver microsomal preparations.[66] These are also in accord with the predictions made with the steric binding model.[71] However, an example is known in 8-MBA metabolism that a methyl substituent can alter the stereoselectivity of cytochrome P-450 isozymes; the 8*S*,9*S*-dihydrodiol is the major enantiomer formed by cytochrome P-450 and the 8*R*,9*R*-dihydrodiol is the major enantiomer formed by cytochrome P-448.[85] So far, it has not been possible to isolate the relatively unstable 1,2-epoxide and 3,4-epoxide as metabolites of BA or MBA. Since epoxide hydrolase catalyzes the addition of water at the nonbenzylic position of a non-K-region epoxide (see below for more discussion), the major 1,2-epoxide enantiomer formed in 8-MBA metabolism can be deduced to have 1*R*,2*S* stereochemistry, since the metabolically formed 1,2-dihydrodiol contained predominantly 1*R*,2*R*-dihydrodiol.[85] Based on the enantiomeric composition of 3,4-dihydrodiol formed (Table 2), 3*S*,4*R*-epoxide is deduced to be the major enantiomer formed in the metabolism at the 3,4-double bonds of 6-MBA and 7-MBA. However, 3*R*,4*R*-dihydrodiol is only slightly favored in the metabolisms of 7,12-DMBA and 4-MBA (Table 2). Both 1,2- and 3,4-double bonds do not fit into the steric binding site model proposed by Jerina et al.[71] Thus, the stereoheterotopic interactions between the 1,2- or 3,4-double bond and cytochrome P-448 (or any other isozymes) cannot yet be predicted by any reliable model.

2. Stereoselective Metabolism of Dihydrodiols

Pathways in the stereoselective metabolism of enantiomeric BaP 7,8-dihydrodiols are

Table 4
ENANTIOMERIC COMPOSITIONS OF THE 3,4-DIHYDRODIOL FORMED IN THE METABOLISM OF BA AND SOME MBAs

Microsomes	3*R*,4*R*-dihydrodiol enantiomer (%)[a]						
	BA[b]	4-MBA[c]	6-MBA[c]	7-MBA[c]	8-MBA[d]	12-MBA[c]	7,12-DMBA[b]
Control	83	—	84	88	~80	—	57
PB	91	—	87	99	~95	~95	62
MC	90	54	92	88	~91	—	64

[a] The enantiomeric compositions were determined either by chiral stationary phase HPLC[66] or estimated from their CD spectral data.
[b] From Reference 86.
[c] Experimental details to be published elsewhere.
[d] Estimated using data in Reference 85.

classic examples indicating the important roles played by the cytochrome P-450 isozymes in the activation of a carcinogenic PAH.[18,38] BaP 7*R*,8*R*-dihydrodiol is stereoselectively metabolized predominantly to BaP *anti*-7*R*,8*S*-dihydrodiol-9*S*,10*R*-epoxide, whereas BaP 7*S*,8*S*-dihydrodiol is metabolized predominantly to BaP *syn*-7*S*,8*R*-dihydrodiol-9*R*,10*S*-epoxide.[18,38] BaP 7*R*,8*R*-dihydrodiol is the predominant enantiomer formed in the metabolism at the 7,8-double bond of BaP.[6,14,15] In the metabolism of BA, MBA and 7,12-DMBA, 3*R*,4*R*-enantiomer constitutes 54 to 99% of the 3,4-dihydrodiol formed (Table 4). Up to now, only metabolism studies of enantiomeric BA 3,4-dihydrodiols have been reported.[81] The results found in BA 3,4-dihydrodiol metabolism are quite different from those observed in the metabolism of the BaP 7,8-dihydrodiol enantiomers. The major products formed in the metabolism of BA 3*R*,4*R*-dihydrodiol were *bis*-dihydrodiols (tetrols) and 3,4-quinone (Figure 3), with only a small percentage (<16% of all the products formed) of bay-region dihydrodiol-epoxides formed. In contrast, no bay-region dihydrodiol-epoxide was detected in the metabolism of BA 3*S*,4*S*-dihydrodiol and only *bis*-dihydrodiol and 3,4-quinone were found.

Enantiomeric BA 8,9-dihydrodiols are metabolized similarly to those found in the metabolism of enantiomeric BaP 7,8-dihydrodiols. BA 8*R*,9*R*-dihydrodiol and 8*S*,9*S*-dihydrodiol are metabolized predominantly to *anti*-8*R*,9*S*-dihydrodiol-10*S*,11*R*-epoxide and *syn*-8*S*,9*R*-dihydrodiol-10*R*,11*S*-epoxide, respectively.[86] Unlike the bay-region dihydrodiol-epoxides of BaP, the 8,9-dihydrodiol-10,11-epoxides formed in the metabolism of the enantiomeric BA 8,9-dihydrodiols are not readily hydrolyzed to tetrols and they can be detected by reversed-phase HPLC using water/methanol as the elution solvent.

In spite of the quasidiaxial conformation, enantiomeric BA 1,2-dihydrodiols are metabolized at the vicinal 3,4-double bonds to form 1,2-dihydrodiol-3,4-epoxides.[52,74,75] The *syn* isomer was greater than 75% of the 1,2-dihydrodiol-3,4-epoxides formed in the metabolism of BA 1*S*,2*S*-dihydrodiol, by three rat-liver microsomal preparations.[73] When BA 1*R*,2*R*-dihydrodiol was used as the substrate, the *anti* isomer was greater than 82% of 1,2-dihydrodiol-3,4-epoxides formed by cytochrome P-448, the major isozyme contained in liver microsomes from 3-methylcholanthrene-treated rats.[52] However, 83% and 86% of the 1,2-dihydrodiol-3,4-epoxides formed by liver microsomes from untreated and phenobarbital-treated rats are the *syn* isomer. Unlike the BA 8,9-dihydrodiol-10,11-epoxides, BA 1,2-dihydrodiol-3,4-epoxides are hydrated by epoxide hydrolase to tetrols.[52]

Metabolism of 7,12-DMBA 3,4-dihydrodiol by rat-liver microsomes has been studied in our laboratory.[100] Only 3,4-dihydrodiol products formed by hydroxylations at the methyl groups of 7,12-DMBA 3,4-dihydrodiol were found;[86,87] neither 3,4-dihydrodiol-1,2-epoxide

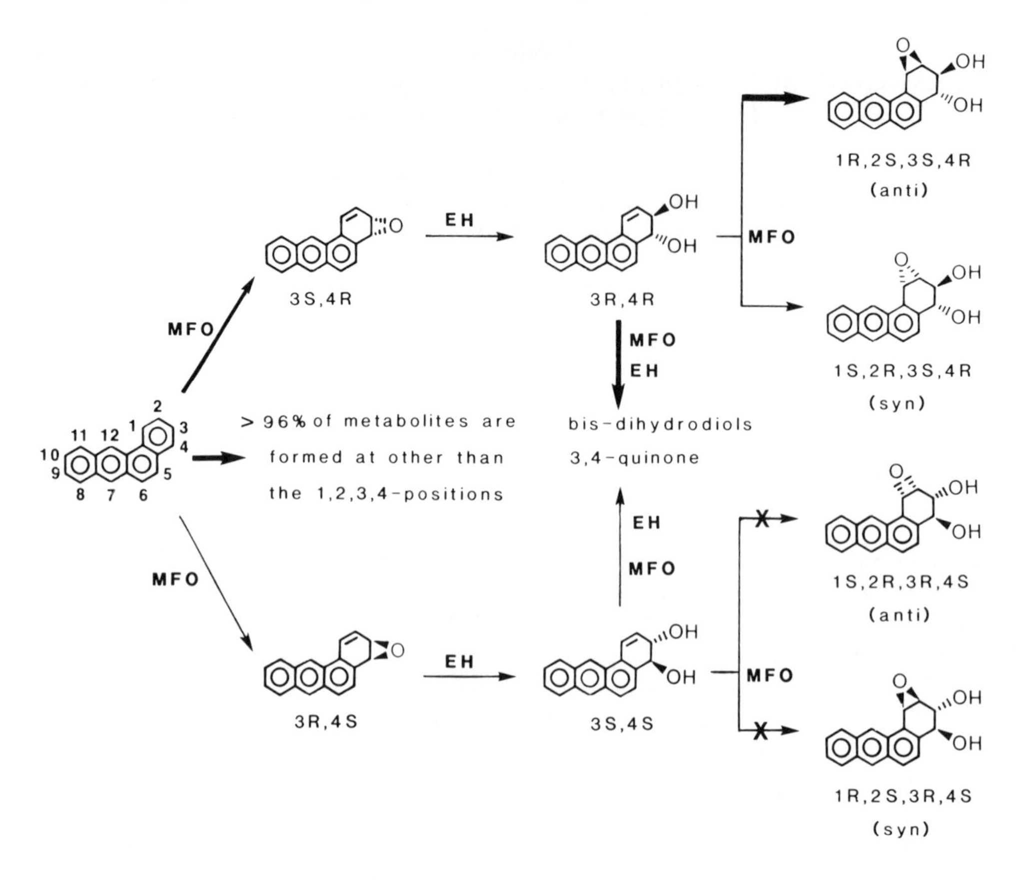

FIGURE 3. Metabolic activation and detoxification pathways of BA. MFO and EH abbreviate for the cytochrome P-450-containing mixed-function oxidases and epoxide hydrolase, respectively. The absolute configurations of the metabolites are as shown. (From Yang, S. K., Mushtaq, M., and Chiu, P.-L., in *Polycyclic Hydrocarbons and Carcinogenesis*, Harvey, R. G., Ed., ACS Symp. Ser. 283, American Chemical Society, Washington, D.C., 1985, 19.)

nor 1,2,3,4-tetrols were found as products. These results suggest that the metabolically formed bay-region 3,4-dihydrodiol-1,2-epoxides of 7,12-DMBA are extremely reactive and bind readily to microsomal proteins (or DNA, if present in the incubation mixture).[88]

C. Epoxide Hydrolase

Epoxide hydrolase catalyzes the regioselective hydration of arene epoxides to *trans*-dihydrodiols.[48,49,54] The mechanisms in the hydration of K-region and non-K-region arene epoxides are different. For non-K-region epoxides, water attack occurs regioselectively, predominantly at the allylic carbon of both epoxide enantiomers.[18,66] The hydration reaction is essentially product stereo-specific, i.e., one epoxide enantiomer is hydrated only to a *R,R*-dihydrodiol, whereas the other epoxide enantiomer is hydrated only to an *S,S*-dihydrodiol.[18,64] However, the rates of hydration of the two enantiomeric epoxides are different. Consequently the epoxide enantiomer that has the slower rate of hydration has a greater chance to be converted nonenzymically to phenolic products. Thus a racemic epoxide is hydrated to a dihydrodiol enriched in one enantiomer (whose precursor epoxide has a faster rate of hydration). This has actually been observed in the hydration of racemic BA 3,4-epoxide[89] and BaP 7,8-epoxide.[18,90,91] Some dihydrodiol-epoxides are also substrates of epoxide hydrolase.[50,52]

The enantiomeric K-region epoxides not only have different mechanisms, but also different rates, of hydration[65,91,92] (Table 5). Both K-region epoxide enantiomers of BaP and chrysene

Table 5
REGIOSELECTIVE HYDRATION OF K-REGION 5,6-EPOXIDE ENANTIOMERS OF SOME PAHs

Arene epoxide	Microsomal protein (mg/mℓ)[a]	Dihydrodiol enantiomer (%) 5*S*,6*S* (4*S*,5*S*)	5*R*,6*R* (4*R*,5*R*)	Ref.
BA 5*S*,6*R*-epoxide	1	10 (*R*)[b]	90 (*S*)	65
Chrysene 5*S*,6*R*-epoxide	3	31 (*R*)	69 (*S*)	92
BaP 4*S*,5*R*-epoxide	1	0	100 (*S*)	65
BcPh 5*S*,6*R*-epoxide	1	100 (*R*)	0	99
12-MBA 5*S*,6*R*-epoxide	1	97 (*R*)	3 (*S*)	84
7,12-DMBA 5*S*,6*R*-epoxide	3	95 (*R*)	5 (*S*)	98[c]
BA 5*R*,6*S*-epoxide	1	65(*R*)	35 (*S*)	65
Chrysene 5*R*,6*S*-epoxide	3	8 (*R*)	92 (*S*)	92
BaP 4*R*,5*S*-epoxide	1	11 (*R*)	89 (*S*)	65
BcPh 5*R*,6*S*-epoxide	1	32 (*R*)	68 (*S*)	99
12-MBA 5*R*,6*S*-epoxide	1	57 (*R*)	43 (*S*)	84
7,12-DMBA 5*R*,6*S*-epoxide	3	46 (*R*)	54 (*S*)	98[c]

[a] Liver microsomes from phenobarbital-treated male Sprague-Dawley rats were used.
[b] Position of water attack at chiral center of K-region epoxide.
[c] The assignments of absolute configuration in Reference 98 were reversed.

From Yang, S. K., Mushtaq, M., and Weems, H. B., *Arch. Biochem. Biophys.*, 255, 48, 1987.

are hydrated predominantly (≥89%) at the *S*-center to form the *R,R*-dihydrodiol. The 5*S*,6*R*-epoxides of BA and chrysene are also hydrated predominantly (~90%) at the *S*-center to form the 5*R*,6*R*-dihydrodiol. In contrast, the 5*S*,6*R*-epoxide enantiomer of BcPh, 12-MBA, and 7,12-DMBA are both hydrated (≥95%) at the *R*-center to form 5*S*,6*S*-dihydrodiol (Table 5), which is the major 5,6-dihydrodiol enantiomer formed in the metabolism of the parent hydrocarbons (Table 3). The epoxide hydrolase-catalyzed water attack in the hydration of the 5*R*,6*S*-epoxide enantiomers of BA, 12-MBA and 7,12-DMBA is considerably less regioselective (Table 5). It is apparent that the K-region *S,R*-epoxides derived from planar PAHs such as BA, BaP, and chrysene are all hydrated by water attack at the *S*-center (Table 5). In contrast, the K-region *S,R*-epoxides of nonplanar PAHs such as BcPh, 12-MBA, and 7,12-DMBA are hydrated at the *R*-center (Table 5). It is also apparent from the results in Table 5 that there is no simple rule to predict the major site of water attack in the epoxide hydrolase-catalyzed hydration reactions of the K-region *R,S*-epoxides.

D. Reactivity of Dihydrodiol-Epoxides

According to the bay-region theory,[10,11] the bay-region 3,4-dihydrodiol-1,2-epoxide, if formed metabolically from BA or any of the MBAs, is predicted to be the most reactive metabolite of that parent hydrocarbon. The chemically synthesized BA *anti*-3,4-dihydrodiol-1,2-epoxide and *syn*-3,4-dihydrodiol-1,2-epoxide are both substantially more tumorigenic than BA.[9] However, 7-MBA *anti*-3,4-dihydrodiol-1,2-epoxide has lower tumor-initiating activity than 7-MBA or 7-MBA 3,4-dihydrodiol.[93] The reason for the low tumor-initiating activity of 7-MBA *anti*-3,4-dihydrodiol-1,2-epoxide may be due to its intrinsic chemical reactivity with cellular macromolecules before reaching the critical target site (presumably DNA). Bay-region dihydrodiol-epoxides for the 11 other MBAs have not been tested, thus the structure-activity relationship for the bay-region dihydrodiol-epoxides of the 12 MBAs is not known. Due to their intrinsic chemical reactivity, it would be very difficult, if not impossible, to obtain a reliable structure-activity relationship by testing 12 MBA bay-region dihydrodiol-epoxides. Silverman and Lowe[41] have used the simple Hückel theory to predict the reactivities of bay-region dihydrodiol-epoxides with a methyl group at various positions. They found that a methyl group at C_5, C_9, and C_{11} position of BA will not enhance the ease of triol carbocation formation at C_1. Thus, the bay-region 3,4-dihydrodiol-1,2-epoxides of 5-MBA, 8-MBA, 9-MBA, and 11-MBA have similar reactivity to BA 3,4-dihydrodiol-1,2-epoxide. On the other hand, the 3,4-dihydrodiol-1,2-epoxides are stabilized due to methyl substitution at either C_6 or C_7 of BA, hence they are more reactive than that of BA. Although the reactivity of 12-MBA 3,4-dihydrodiol-1,2-epoxide could not be predicted, it is believed to be due to its unusual conformation.[41]

E. Conjugation Reactions

Conjugation of glutathione with an M-region epoxide and conjugation of uridine diphosphate glucuronic acid (UDPGA) and 3-phosphoadenosine-5-phosphosulfate (PAPS) with an M-region dihydrodiol are both reactions that can decrease the effective concentration of the metabolically formed M-region dihydrodiol. Interaction between glutathione and bay-region dihydrodiol-epoxides also decreases the effective concentration of metabolically formed bay-region dihydrodiol-epoxides. Presumably, these conjugation reactions are all detoxification pathways. However, as discussed in Section VII.A above, sulfate conjugates of hydroxymethylbenz[a]anthracenes may play a role in the activation of MBAs. Glucuronide of 3-hydroxy-BaP was found to bind to DNA in the presence of β-glucuronidase.[55] To date, there has not been any evidence to implicate sulfate and glucuronide conjugates in the metabolic activation of MBAs.

F. Metabolism in Target Organs and Nontarget Organs

Metabolism studies of BA and MBAs have mostly been carried out using rat-liver homogenates and microsomal fractions.[94] The metabolisms of BA and 7-MBA has also been studied in human and rat lung tissues, in cultured human lymphocytes, and on mouse skin in vivo.[94] Metabolites formed in rat-liver homogenates and microsomal fractions are also found to be formed in mouse skin in vivo. However, the quantitative distributions of various metabolites are substantially different. Thus, pathways of metabolism in liver (nontarget organ) and in mouse skin (target organ) or other tissues may be quite different; a major pathway in liver enzymes may be a minor pathway in mouse skin, and vice versa. Our understanding of the differences in metabolism between target and nontarget organs is very limited. Metabolism studies of MBAs in mouse skin, other than 7-MBA, have not been carried out.

Cytochrome P-450 isozymes and epoxide hydrolase, the two most important enzyme systems in the drug-metabolizing enzyme complex, in liver and in mouse skin, may be quite different in their amino acid composition and conformation. However, both enzyme systems in liver and in mouse skin catalyze the same oxidative reactions (epoxidation and hydration) in PAH metabolism, and they apparently differ in their regioselectivity and stereoselectivity. Due to abundance of cytochrome P-450 isozymes and epoxide hydrolase present in the livers of rats, a relatively large amount of metabolites (μg to mg range) can be obtained relatively easily by incubation of the parent hydrocarbon with rat liver microsomes. Thus, metabolism studies using rat liver microsomes generally provide a clearer understanding of the possibilities of metabolic pathways. The availability of metabolites formed by oxidative metabolism greatly facilitates the development of separation systems (GC, HPLC, TLC, etc.) and the structural identification of the metabolites. Once an analytical system is established and the structures of metabolites are identified, it is rather easy to analyze the metabolites formed in any other in vitro and in vivo systems.

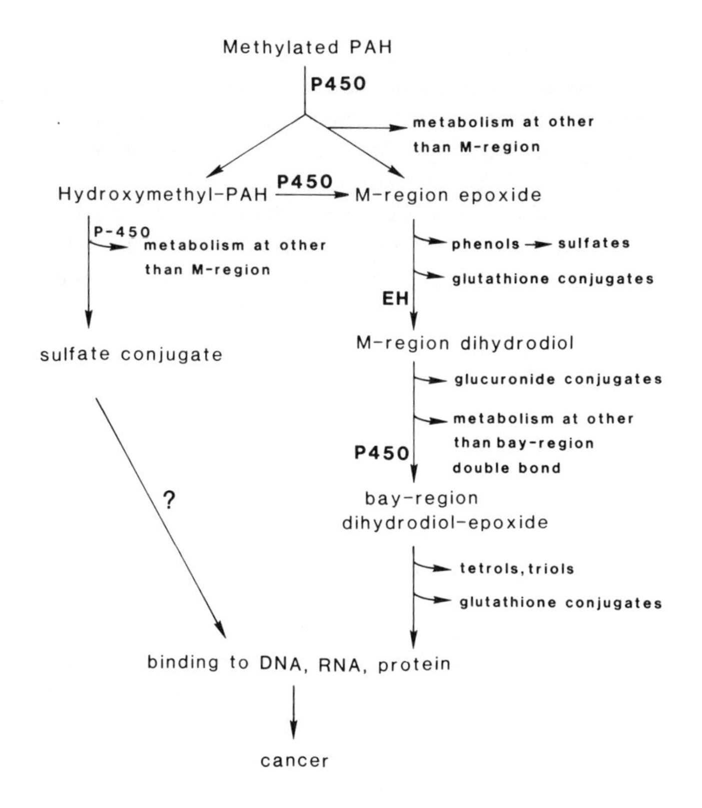

FIGURE 4. Possible metabolic activation and detoxification pathways of methylated polycyclic aromatic hydrocarbons.

VIII. OVERALL CONSIDERATIONS

In BaP metabolism, the 7,8-double bond is not the most favored site of metabolism. In fact, cytochrome P-450 isozymes are not regioselective toward the 7,8-double bond, but rather toward the 1,2-, 2,3-, and 9,10-double bonds. Due to the stereoselectivity of cytochrome P-450 isozymes and the regioselectivity of epoxide hydrolase, the 7*R*,8*S*-epoxide and 7*R*,8*R*-dihydrodiol are preferentially formed, which leads to the formation of *anti*-7*R*,8*S*-dihydrodiol-9*S*,10*R*-epoxide as the ultimate carcinogen.[93] The amount and reactivity of metabolically formed bay-region dihydrodiol-epoxide are important factors in determining the carcinogenicity of the parent PAH. Along the PAH → M-region epoxide → M-region dihydrodiol → bay-region dihydrodiol-epoxide pathway, other metabolic reactions can decrease the effective concentration of bay-region dihydrodiol-epoxide eventually formed at the target site in vivo (Figure 4).

Unlike in the metabolism of carcinogenic BaP, metabolic reactions do not favor the formation of the bay-region dihydrodiol-epoxides in the metabolism of BA, MBAs, and 7,12-DMBA. The 3,4-double bond is a very minor site in BA metabolism by rat-liver microsomal enzymes. Metabolism of BA in mouse skin maintained in short-term organ culture produced higher amount of 3,4-dihydrodiol than those of 5,6- and 8,9-dihydrodiols.[96] However, it is not known if the noncarcinogenic dihydrodiols are preferentially removed by conjugation reactions to yield water soluble products. Another important factor in BA metabolism is that the vicinal 1,2-double bond is not the major metabolic site in the metabolism of both enantiomeric BA 3,4-dihydrodiols (Figure 3). Together with the deactivation metabolic reactions, the effective concentration of bay-region 3,4-dihydrodiol-1,2-epoxide, a known carcinogenic compound,[9] is extremely low. This is probably one of the reasons why BA is a very weak carcinogen.

The 3,4-dihydrodiols of 6-MBA, 7-MBA, 8-MBA, and 7,12-DMBA are all known to be either more mutagenic or tumorigenic than the corresponding parent hydrocarbon.[9] Unfortunately, the 3,4-dihydrodiols of the noncarcinogenic MBAs have not been available for mutagenicity or carcinogenicity testings. Also, in vivo metabolism studies of the majority of the MBAs have not been carried out. The 3,4-dihydrodiol is a relatively minor product formed in the metabolism of all MBAs except 8-MBA (Table 2) by rat-liver microsomes. Studies of the metabolic pathways of 3,4-dihydrodiols of various MBAs will also be of great value in the understanding of the differences of carcinogenic potencies of the twelve MBAs. Presumably, the major metabolic pathway of these 3,4-dihydrodiols is the formation of the highly reactive bay-region 3,4-dihydroidol-1,2-epoxide. Except for BA, 7-MBA, and 7,12-DMBA which have been studied in a short-term mouse skin organ culture, no other MBAs have been studied by using enzyme systems other than rat-liver enzymes.

IX. CONCLUDING REMARKS

We have known, for more than 20 years, that methyl substitutions at some positions (6, 7, 8, and 12) of BA, a weak carcinogen, result in potent carcinogens. To date, in spite of the success of the bay-region theory in helping to understand metabolic activation pathways of many unsubstituted PAHs, we still do not understand why, for example, 7-MBA is a more potent carcinogen than 6-MBA. It is rather straightforward to understand why methyl substitutions at C_1, C_2, C_3, C_4 positions do not convert BA to a potent carcinogen. It is also easy to understand why 5-MBA is as weak a carcinogen as BA, since the quasiaxial hydroxyl groups of 5-MBA 3,4-dihydrodiol, if formed metabolically from 5-MBA, inhibit the metabolic formation of bay-region 3,4-dihydrodiol-1,2-epoxides. But why do methyl substitutions at C_9, C_{10}, and C_{11} positions not enhance the carcinogenicity of BA? A possible explanation for the very weak carcinogenicity of 9-MBA, 10-MBA, and 11-MBA may be due to a combination of the following factors: (1) low metabolic oxidation at the M-region 3,4-double bond of the parent PAH, (2) unstable 3,4-epoxide, (3) low metabolic oxidation at the bay-region 1,2-double bond of the 3,4-dihydrodiol, and (4) the intrinsic chemical reactivity of the resulting 3,4-dihydrodiol-1,2-epoxide is not enhanced by the methyl substituent.

The following experiments, when carried out under in vivo conditions on mouse skin, should enhance our understanding of the structure-activity relationship of BA and MBA isomers; (1) the relative amounts of 3,4-dihydrodiols formed from 12 MBAs, (2) the metabolic fates of the 12 MBA 3,4-dihydrodiols, (3) DNA binding activities of the 12 MBA 3,4-dihydrodiols, (4) relative tumor-initiating activities of the 12 MBA 3,4-dihydrodiols, (5) the metabolic fates of the 12 hydroxymethylbenz[a]anthracenes, (6) relative tumor-initiating activities of the 12 hydroxymethylbenz[a]anthracenes. Availability of [^{14}C] and [^{3}H]-labeled compounds with high specific activity will greatly facilitate these experiments. It is important to carry out each of the experiments with a complete set of isomers. To date, except for the complete set of MBA isomers that has been tested for its carcinogenic and tumor-initiating activities, other experiments such as those mentioned in this section using a complete set of oxidative derivatives have not been carried out.

ACKNOWLEDGMENTS

The author acknowledges Dr. Mohammad Mushtaq's contribution of unpublished experimental results cited in Tables 3 and 4. This work was supported by U.S. Public Health Service grant CA29133. The opinions or assertions contained herein are the private ones of the author and are not to be construed as official or reflecting the views of the Department of Defense of the Uniformed Services University of the Health Sciences.

REFERENCES

1. **Baird, W. M., Dipple, A., Grover, P. L., Sims, P., and Brookes, P.,** Studies on the formation of hydrocarbon-deoxyribonucleoside products by the binding of derivatives of 7-methylbenz[a]anthracene to DNA in aqueous solution and in mouse embryo cells in culture, *Cancer Res.*, 33, 2386, 1973.
2. **Baird, W. M., Harvey, R. G., and Brookes, P.,** Comparison of the cellular DNA-binding products of benzo[a]pyrene with the products formed by the reaction of benzo[a]pyrene-4,5-oxide with DNA, *Cancer Res.*, 35, 54, 1975.
3. **Borgen, A., Darvey, H., Castagnoli, N., Crocker, T. T., Rasmussen, R. E., and Wang, I. Y.,** Metabolic conversion of benzo[a]pyrene by Syrian hamster liver microsomes and binding of metabolites to deoxyribonucleic acid, *J. Med. Chem.*, 16, 502, 1973.
4. **Sims, P., Grover, P. L., Swaisland, A., Pal, A., and Hewer, A.,** Metabolic activation of benzo[a]pyrene proceeds by a diol-epoxide, *Nature*, 252, 236, 1974.
5. **Daudel, P., Duquesne, M., Vigny, P., Grover, P. L., and Sims, P.,** Fluorescence spectral evidence that benzo[a]pyrene-DNA products in mouse skin arise from diol-epoxides, *FEBS Lett.*, 57, 250, 1975.
6. **Huberman, E., Sachs, L., Yang, S. K., and Gelboin, H. V.,** Identification of mutagenic metabolites of benzo[a]pyrene in mammalian cells, *Proc. Natl. Acad. Sci. U.S.A.*, 73, 607, 1976.
7. **Newbold, R. F. and Brookes, P.,** Exceptional mutagenicity of a benzo[a]pyrene diol epoxide in cultured mammalian cells, *Nature*, 261, 52, 1976.
8. **Wood, A. W., Wislocki, P. G., Chang, R. L., Levin, W., Lu, A. Y. H., Yagi, H., Hernandez, O., Jerina, D. M., and Conney, A. H.,** Mutagenicity and cytotoxicity of benzo(a)pyrene benzo-ring epoxides, *Cancer Res.*, 36, 3358, 1976.
9. **Wislocki, P. G. and Lu, A. Y. H.,** Carcinogenicity and mutagenicity of proximate and ultimate carcinogens of polycyclic aromatic hydrocarbons, chap. 1, this volume.
10. **Jerina, D. M. and Daly, J. W.,** Oxidation at carbon, in *Drug Metabolism — from Microbe to Man*, Parke, D. V. and Smith, R. L., Eds., Taylor and Frances Ltd., London, 1976, 13.
11. **Jerina, D. M., Lehr, R. E., Akagi, H., Hernandez, O., Dansette, P. M., Wood, A. W., Wislocki, P. G., Chang, R. L., Levin, W., and Conney, A. H.,** Mutagenicity of benzo(a)pyrene derivatives and the description of a quantum mechanical model which predicts the ease of carbonium ion formation from diol epoxides, in *In Vitro Metabolic Activation and Mutagenesis Testing*, deSerres, F. J., Fouts, J. R., Bend, J. R. and Philpot, R. M., Eds., Elsevier/North-Holland Biomedical Press, Amsterdam, 1976, 159.
12. **Yagi, H., Hernandez, O., and Jerina, D. M.,** Synthesis of (±)-7β,8α-dihydrozy-9β,10β-epoxy-7,8,9,10-tetrahydrobenzo[a]pyrene with stereochemistry related to the anileukemic triptolides, *J. Am. Chem. Soc.*, 97, 6881, 1975.
13. **Beland, F. A. and Harvey, R. G.,** The isomeric 9,10-epoxides of *trans*-7,8-dihydroxy-7,8-dihydrobenzo[a]pyrene, *J. Chem. Soc. Chem. Commun.*, 85, 1976.
14. **Yang, S. K., McCourt, D. W., Roller, P. P., and Gelboin, H. V.,** Enzymatic conversion of benzo[a]pyrene leading predominantly to the diol-epoxide *r*-7,*t*-8-dihydroxy-*t*-9,10-oxy-7,8,9,10-tetrahydrobenzo[a]pyrene through a single enantiomer of *r*-7,*t*-8-dihydroxy-7,8-dihydrobenzo[a]pyrene, *Proc. Natl. Acad. Sci. U.S.A.*, 73, 2594, 1976.
15. **Thakker, D. R., Yagi, H., Lu, A. Y. H., Levin, W., Conney, A. H., and Jerina, D. M.,** Metabolism of benzo[a]pyrene: conversion of (±)-*trans*-7,8-dihydroxy-7,8-dihydrobenzo[a]pyrene to highly mutagenic 7,8-diol-9,10-epoxides, *Proc. Natl. Acad. Sci. U.S.A.*, 73, 3381, 1976.
16. **Nakanishi, K., Kasai, H., Cho, H., Harvey, R. G., Jeffrey, A. M., Jennette, K. W., and Weinstein, I. B.,** Absolute configuration of a ribonucleic acid adduct formed *in vivo* by metabolism of benzo[a]pyrene, *J. Am. Chem. Soc.*, 99, 258, 1977.
17. **Koreeda, M., Moore, P. D., Wislocki, P. G., Levin, W., Conney, A. H., Yagi, H., and Jerina, D. M.,** Binding of benzo[a]pyrene 7,8-diol-9,10-epoxides to DNA, RNA, and protein of mouse skin occurs with high stereoselectivity, *Science*, 199, 778, 1978.
18. **Yang, S. K., McCourt, D. W., Leutz, J. C., and Gelboin, H. V.,** Benzo[a]pyrene diol epoxides: mechanism of enzymatic formation and optically active intermediates, *Science*, 196, 1199, 1977.
19. **Weinstein, I. B., Jeffrey, A. M., Jennette, K. W., Blobstein, S. H., Harris, C., Harvey, R. G., Autrup, H., Kasai, H., and Nakanishi, K.,** Benzo[a]pyrene diol epoxides as intermediates in nucleic acid binding in vitro and in vivo, *Science*, 193, 592, 1976.
20. **Jeffrey, A. M., Jennette, K. W., Blobstein, S. H., Weinstein, I. B., Kasai, H., Beland, F. A., Harvey, R. G., Miura, I., and Nakanishi, K.,** Benzo[a]pyrene-nucleic acid derivatives found in vivo: structure of a benzo[a]pyrenetetrahydrodiol epoxide-guanosine adduct, *J. Am. Chem. Soc.*, 98, 5714, 1976.
21. **Koreeda, M., Moore, P. D., Yagi, H., Yeh, H. J. C., and Jerina, D. M.,** Alkylation of polyguanylic acid at the 2-amino group and phosphate by the potent mutagen (±)-7β,8α-dihyroxy-9β,10β-epoxy-7,8,9,10-tetrahydrobenzo[a]pyrene, *J. Am. Chem. Soc.*, 98, 6720, 1976.

22. **Meehan, T., Strau, K., and Calvin, M.,** Benzo[a]pyrene diol epoxide covalently binds to deoxyguanosine and deoxyadenosine in DNA, *Nature,* 269, 725, 1977.
23. **Jeffrey, A. M., Weinstein, I. B., Jennette, K. W., Grzeskowiak, K., Autrup, H., Nakanishi, K., Harvey, R. G., and Harris, C.,** Structures of benzo(a)pyrene-nucleic acid adducts in human and bovine explants, *Nature,* 269, 348, 1977.
24. **Huberman, E., Yang, S. K., McCourt, D. W., and Gelboin, H. V.,** Mutagenicity to mammalian cells in culture by (+) and (−) *trans*-7,8-dihdroxy-7,8-dihydrobenzo(a)pyrenes and the hydrolysis and reduction products of two stereoisomeric benzo(a) pyrene 7,8-diol-9,10-epoxides, *Cancer Lett.,* 4, 35, 1977.
25. **Levin, W., Wood, A. W., Yagi, H., Dansette, P. M., Jerina, D. M., and Conney, A. H.,** Carcinogenicity of benzo[a]pyrene 4,5-, 7,8-, and 9,10-oxides on mouse skin, *Proc. Natl. Acad. Sci. U.S.A.,* 73, 243, 1976.
26. **Kapitulnik, J., Wislocki, P. G., Levin, W., Yagi, H., Thakker, D. R., Akagi, H., Koreeda, M., Jerina, D. M., and Conney, A. H.,** Marked differences in the carcinogenic activity of optically pure (+)- and (−)-trans-7,8-dihydroxy-7,8-dihydrobenzo(a)pyrene in newborn mice, *Cancer Res.,* 38, 2661, 1978.
27. **Levin, W., Wood, A. W., Chang, R. L., Slaga, T. J., Yagi, H., Jerina, D. M., and Cooney, A. H.,** Marked differences in the tumor-initiating activity of optically pure (+)- and (−)-*trans*-7,8-dihydroxy-7,8-dihydrobenzo(a)pyrene on mouse skin, *Cancer Res.,* 37, 2721, 1977.
28. **Buening, M. K., Wislocki, P. G., Levin, W., Yagi, H., Thakker, D. R., Akagi, H., Koreeda, M., Jerina, D. M., and Conney, A. H.,** Tumorigenicity of the optical enantiomers of the diastereomeric benzo[a]pyrene 7,8-diol-9,10-epoxides in newborn mice: exceptional activity of (±)-7β,8α-dihydroxy-9a,10a-epoxy-7,8,9,10-tetrahydrobenzo[a]pyrene, *Proc. Natl. Acad. Sci. U.S.A.,* 75, 5358, 1978.
29. **Slaga, T. J., Bracken, W. J., Gleason, G., Levin, W., Yagi, H., Jerina, D. M., and Conney, A. H.,** Marked differences in the skin tumor-initiating activities of the optical enantiomers of the diastereomeric 7,8-diol-9,10-epoxides, *Cancer Res.,* 39, 67, 1979.
30. **Pullman, A. and Pullman, B.,** Electronic structure and carcinogenic activity of aromatic molecules, New Developments, *Adv. Cancer, Res.,* 3, 117, 1955.
31. **Pullman, B.,** Recent developments on the mechanism of chemical carcinogenesis by aromatic hydrocarbons, *Int. J. Quantum Chem.* 16, 669, 1979.
32. **Jones, P. W. and Freudenthal, R. I., Eds.,** *Carcinogenesis — A Comprehensive Survey.* Vol. 3, *Polynuclear Aromatic Hydrocarbons,* 2nd Int. Symp. Analysis, Chemistry, and Biology, Raven Press, New York, 1978.
33. **Gelboin, H. V. and Ts'o, P. O. P., Eds.,** *Polycyclic Hydrocarbons and Cancer,* Vol. 1 & 2, Academic Press, New York, 1978.
34. **Gelboin, H. V. and Ts'o, P. O. P., Eds.** *Polycyclic Hydrocarbons and Cancer,* Vol. 3, Academic Press, New York, 1981.
35. **Harvey, R. G.,** Activated metabolites of carcinogenic hydrocarbons, *Acc. Chem. Res.,* 14, 218, 1981.
36. **Harvey, R. G., Ed.,** *Polycyclic Hydrocarbons and Carcinogenesis,* ACS Symp. Ser. 283, American Chemical Society, Washington, D.C., 1985.
37. **Gelboin, H. V.,** Benzo[a]pyrene metabolism, activation, and carcinogenesis: role and regulation of mixed-function oxidases and related enzymes, *Physiol. Rev.,* 60, 1107, 1980.
38. **Conney, A. H.,** Induction of microsomal enzymes by foreign chemicals and carcinogenesis by polycyclic aromatic hydrocarbons: G. H. A. Clowes memorial lecture, *Cancer Res.,* 42, 4875, 1982.
39. **Pelkonen, O. and Nebert, D. W.,** Metabolism of polycyclic aromatic hydrocarbons: etiologic role in carcinogenesis, *Pharmacol. Rev.,* 34, 190, 1982.
40. **Dipple, A., Moschel, R. C., and Bigger, C. A. H.,** Polynuclear aromatic carcinogens, in *Chemical Carcinogens,* Vol. 2, 2nd ed., Searle, C. E., Ed., ACS Monograph 182, American Chemical Society, Washington, D.C., 1984, 41.
41. **Silverman, B. D. and Lowe, J. P.,** Diol-epoxide reactivity of methylated polycyclic aromatic hydrocarbons (PAH): ranking the reactivity of the positional monomethyl isomers, in *Polynuclear Aromatic Hydrocarbons: Physical and Biological Chemistry,* Cooke, M., Dennis, A. J., and Fisher, G. L., Eds., Battelle Press, Columbus, Ohio, 1982, 743.
42. **Smith, I. A. and Seybold, P. G.,** Methylbenz[a]anthracenes: correlations between theoretical reactivity indices and carcinogenicity, *Int. J. Quantum Chem: Quantum Biol.,* Symp. 5, 311, 1978.
43. **Loew, G. Poulsen, M., Ferrell, J., and Chaet, D.,** Quantum chemical studies of methylbenz[a]anthracenes: metabolism and correlations with carcinogenicity, *Chem. Biol. Interact.,* 31, 319, 1980.
44. **Cavalieri, E. L. and Rogan, E. G.,** One-electron oxidation in aromatic hydrocarbon carcinogenesis, in *Polycyclic Hydrocarbons and Carcinogenesis,* Harvey, R. G., Ed., ACS Symp. Ser. 283, American Chemical Society, Washington, D.C., 1985, 289.
45. **Flesher, J. W. and Myers, S. R.,** Oxidative metabolism of 7-methylbenz[a]anthracene, 12-methylbenz[a]anthracene and 7,12-dimethylbenz[a]anthracene by rat liver cytosol, *Cancer Lett.,* 26, 83, 1985.
46. **Watabe, T., Hakamata, Y., Hiratsuka, A., and Ogura, K.,** A 7-hydroxymethyl sulphate ester as an active metabolite of the carcinogen, 7-hydroxymethylbenz[a]anthracene, *Carcinogenesis,* 7, 207, 1986.

47. **Hecht, S. S., Melikian, A. A., and Amin, S.,** Effects of methyl substitution on the tumorigenicity and metabolic activation of polycyclic aromatic hydrocarbons, chapter 4, this volume.
48. **Jerina, D. M. and Daly, J. W.,** Arene oxides: a new aspect of drug metabolism, *Science,* Washington, D.C., 185, 573, 1974.
49. **Sims, P. and Grover, P. L.,** Epoxides in polycyclic aromatic hydrocarbon metabolism and carcinogenesis, *Adv. Cancer Res.,* 20, 165, 1974.
50. **Sayer, J. M., Yagi, H., van Bladeren, P. J., Levin, W., and Jerina, D. M.,** Stereoselectivity of microsomal epoxide hydrolase toward diol epoxides and tetrahydroepoxides derived from benz[a]anthracene, *J. Biol. Chem.,* 260, 1630, 1985.
51. **Robertson, I. G. C. and Jernström, B.,** The enzymatic conjugation of glutathione with bay-region diol-epoxides of benzo[a]pyrene, benz[a]anthracene, and chrysene, *Carcinogenesis,* 7, 1633, 1986.
52. **Chou, M. W., Chiu, P.-L., Fu, P. P., and Yang, S. K.,** Effect of enzyme induction on the stereoselective metabolism of optically pure (−)1*R*,2*R*- and (+)1*S*,2*S*-dihydroxy-1,2-dihydrobenz[a]anthracenes to vicinal 1,2-dihydrodiol 3,4-epoxides by rat liver microsomes, *Carcinogenesis,* 4, 629, 1983.
53. **Lu, A. Y. H. and West, S.,** Multiplicity of mammalian microsomal cytochromes P-450, *Pharmacol. Rev.,* 31, 277, 1980.
54. **Lu, A. Y. H. and Miwa, G. T.,** Molecular properties and biological functions of microsomal epoxide hydrase, *Ann. Rev. Pharmacol. Toxicol.,* 20, 513, 1980.
55. **Kinoshita, N. and Gelboin, H. V.,** β-Glucuronidase catalyzed hydrolysis of benzo[a]pyrene-3-glucuronide and binding to DNA, *Science,* Washington, D.C., 199, 307, 1978.
56. **Yang, S. K., Chou, M. W., and Fu, P. P.,** Metabolic and structural requirements for the carcinogenic potencies of unsubstituted and methyl-substituted polycyclic aromatic hydrocarbons, in *Carcinogenesis: Fundamental Mechanisms and Environmental Effects,* Pullman, B., Ts'o, P. O. P, and Gelboin, H. V., Eds., Dordrecht/Holland, 1980, 143.
57. **Sayer, J. M., Yagi, H., Croisy-Delcey, M., and Jerina, D. M.,** Novel bay-region diol epoxides from benzo[c]phenanthrene, *J. Am. Chem. Soc.,* 103, 4970, 1981.
58. **Silverman, B. D.,** Steric crowding and conformer stability of the diolepoxides of benzo[c]phenanthrene, *Cancer Biochem. Biophys.,* 6, 23, 1982.
59. **Silverman, B. D.,** Diol-epoxide conformer stability of polycyclic aromatic hydrocarbons (PAH): the effect of a bay-region methyl group, in *Polynuclear Aromatic Hydrocarbons: Formation, Metabolism and Measurement,* Battelle Press, Columbus, Ohio, 1983, 1113.
60. **Bao, Z. and Yang, S. K.,** Two K-regions of 5-methylchrysene are sites of oxidative metabolism, *Biochem. Biophys. Res. Commun.,* 141, 734, 1986.
61. **Briant, C. E., Jones, D. W., and Shaw, J. D.,** Molecular dimensions in crystals of methyl-substituted and other derivatives of benz[*a*]anthracene, *J. Mol. Struct.,* 130, 167, 1985.
62. **Yang, S. K., Chou, M. W., and Fu, P. P.,** Microsomal oxidations of methyl-substituted and unsubstituted aromatic carbons of monomethylbenz[*a*]anthracenes, in *Polynuclear Aromatic Hydrocarbons: Chemical Analysis and Biological Fate,* Cook, M. and Dennis, A. J., Eds., Battelle Press, Columbus, Ohio, 1981, 253.
63. **Chiu, P.-L. and Yang, S. K.,** Liquid chromatographic separation of five *trans*-dihydrodiols of benz[*a*]anthracene, *J. Liq. Chromatogr.,* 9, 701, 1986.
64. **Mushtaq, M., Bao, Z., and Yang, S. K.,** Reversed-phase HPLC separation of phenolic derivatives of benzo[*a*]pyrene, benz[*a*]anthracene, and chrysene with monomeric and polymeric C18 columns, *J. Chromatogr.,* 385, 293, 1987.
65. **Yang, S. K. and Chiu, P.-L.,** Cytochrome P-450-catalyzed stereoselective epoxidation at the K-region of benz[*a*]anthracene and benzo[*a*]pyrene, *Arch. Biochem. Biophys.,* 240, 546, 1985.
66. **Mushtaq, M., Weems, H. B., and Yang, S. K.,** Metabolic and stereoselective formation of non-K-region benz[*a*]anthracene 8,9- and 10,11-epoxides, *Arch. Biochem. Biophys.,* 246, 478, 1986.
67. **Thakker, D. R., Levin, W., Yagi, H., Ryan, D., Thomas, P. E., Karle, J. M., Lehr, R. E., Jerina, D. M., and Conney, A. H.,** Metabolism of benzo[*a*]anthracene to its tumorigenic 3,4-dihydrodiol, *Mol. Pharmacol.,* 15, 138, 1979.
68. **Fu, P. P., Chou, M. W., and Yang, S. K.,** In vitro metabolism of 12-methylbenz[*a*]anthracene: effect of the methyl group on the stereochemistry of a 5,6-dihydrodiol metabolite, *Biochem. Biophys. Res. Commun.,* 106, 940, 1982.
69. **Yang, S. K., Chou, M. W., Weems, H. B., and Fu, P. P.,** Enzymatic formation of an 8,9-diol from 8-methylbenz[*a*]anthracene, *Biochem. Biophys. Res. Commun.,* 90, 1136, 1979.
70. **Thakker, D. R., Levin, W., Yagi, H., Yeh, H. J. C., Ryan, D. E., Thomas, P. E., Conney, A. H., and Jerina, D. M.,** Stereoselective metabolism of the (+)-(S,S)- and (−)-(R,R)- Enantiomers of *trans*-3,4-dihydroxy-3,4-dihydrobenzo[c]phenanthrene by rat and mouse liver microsomes and by a purified and reconstituted cytochrome P-450 system, *J. Biol. Chem.,* 261, 5404, 1986.

71. **Jerina, D. M., Michaud, D. P., Feldman, R. J., Armstrong, R. N., Vyas, K. P., Thakker, D. R., Yagi, H., Thomas, P. E., Ryan, D. E., and Levin, W.,** Stereochemical modeling of the catalytic site of cytochrome P-450c, in *Microsomes, Drug Oxidations and Drug Toxicity,* Sato, R. and Kato, R., Eds., Japan Scientific Societies Press, Tokyo, 1982, 195.
72. **Chou, M. W., Yang, S. K., Sydor, W., and Yang, C. S.,** Metabolism of 7,12-dimethylbenz[*a*]anthracene and 7-hydroxymethyl-12-methylbenz[a]anthracene by rat liver nuclei and microsomes, *Cancer Res.,* 41, 1559, 1981.
73. **Mushtaq, M., Fu, P. P., Miller, D. W., and Yang, S. K.,** Metabolism of 6-methylbenz[*a*]anthracene by rat liver microsomes and mutagenicity of metabolites, *Cancer Res.,* 45, 4006, 1985.
74. **Yang, S. K. and Chou, M. W.,** Metabolism of the bay-region *trans*-1,2-dihydrodiol of benz[*a*]anthracene in rat liver microsomes occurs primarily at the 3,4-double bond, *Carcinogenesis,* 1, 803, 1980.
75. **Vyas, K. P., van Bladeren, P. J., Thakker, D. R., Yagi, H., Sayer, J. M., Levin, W., and Jerina, D. M.,** Regioselectivity and stereoselectivity in the metabolism of *trans*-1,2-dihydroxy-1,2-dihydrobenz[*a*]anthracene by rat liver microsomes, *Mol. Pharmacol.,* 24, 115, 1983.
76. **Chou, M. W., Fu, P. P., and Yang, S. K.,** Metabolic conversion of dibenz[a,h]anthracene (±)*trans*-1,2-dihydrodiol and chrysene (±)*trans*-1,2-dihydrodiol to vicinal dihydrodiol-epoxides, *Proc. Natl. Acad. Sci. U.S.A.,* 78, 4270, 1981.
77. **Vyas, K. P., Yagi, H., Levin, W., Conney, A. H., and Jerina, D. M.,** Metabolism of (−)-*trans*-(3R,4R)-dihydroxy-3,4-dihydrochrysene to diol epoxides by liver microsomes, *Biochem. Biophys. Res. Commun.,* 98, 961, 1981.
78. **Nordqvist, M., Thakker, D. R., Vyas, K. P., Yagi, H., Levin, W., Ryan, D. E., Thomas, P. E., Conney, A. H., and Jerina, D. M.,** Metabolism of chrysene and phenanthrene to bay-region diol epoxides by rat liver enzymes, *Mol. Pharmacol.,* 19, 168, 1981.
79. **Thakker, D. R., Yagi, H., Lehr, R. E., Levin, W., Buening, M., Lu, A. Y. H., Chang, R. L., Wood, A. W., Conney, A. H., and Jerina, D. M.,** Metabolism of *trans*-9,10-dihydroxy-9,10-dihydrobenzo[*a*]pyrene occurs primarily by arylhydroxylation rather than formation of a diol epoxide, *Mol. Pharmacol.,* 14, 502, 1978.
80. **Wood, A. W., Levin, W., Thakker, D. R., Yagi, H., Chang, R. L., Ryan, D. E., Thomas, P. E., Dansette, P. M., Whittaker, N., Turujman, S., Lehr, R. E., Jerina, D. M., and Conney, A. H.,** Biological activity of benzo[*e*]pyrene. An assessment based on mutagenic activities and metabolic profiles of the polycyclic hydrocarbon and its derivatives, *J. Biol. Chem.,* 254, 4408, 1979.
81. **Thakker, D. R., Levin, W., Yagi, H., Tada, M., Ryan, D. E., Thomas, P. E., Conney, A. H., and Jerina, D. M.,** Stereoselective metabolism of the (+)- and (−)-enantiomers of *trans*-3,4-dihydroxy-3,4-dihydrobenz[*a*]anthracene by rat liver microsomes and by a purified and reconstituted cytochrome P-450 system, *J. Biol. Chem.,* 257, 5103, 1982.
82. **Baird, W. M. and Pruess-Schartz, D.,** Polycyclic aromatic hydrocarbon-DNA adducts and their analysis: a powerful technique for characterization of pathways of metabolic activation of hydrocarbons to ultimate carcinogenic metabolites, chap. 13, Vol. 2, this work.
83. **Weems, H. B., Mushtaq, M., and Yang, S. K.,** Resolution of enantiomeric epoxides of polycyclic aromatic hydrocarbons by chiral stationary phase high-performance liquid chromatography, *Anal. Biochem.,* 148, 328, 1985.
84. **Yang, S. K., Mushtaq, M., Weems, H. B., Miller, D. W., and Fu, P. P.,** Stereoselective formation and hydration of 12-methylbenz[*a*]anthracene 5,6-epoxide enantiomers by rat liver microsomal enzymes, *Biochem. J.,* 245, 191, 1987.
85. **Yang, S. K., Chou, M. W., Fu, P. P., Wislocki, P. G., and Lu, A. Y. H.,** Epoxidation reactions catalyzed by rat liver cytochrome P-450 and P-448 occur at different faces of the 8,9-double bond of 8-methylbenz[*a*]anthracene, *Proc. Natl. Acad. Sci. U.S.A.,* 79, 6802, 1982.
86. **Yang, S. K.,** Structural factors that enhance or attenuate the carcinogenicity of polycyclic aromatic hydrocarbons, in *Biology of Cancer,* Vol. 1, 13th Internat. Cancer Congress, Part B., Mirand, E. A., Hutchinson, W. B., and Mihich, E., Eds., Alan R. Liss, New York, 1983, 151.
87. **Chou, M. W. and Yang, S. K.,** Combined reversed-phase and normal-phase high performance liquid chromatography in the purification and identification of 7,12-dimethylbenz[*a*]anthracene metabolites, *J. Chromatogr.,* 185, 635, 1979.
88. **Chou, M. W. and Yang, S. K.,** Identification of four *trans*-3,4-dihydrodiol metabolites of 7,12-dimethylbenz[*a*]anthracene and their in vitro DNA binding activities upon further metabolism, *Proc. Natl. Acad. Sci. U.S.A.,* 75, 5466, 1978.
89. **Thakker, D. R., Levin, W., Yagi, H., Turujman, S., Kapadia, D., Conney, A. H., and Jerina, D. M.,** Absolute stereochemistry of the *trans*-dihydrodiols formed from benzo[*a*]anthracene by liver microsomes, *Chem. Biol. Interact.,* 27, 145, 1979.
90. **Thakker, D. R., Yagi, H., Levin, W., Lu, A. Y. H., Conney, A. H., and Jerina, D. M.,** Stereospecificity of microsomal and purified epoxide hydrase from rat liver, *J. Biol. Chem.,* 252, 6328, 1977.

91. **Armstrong, R. N., Kedzierski, B., Levin, W., and Jerina, D. M.,** Enantioselectivity of microsomal epoxide hydrolase toward arene oxide substrates, *J. Biol. Chem.*, 256, 4726, 1981.
92. **Weems, H. B., Fu, P. P., and Yang, S. K.,** Stereoselective metabolism of chrysene by rat liver microsomes. Direct separation of diol enantiomers by chiral stationary phase h.p.l.c., *Carcinogenesis,* 7, 1221, 1986.
93. **Slaga, T. J., Gleason, G. L., Mills, G., Ewald, L., Fu, P. P., Lee, H. M., and Harvey, R. G.,** Comparison of the skin tumor-initiating activities of dihydrodiols and diol-epoxides of various polycyclic aromatic hydrocarbons, *Cancer Res.*, 40, 1981, 1980.
94. **Sims, P. and Grover, P. L.,** Involvement of dihydrodiols and diol epoxides in the metabolic activation of polycyclic hydrocarbons other than benzo[*a*]pyrene, in *Polycyclic Hydrocarbons and Cancer,* Vol. 3, Gelboin, H. V. and Ts'o, P. O. P., Eds., Academic Press, New York, 1981, 117.
95. **Yang, S. K., Mushtaq, M., and Chiu, P.-L.,** Stereoselective metabolism and activations of polycyclic aromatic hydrocarbons, in *Polycyclic Hydrocarbons and Carcinogenesis,* Harvey, R. G., Ed., ACS Symp. Ser. 283, American Chemical Society, Washington, D.C., 1985, 19.
96. **MacNicoll, A. D., Grover, P. L., and Sims, P.,** The metabolism of a series of polycyclic aromatic hydrocarbons by mouse skin maintained in short-term organ culture, *Chem. Biol. Interact.*, 29, 169, 1980.
97. **Yang, S. K., Chou, M. W., and Fu, P. P.,** Metabolism of 6-, 7-, 8-, and 12-methylbenz[*a*]anthracenes and hydroxymethylbenz[a]anthracenes, in *Polynuclear Aromatic Hydrocarbons: Chemistry and Biological Effects,* Bjørseth, A. and Dennis, A. J., Eds., Battelle Press, Columbus, Ohio, 1980, 645.
98. **Mushtaq,, M., Weems, H. B., and Yang, S. K.,** Resolution and absolute configuration of 7,12-dimethylbenz[*a*]anthracene 5,6-epoxide enantiomers, *Biochem. Biophys. Res. Commun.*, 125, 539, 1984.
99. **Yang, S. K., Mushtaq, M., and Weems, H. B.,** Stereoselective formation and hydration of benzo[*c*]phenanthrene of 3,4- and 5,6-epoxide enantiomers by rat liver microsomal enzymes, *Arch. Biochm. Biophys.*, 255, 48, 1987.
100. **Yang, S. K.,** Unpublished results.

Chapter 6

STRUCTURE-ACTIVITY RELATIONSHIPS AMONG TRICYCLIC POLYNUCLEAR AROMATIC HYDROCARBONS

Edmond J. LaVoie and Joseph E. Rice

TABLE OF CONTENTS

I. INTRODUCTION

The environmental occurrence and carcinogenic activity of polynuclear aromatic hydrocarbons (PAH) have been frequently reviewed.[1-6] For several decades, investigations have been conducted on the structure-carcinogenicity relationships of PAH. The objective of many of these studies was directed toward a determination of the relative carcinogenic potency of PAH. In a comparison of the carcinogenic activities of several unsubstituted PAH, it was suggested that only those PAH containing four or more rings in their structure would have significant activity. The influence of methyl substituents on the biological activity of several of the more potent carcinogenic PAH was also investigated. In these extensive studies, it was shown that methyl substitution at specific sites could further enhance the carcinogenic activity of several PAH. Studies outlining the biological activity of methylated PAH containing four and five rings are reviewed in Chapter 4. In a limited number of cases, methylated derivatives of an inactive parent hydrocarbon were carcinogenic. While several of the methylated derivatives of tricyclic PAH were found to be inactive as tumor initiators, there are specific methylated analogues of anthracene and phenanthrene which have been shown to be tumorigenic.

The focus of this chapter is on the structure-activity relationships among tricyclic PAH. It is difficult to obtain quantitative data on the levels of tricyclic PAH and their methylated derivatives in the environment because of their relative volatility. Despite this fact, the widespread occurrence of tricyclic PAH and their methylated derivatives has been documented in recent years.[4-7] In view of their environmental prevalence, there has been increased interest in the biological activities of tricyclic PAH. This chapter will review the structure-activity relationships of tricyclic PAH as it pertains to their mutagenic, tumorigenic, and tumor-promoting activities. The biological activity of these PAH, as well as their parent hydrocarbons, will be discussed. Historical data on the carcinogenic activity of these PAH was obtained from the *Survey of Compounds Which Have Been Evaluated for Carcinogenic Activity* as published by the Public Health Service.[11a-l] Data on the mutagenic activity of tricyclic PAH were compiled through the assistance of the Environmental Mutagen Information Service (EMIS) located in Oak Ridge, Tennessee.

Studies on the structure-activity of tricyclic PAH can provide valuable information with regard to the mechanism of action and the molecular basis for the biological activity of PAH. The structure-activity relationships associated with these PAH are frequently more readily interpretable in view of their less complex structure. Interest in the biological activity of tricyclic PAH is related not only to their environmental occurrence but also to the possibility that their small molecular size might be associated with unique genotoxic effects. Within this subclass of PAH, the potential for mutagenic, tumorigenic, and tumor-promoting activity has been documented. The effect of varying structure on the biological activities of tricyclic PAH will be discussed based upon currently available data. In those instances where additional experimental data are available, the suggested molecular basis which underlies an observed structure-activity relationship will be presented.

II. FLUORENE AND FLUORENE DERIVATIVES

A. Mutagenic Activity

Studies on used crankcase oil and coal-liquefaction fractions demonstrated that low-molecular weight fractions had significant mutagenic activity.[12,13] These mutagenic subfractions largely consisted of methylated derivatives of tricyclic PAH. Based upon these observations, a systematic study on the mutagenic activity of methylated fluorenes was performed.

Comparison of the mutagenic activity of an extensive series of methylated fluorene derivatives indicated that there were rigid structural requirements for mutagenic activity within

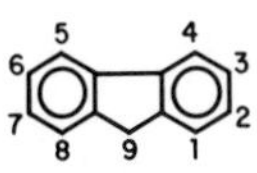

FIGURE 1. Structure and numbering of fluorene.

Table 1
MUTAGENIC ACTIVITY OF FLUORENE AND FLUORENE DERIVATIVES IN *S. TYPHIMURIUM*

	Mutagenic activity			Mutagenic activity	
	TA98	**TA100**		**TA98**	**TA100**
Fluorene and methylfluorenes[14,15,17]			**Dimethyl- and tri-methylfluorenes**[14,16,17]		
Fluorene	−	−	1,9-Dimethylfluorene	+	+
1-Methylfluorene	−	−	2,3-Dimethylfluorene	−	−
2-Methylfluorene	−	−	2,7-Dimethylfluorene	−	−
3-Methylfluorene	−	−	2,9-Dimethylfluorene	+	+
4-Methylfluorene	−	−	3,9-Dimethylfluorene	+	+
9-Methylfluorene	+	+	4,9-Dimethylfluorene	+	+
			9,9-Dimethylfluorene	−	−
Derivatives of methylfluorenes[14,16,17]			2,3,9-Trimethylfluorene	+	+
			2,7,9-Trimethylfluorene	−	−
2-Fluoro-9-methylfluorene	+	+	**9-Alkylfluorenes**[15]		
2,7-Difluoro-9-methylfluorene	+	+			
9-Hydoxymethylfluorene	−	−	9-Ethylfluorene	+	+
9-Hydroxy-9-methylfluorene	−	−	9-*n*-Propylfluorene	+	+
			9-*i*-Propylfluorene	−	−

Note: The data listed provide a qualitative assessment of mutagenic activity in *S. typhimurium* TA98 and TA100. In all instances compounds were assayed at doses of 5 to 200 μg per plate. A (+) response is indicated for any compound which produced a twofold increase in His^+ revertants/plate.

this group of PAH.[14-16] In a comparative mutagenicity assay of all the positional isomers of methyl fluorene (five isomers, see Figure 1), only 9-methylfluorene was found to be mutagenic in *Salmonella typhimurium* TA98 and TA100 in the presence of liver homogenate from Aroclor®-pretreated rats. In the absence of a metabolic activation system, none of the monomethylated fluorenes was active as a mutagen in either of these tester strains.[14,17]

The association of mutagenic activity with the presence of a single methyl substituent at the 9 position of fluorene was also evident from studies performed on a series of polymethylated derivatives of fluorene. Those fluorene derivatives which have been assayed as mutagens in *S. typhimurium* are listed in Table 1. The lack of mutagenic potency observed in assays with 2,3-, 2,7-, and 9,9-dimethylfluorene is consistent with the requirement of a single alkyl substituent at the 9 position. In a comparison of the mutagenic dimethylfluorenes, both 1,9- and 4,9-dimethylfluorene were more potent mutagens than either 2,9- or 3,9-dimethylfluorene. The only polymethylated fluorene derivative that had a single methyl substituent at the 9 position and was not mutagenic under similar assay conditions was 2,7,9-trimethylfluorene. The inactivity of 1,1-diphenylethane indicates that the "biphenyl-character" and/or rigidity of the fluorene ring system is an essential requirement for retention of activity.[17]

The effect of increasing the length of the alkyl group at the 9 position on the mutagenic activity of substituted fluorenes was also determined by comparing the relative mutagenic

FIGURE 2. Hydroperoxides of 9-methylfluorene and 1,9-dimethylfluorene assayed for mutagenic activity.

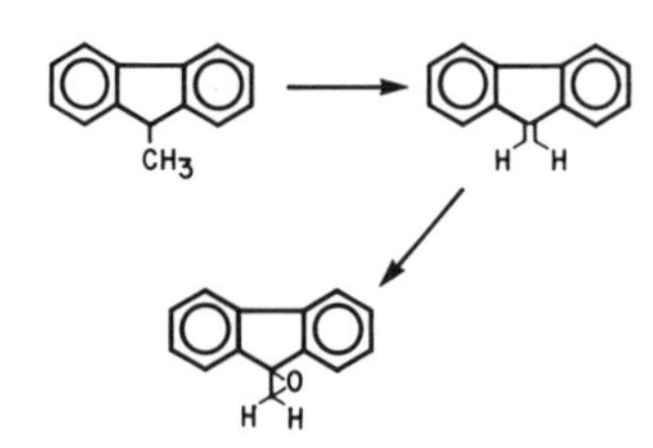

FIGURE 3. Proposed mechanism for the formation of an ultimate mutagen of 9-methylfluorene involving a 9-methylidene intermediate.

potency of 9-methylfluorene vs. 9-ethyl-, 9-*n*-propyl-, and 9-*i*-propylfluorene.[17] While 9-ethylfluorene was as active as 9-methylfluorene, 9-*n*-propylfluorene was much less active. In the case of 9-*i*-propylfluorene, the branching of this alkyl group resulted in the loss of mutagenic activity for this 9-substituted fluorene.

These data suggest that the 9 position of fluorene is directly associated with, or is in close proximity to, the site which is ultimately responsible for the activity of mutagenic fluorene derivatives. The observation that fluorine substitution, as in the case of 2-fluoro-9-methylfluorene or 2,7-difluoro-9-methylfluorene, did not inhibit mutagenic activity suggests that epoxide formation at either the 1,2- or 7,8-positions of 9-methylfluorene is not associated with its ultimate activation to a mutagen.[17]

Hydroxylation of either the 9-methyl group or the 9 position of 9-methylfluorene resulted in the loss of mutagenic activity in *S. typhimurium*. It has been speculated that for certain methylated PAH, formation of a reactive ester at a benzylic position may be related to their exceptional biological activity.[18,19] The mutagenicity data on 9-methyl-9-fluorenol suggests that the subsequent formation of such reactive esters in this series of compounds is not a likely mechanism by which methylated fluorenes are ultimately converted to mutagens. 9-Alkylfluorenes are known to form stable hydroperoxides at the 9 position.[20] If such hydroperoxides could be responsible for the mutagenic activity of methylated fluorenes, their involvement as ultimate mutagens would be consistent with the observed structure-mutagenicity relationships. Both 9-hydroperoxy-9-methylfluorene and 9-hydroperoxy-1,9-dimethylfluorene (Figure 2) were evaluated as mutagens in *S. typhimurium* TA98 and TA100 in the presence and absence of metabolic activation. At the highest doses employed, 9-hydroperoxy-9-methylfluorene was weakly mutagenic. While the absence of mutagenic activity for this peroxide when assayed without metabolic activation indicated it was not an ultimate mutagen, its weak mutagenic activity in the presence of metabolic activation indicated that it could contribute to the mutagenic activity observed with 9-methylfluorene.[17] In the case of 9-hydroperoxy-1,9-dimethylfluorene, the weak mutagenic response of the peroxide relative to 1,9-dimethylfluorene clearly indicated that these hydroperoxides are not likely to contribute in any significant degree to the mutagenic activity of methylated fluorenes.[16]

An alternate mechanism, involving the formation of a 9-methylidene intermediate with the ultimate formation of fluorene-9,2′-oxirane (Figure 3), has been proposed.[15] While 9-methylidenefluorene is somewhat unstable, 9-ethylidenefluorene can be synthesized and isolated as a pure compound. The mutagenic activity of 9-ethylidenefluorene in the presence of metabolic activation as well as the potent mutagenic activity of fluorene-9,2′-oxirane in the absence of metabolic activation support this hypothesis.[15] Studies performed to isolate 9-methylidenefluorene or trap fluorene-9,2′-oxirane from incubation mixtures of 9-methylfluorene with similar rat-liver preparations as employed in assays of its mutagenic activity were not successful.

The fluorene nucleus is common to several nonalternant PAH. Studies have already shown that a similar structure-mutagenicity relationship exists for methylated benzofluorenes.[17] Determining the mechanisms by which methylated fluorenes are ultimately activated to mutagens will be a significant contribution to our understanding of the molecular basis for the biological activity of PAH.

B. Tumorigenic Activity

Fluorene and several derivatives of fluorene, because of their prevalence in the environment, were among the first pure hydrocarbons assayed for carcinogenic activity. The bioassays performed to assess the tumorigenic activity of fluorene, fluorenone, and methylated derivatives of fluorene are outlined in Table 2. Neither fluorene nor fluorenone were found to have significant carcinogenic activity in these bioassays. The International Agency for Research on Cancer (IARC) considers the available bioassay data on fluorene to be "inadequate to permit an evaluation of carcinogenicity . . . to experimental animals."[6] Bioassays on the tumor-initiating activity of those methylated fluorenes which are mutagenic in *S. typhimurium* indicated that none of these fluorene derivatives is active on mouse skin. In view of these data, the potential health effects of mutagenic fluorene derivatives remain uncertain. Studies in vitro using the hepatocyte primary culture/DNA repair test have been performed on fluorene, 9-methylfluorene and 1,9-dimethylfluorene.[30,31] The results of these bioassays indicate that in these intact cells no unscheduled DNA synthesis was detected at any of the concentrations employed.[31] Further studies in mammalian cells and whole animals, however, are needed to more accurately determine the potential adverse health effects of these fluorene derivatives.

III. ACENAPHTHYLENE AND ACENAPHTHENE

Acenaphthylene and acenaphthene (Figure 4) were assayed for carcinogenic activity at relatively high doses as outlined in Table 2. Under these experimental conditions, neither of these tricyclic PAH were active as carcinogens. Despite their prevalence in the environment, there are no recent studies on their potential carcinogenic activity in animals.[32] Studies on the biological activity of these tricyclic PAH in vitro have indicated that these parent hydrocarbons are not genotoxic. These compounds were inactive as mutagens as measured by forward mutation to azaguanine resistance in *S. typhimurium* TM677.[33,34] Similar results have been observed for these parent hydrocarbons and their 3-methyl derivatives.[35] Studies on the mutagenic activity of acenaphthene in *S. typhimurium* TA1535, TA1537, TA1538, and TA98 revealed that in the absence of metabolic activation, this compound was toxic at a dose level of 200 μg per plate.[36] Acenaphthene was also shown not to exhibit significant noncovalent interaction with nucleosides.[37]

The limited studies on the biological activity of acenaphthylene and acenaphthene are in marked contrast to the detailed studies on the mutagenic and carcinogenic activity of their 5-nitro derivatives.[38-42] These studies have shown that both of these nitro derivatives are potent mutagenic and carcinogenic agents. A detailed review of all of the biological activities of these derivatives of acenaphthylene and acenaphthene, however, is beyond the scope of this review.

The limited data on the biological activity of acenaphthylene and acenaphthene suggest that these hydrocarbons are not genotoxic. Among their methylated analogues, only 3-methylacenaphthylene and 3-methylacenaphthene have been evaluated as mutagens. Neither

Table 2
BIOASSAYS ON THE TUMORIGENIC ACTIVITY OF FLUORENE, FLUORENONE, METHYLATED DERIVATIVES OF FLUORENE, ACENAPHTHYLENE, AND ACENAPHTHENE

Compound	Species/strain	Est. total dose animal	Route and site of application	Comments[a]	Ref.
Fluorene	Mice; CF1	3.0—4.0 mg	Skin; topical	0% TBA	21
Fluorene	Mice	90 mg[b]	Skin; topical	Coapplication with croton oil; 0% TBA	22
Fluorene	Mice; CD1	1.0 mg	Skin; topical	As a tumor initiator with TPA; 5% TBA	14, 17
Fluorene	Mice	25 mg	i.p.	0% TBA	23
Fluorene	Rats	2.0 g	p.o.; in the diet	0% TBA	24
Fluorenone	Rats	30 g	p.o.; in the diet, 600 days	No significant lesions	25
Fluorenone	Rats	60 mg	s.c.	Viewed as noncarcinogenic	25
9-Methylfluorene	Mice; CD1	1.0 mg	Skin; topical	As a tumor initiator with TPA; 5% TBA, 0.05 T/A	14, 17
1,9-Dimethylfluorene	Mice; CD1	1.0 mg	Skin; topical	As a tumor initiator with TPA; 14% TBA, 0.16 T/A	14, 15, 16
2,9-Dimethylfluorene	Mice; CD1	1.0 mg	Skin; topical	As a tumor initiator with TPA; 4.2% TBA, 0.04 T/A	16
3,9-Dimethylfluorene	Mice; CD1	1.0 mg	Skin; topical	As a tumor initiator with TPA; 12% TBA, 0.12 T/A	16
4,9-Dimethylfluorene	Mice; CD1	1.0 mg	Skin; topical	As a tumor initiator with TPA as promoter, 8% TBA, 0.08 T/A	16
2,3,9-Trimethylfluorene	Mice; CD1	1.0 mg	Skin; topical	Evaluated as a tumor initiator with TPA as promoter; 0% TBA	16
Acenaphthene	Mice	(1.0% sol)	Skin; topical	No tumors	26, 27
Acenaphthene	Mice	9.0 mg[b]	Skin; topical	Coapplication of croton oil, 2.2% TBA	22
Acenaphthylene	Rats; SD	300 mg	Intragastric	No tumors observed	28
Acenaphthylene	Mice	2.5 g	Skin; topical	No tumors	29

[a] Percent tumor-bearing animals is listed as % TBA. The average number of tumors of all animals in the experimental group is given as T/A in these tables.

[b] Calculation of estimated total dose assumes that approximately 100 $\mu\ell$ of solution was painted on the skin of mice. These values are only crude approximations in view of the limited experimental detail available.

of these methyl derivatives was active as a mutagen in the presence of metabolic activation. The genotoxic potential of the 4- and 5-methyl derivatives of these PAH has not been evaluated.

IV. ANTHRACENE AND ANTHRACENE DERIVATIVES

A. Mutagenic Activity

Anthracene and several derivatives of anthracene have been assayed for mutagenic activity in a variety of test systems. In general, anthracene and each isomer of methylanthracene

ACENAPHTHENE ACENAPHTHYLENE

FIGURE 4. Structure and numbering of acenaphthene and acenaphthylene.

Table 3
MUTAGENIC ACTIVITY OF ANTHRACENE AND ANTHRACENE DERIVATIVES IN *SALMONELLA TYPHIMURIUM*

Anthracene and methylated anthracenes	Mutagenic activity[a]					
	TA98[b]	**Ref.**	**TA100[b]**	**Ref.**	**8AGr**	**Ref.**
Anthracene	−	14, 45, 46, 48	−	14, 45, 48, 51	−	33
1-Methylanthracene	−	14, 43, 44	−	14, 43, 44	N.D.[c]	
2-Methylanthracene	−	14, 43, 44	−	14, 43, 44	+	33
9-Methylanthracene	−	14, 42—45	−	14, 42—45	+	33
2,9-Dimethylanthracene	−	44	−	44	N.D.	
9,10-Dimethylanthracene	+	43—47, 48	+	43—45, 48	−	33
2,9,10-Trimethylanthracene	+	44	+	44	N.D.	
2,3,9,10-Tetramethylanthracene	+	44	+	44	N.D.	
Oxygenated derivatives						
Anthrone	−	47, 49	−	47, 49	N.D.	
Anthraquinone	−	47, 49, 50	−	47, 49, 50	N.D.	
2-Methylanthraquinone	−	47, 50	−	47, 50	N.D.	
Anthralin	−	49, 51	−	49, 51	N.D.	

[a] Qualitative data are presented in this table for comparative purposes.

[b] The data listed provide a qualitative assessment of mutagenic activity in *S. typhimurium* TA98 and TA100. In all instances compounds were assayed at doses of 5—200 μg per plate. A (+) response is indicated for any compound which produced a twofold increase in His^+ revertants/plate.

[c] N.D., not determined.

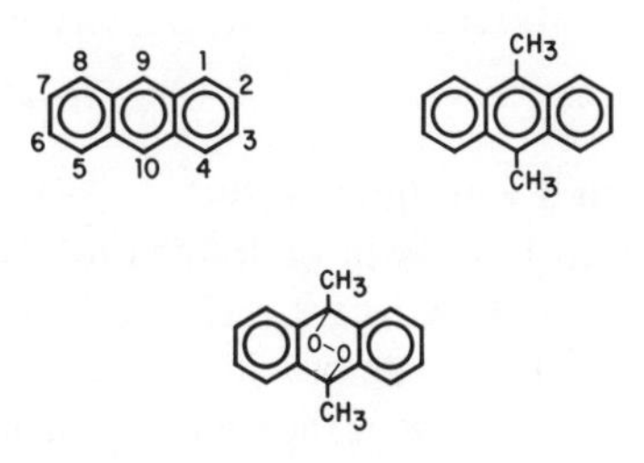

FIGURE 5. Structure and numbering of anthracene and the structure of 9,10-dimethylanthracene and its photo-oxide.

were not mutagenic. Table 3 summarizes some of the results obtained from assays on the mutagenic activity of anthracene, methylated derivatives of anthracene, and oxygenated derivatives of anthracene. Anthracene was frequently selected as a negative control compound in many of the in vitro mutagenicity assays. 9,10-Dimethylanthracene, which is weakly carcinogenic (see Section IV.B), was often employed to determine the effectiveness of these test systems to detect weak carcinogenic agents. Both anthracene and 9,10-dimethylanthracene (Figure 5) were among the 42 coded compounds selected by the International Collaborative Program evaluation of short-term tests for carcinogens.[52] Anthracene did not increase mitotic recombination in *Saccharomycis* D_3[53] and was not more toxic in DNA repair-deficient strains of bacteria.[54] While 9,10-dimethylanthracene was also not more toxic in DNA repair-deficient strains of bacteria, 9,10-dimethylanthracene was mutagenic in *Drosophila melanogaster*[55,56] and was genotoxic in the SOS Chromotest.[57] In contrast to anthracene, 9,10-dimethylanthracene induces unscheduled DNA synthesis in primary rat hepatocyte cultures.[30]

9,10-Dimethylanthracene was most commonly associated with unique genotoxicity relative to anthracene or the monomethylated derivatives of anthracene. An exception to this general trend was observed in assays performed with the forward mutation to azaguanine resistance in *S. typhimurium* TM677. In addition, studies with human lymphoblasts indicated that 2- and 9-methylanthracene are active as mutagens; anthracene and 9,10-dimethylanthracene were not mutagenic in this system.[58] The association of increased genotoxicity in various in vitro bioassays with the presence of methyl substituents at both the 9- and 10-positions of anthracene, however, is consistent with bioassays on the relative tumorigenic activity of substituted anthracenes.

It is known that, in contrast to other anthracene derivatives, photo-oxidation of 9,10-dimethylanthracene results in the formation of a stable peroxide (see Figure 5). The involvement of such a peroxide in the ultimate activation of anthracene derivatives substituted at both the 9 and 10 positions to mutagens would be consistent with the observed structure-mutagenicity data. Bioassays on this photo-oxide, however, have clearly shown that it is not the major ultimate mutagen of 9,10-dimethylanthracene.

Anthrone, anthraquinone, and 2-methylanthraquinone were inactive as mutagens in *S. typhimurium* (Figure 6). While anthralin is active as a tumor promoter in bioassays performed on mouse skin (see IV.D), it is not active as a carcinogen and is not mutagenic in *S. typhimurium*.[59] While anthralin did induce a lower number of sister chromatid exchanges, this activity did not correlate with the potency of various other tumor promoters.[60] There are only a limited number of chemical agents which have shown significant activity as tumor promoters in the two-stage mouse skin carcinogenicity bioassay. Thus, there has been intensive interest in the biochemical effects of anthralin. Anthralin has been shown to inhibit DNA repair synthesis while its nonpromoting analogue 1,8-dihydroxyanthraquinone (see IV.D) had no effect.[61] Anthralin was also reported to induce mitotic aneuploidy in cultures of yeast strain D_6.[62] Anthralin was found not to influence the rate of DNA synthesis in human fibroblasts[63] and not to have recombinogenic activity in *Saccharomyces cerevisciae* D_3.[53]

B. Tumor-Initiating Activity and Carcinogenicity

Representative bioassays which have been performed to assess the tumorigenic activity of anthracene are outlined in Table 4. A recent review of the bioassay data on anthracene by the IARC concluded that "available data provide no evidence that anthracene is carcinogenic to experimental animals".[6] In view of their environmental occurrence, several methylated derivatives of anthracene have also been assayed for tumorigenic activity. These bioassays are summarized in Table 5. The monomethylated derivatives of anthracene have not been shown to have significant tumorigenic activity in the various bioassays performed in mice. Among the various dimethylanthracenes which have been bioassayed, 9,10-dimethylanthracene was the only isomer which was active as a complete carcinogen and tumor initiator. These results parallel the frequently observed genotoxicity of 9,10-dimethylanthracene relative to anthracene, methylanthracene, and several isomeric dimethylanthracenes (see Section IV.A). 9,10-Dimethylanthracene was the first tricyclic PAH which was shown to be tumorigenic. In contrast to frequent generalization regarding the requirement for a bay region for tumorigenic activity among alternate PAH, this methylated anthracene is devoid of a bay region. Other methylated anthracenes which were active as tumor initiators include 2,9,10-trimethylanthracene and 2,3,9,10-tetramethylanthracene.

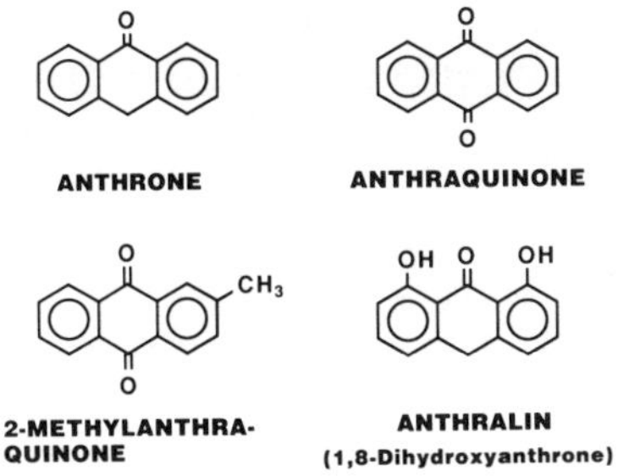

ANTHRONE

ANTHRAQUINONE

2-METHYLANTHRA-QUINONE

ANTHRALIN (1,8-Dihydroxyanthrone)

FIGURE 6. Structure of anthrone, anthraquinone, 2-methylanthraquinone, and anthralin.

Table 4
REPRESENTATIVE BIOASSAYS ON THE TUMORIGENIC ACTIVITY OF ANTHRACENE

Species/strain	No. of animals	Est. total dose/animal	Route and site of administration	Comments[a]	Ref.
Mice	100	1% Sol.	Skin; topical	No tumors	26, 27
Mice	50	1% Sol.	Skin; topical	No tumors	20, 21
Mice; "S" strain	20	30 mg	Skin; topical	As a tumor initiator with croton oil as promoter; 15% TBA; 0.2 T/A	64
Mice; CD1	20	1.0 mg	Skin; topical	As a tumor initiator using TPA as promoter; 15% TBA; 0.15 T/A	43, 44
Mice; Swiss Millerton	5	150 mg	Skin; topical	Complete carcinogen, 0% TBA	65
Mice; CD1	30	1.78 mg	Skin; topical	As a tumor initiator using TPA as promoter; 14% TBA, 0.14 T/A	66
Mice; CC57	25	2.5 mg	s.c.	No tumors	67
Mice; C57B1	26	5.0 mg	s.c.	No tumors	68
Rats; BDI, BDIII	28	4.5 g	p.o.	Incidence of tumors not significant	69
Rats; BDI, BDIII	10	660 mg	i.p.	1 Spindle cell sarcoma	69
Rats; BDI, BDIII	10	660 mg	s.c.	1 Myxosarcoma after 17 months	69
Rats; Osborne-Mendel	37	0.5 mg	Lung implant	0% TBA	70
Rabbits	9	10—20 mg pellets	Implants in the brain	No tumors	71

[a] Percent tumor-bearing animals is listed as % TBA. The average number of tumors of all animals in the experimental group is given as T/A in these tables.

The one common structural feature of all of the tumorigenic anthracene derivatives is the presence of methyl substituents at both the 9 and 10 positions. The lack of tumor-initiating activity of the endoperoxide of 9,10-dimethylanthracene indicates that this compound is not involved in the activation of 9,10-dimethylanthracene to a tumorigenic agent. While the exact mechanism by which these agents are tumorigenic has not been conclusively established, comparative studies on the metabolism of anthracene, 9-methylanthracene, and 9,10-dimethylanthracene have suggested that the requisite formation of an epoxide adjacent to the *peri*-methyl group of 9,10-dimethylanthracenes may be related to their genotoxic activities.

Table 5
BIOASSAYS ON METHYLATED ANTHRACENES

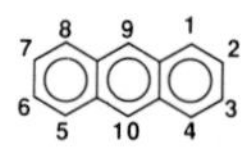

Methyl isomer	Species/strain	Est. total dose	Route and site of administration	Comments[a]	Ref.
1	Mice; CD1/F	1.0 mg	Skin; topical	As a tumor initiator using TPA as promoter; 15% TBA; 0.15 T/A	43, 44
2	Not specified		Skin; topical	0 tumors	12
	Mice; CD1	1.0 mg	Skin; topical	10% TBA, 0.15 T/A	43, 44
9	Mice	?	Skin; topical	0 tumors	73
9	Mice	5.52 mg	Skin; topical	0% TBA	74
9	Mice; Str. A	7.0 mg	s.c.	0% TBA	75
9	Newborn mice	30—40 μg	s.c.	5.5% TBA	76
9	Newborn mice	30—40 μg	i.p.	9.0% TBA	76
9	Mice	75 μg	Skin; topical	As a tumor initiator with extract from croton seeds used for promotion	70
9	Mice; CD1	1.0 mg	Skin; topical	As a tumor initiator with TPA as promoter; 20% TBA; 0.25 T/A	43, 44
1,2	Mice; C_3H	30 mg	Skin; topical	0% TBA	77
1,2	Mice; C_3H	60 mg	s.c.	0% TBA	77
1,3	Mice; C_3H	30 mg	Skin; topical	20% lung adenoma	77
1,3	Mice; C_3H	60 mg	s.c.	0% TBA	77
1,3	Mice; C_3H	20 mg	Pellet, implant, s.c.	0% TBA	77
1,4	Mice; C_3H	30 mg	Skin; topical	0% TBA	77
1,4	Mice; C_3H	60 mg	s.c.	0% TBA	77
1,4	Mice; C_3H	20 mg	Pellet, implant, s.c.	10% lung adenomas 0% skin tumors	77
2,3	Mice; C_3H	30 mg	Skin; topical	0% TBA	77
2,3	Mice; C_3H	60 mg	s.c.	10% lung adenomas 0% skin tumors	77
2,3	Mice; C_3H	20 mg	Pellet, implant, s.c.	10% lung adenomas 0% skin tumors	77
2,9	Mice; CD1	1.0 mg	Skin; topical	As a tumor initiator using TPA as promoter; 12% TBA; 0.12 T/A	43, 44
9,10	Mice		Skin; topical	15% TBA, 5% mal. carcinoma	78
9,10	Mice; C_3H	30 mg	Skin; topical	20% lung adenomas 10% skin tumors	77
9,10	Mice; C_3H	60 mg	s.c.	0% TBA	77
9,10	Mice; C_3H	20 mg	Pellet, implant, s.c.	40% lung adenoma 0% skin tumors	77
9,10	Mice	4.1 mg	Implant, s.c.	Significant preneoplastic changes	79
9,10	Mice; A/He	15 μmol/kg	i.v.	1 of 6 with lung adenoma	80
		20 μmol/kg	i.v.	0% TBA	80
9,10	Mice	24 mg	Skin; topical	30% TBA; 0.55 T/A 6 carcinomas	81
9,10	Mice	206 μg	Skin; topical	As a tumor initiator using TPA as promoter; 30% TBA; 0.33 T/A	82

Table 5 (continued)
BIOASSAYS ON METHYLATED ANTHRACENES

Methyl isomer	Species/strain	Est. total dose	Route and site of administration	Comments[a]	Ref.
9,10	Mice; Sencar	412 μg	Skin; topical	As a tumor initiator using TPA as promoter; 27% TBA; 0.37 T/A	83
9,10	Mice; CD1	1.0 mg	Skin; topical	As a tumor initiator using TPA as promoter; 35% TBA, 0.58 T/A	43, 44
2,7,9	Mice	40 mg	Skin; topical	0% TBA	130
2,9,10	Mice	2.0 mg	s.c.	0% TBA	84
2,9,10	Mice; CD1	1.0 mg	Skin; topical	As a tumor initiator using TPA as promoter; 56% TBA; 1.58 T/A	43, 44
2,3,6,7	Mice; C_3H	0.3%	Skin	0% TBA	85
2,3,6,7	Mice; C_3H	5 mg	s.c.		85
2,3,9,10	Mice; CD1	1.0 mg	Skin; topical	As a tumor initiator using TPA as promoter; 24% TBA; 0.28 T/A	43, 44

[a] Percent tumor-bearing animals is listed as % TBA. The average number of tumors of all animals in the experimental group is given as T/A in these tables.

C. Metabolism of Anthracene, 9-Methylanthracene, and 9,10-Dimethylanthracene

The major metabolites of anthracene as formed in vitro are *trans*-1,2-dihydro-1,2-dihydroxyanthracene and anthraquinone.[43,44,86-88] Using a similar activation system as employed in assays for mutagenic activity in *S. typhimurium,* the major metabolites of 9-methylanthracene were identified as the 1,2- and 3,4-dihydrodiols of 9-methylanthracene, 9-hydroxymethylanthracene, and anthraquinone.[43,44,89] While the mechanism by which 9-methylanthracene is converted to anthraquinone is not known, this demethylation, under the in vitro incubation conditions employed, does represent an interesting metabolic transformation. The symmetric nature of 9-methylanthracene provides an excellent model compound to determine the influence of a *peri*-methyl group on dihydrodiol formation. While both the 1,2- and 3,4-dihydrodiols of 9-methylanthracene were detected, these dihydrodiols were found to form in a ratio of approximately one to three.

The major metabolites of 9,10-dimethylanthracene identified from similar incubations with rat-liver homogenates were *trans*-1,2-dihydro-1,2-dihydroxy-9,10-dimethylanthracene, 9-hydroxymethyl-10-methylanthracene, and both the 1,2- and 3,4-dihydrodiol of 9-hydroxymethyl-10-methylanthracene.[43,44,90] Bioassays on the mutagenic activity of these metabolites of 9,10-dimethylanthracene did not reveal the presence of a potent proximate mutagen. In comparison with the metabolites of anthracene and 9-methylanthracene, there were no major quantitative differences in overall dihydrodiol formation which could correlate with the differences in their genotoxic activities. These data, taken together, suggest that the epoxide precursor to the 1,2-dihydrodiol of 9,10-dimethylanthracene may be associated with its mutagenic activity. While the extent of dihydrodiol formation is similar in the metabolism of anthracene, 9-methylanthracene, and 9,10-dimethylanthracene, the epoxide precursor to the 1,2-dihydrodiol of 9,10-dimethylanthracene has a *peri*-methyl substituent adjacent to the epoxide ring which may be associated with its biological activity.

The 5,6-oxide of benz[a]anthracene and 7-methylbenz[a]anthracene have been shown to be active as tumor initiators on mouse skin. While these oxides do not represent the major activated form of their parent hydrocarbons, they do possess weak tumorigenic activity (see Chapter 4). At one fourth the total initiation dose, 7-methylbenz[a]anthracene-5,6-oxide produced a similar tumorigenic response as the 5,6-oxide of benz[a]anthracene.[91] The 5,6-oxide of 7-methylbenz[a]anthracene was also more carcinogenic than the 5,6-oxide of benz[a]anthracene after subcutaneous injection in mice.[92] These data support the hypothesis that a *peri*-methyl substituent may enhance the relative tumorigenic potency of such simple arene oxides.

D. Tumor-Promoting Activity of Oxygenated Derivatives of Anthracene

The tumor-promoting activity of anthralin (1,8-dihydroxy-9-anthrone) has been well documented.[93-99] Anthralin has also been shown to be cocarcinogenic when applied together with benzo[a]pyrene on mouse skin.[102] The tumor-promoting activities of several analogues of anthralin are summarized in Table 6. Based upon these data, it was concluded that the structural features of anthralin necessary for maintaining tumor-promoting activity are at least one phenolic hydroxyl group hydrogen bonded to the C-9 carbonyl oxygen and one benzylic proton at the C-10 position. While 1,8-dihydroxyanthraquinone is inactive as a tumor promoter, anthralin analogues substituted at the 10-position by an acetyl or a myristyl group retained significant activity. The requirement for at least one proton at the C-10 position of anthralin is also consistent with the lack of tumor-promoting activity observed for 1-hydroxyfluorenone.[99]

Anthralin and analogues of anthralin represent an interesting class of tumor promoters. Recent studies have shown that chrysarobin (3-methylanthralin; 1,8-dihydroxy-3-methyl-9-anthrone) is a more potent tumor promoter than anthralin.[101] Since tumor promotion is likely to play an important role in the etiology of human cancer, the mechanism of action and structure-activity of these oxidized derivatives of anthracene are actively being investigated. No other oxidized PAH derivatives to date have exhibited such reproducible and significant tumor-promoting activity as these tricyclic analogues of anthracene.

V. PHENANTHRENE AND PHENANTHRENE DERIVATIVES

A. Mutagenic Activity

Methylated derivatives of phenanthrene have been detected as environmental pollutants in urban air, cigarette smoke, diesel engine exhaust, and coal liquefaction products.[13,103-105] Among the tricyclic PAH which have been detected as environmental pollutants, phenanthrene and its derivatives are among the more prevalent components.

The mutagenic activity of phenanthrene and several methylated derivatives of phenanthrene has been extensively investigated (Table 7). In a comparative study of the mutagenic activity of phenanthrene and of each of the five possible isomers of methylphenanthrene in *S. typhimurium* (Figure 7), it was shown that only 1- and 9-methylphenanthrene were active as mutagens.[14,106,107] For both 1- and 9-methylphenanthrene, the presence of a metabolic activation system consisting of the liver homogenate from Aroclor®-pretreated Fischer rats was required for mutagenic activity.[14,106,107] The exceptional mutagenic activity of both 1- and 9-methylphenanthrene relative to phenanthrene itself was also noted in the forward mutation assay for 8-azaguanine resistance in *S. typhimurium* TM677.[33,58] In this forward mutagenesis assay, 2-methylphenanthrene was also reported to be mutagenic.

The observed structure-mutagenicity relationships observed for methylated phenanthrenes indicated that the presence of a methyl substituent at or adjacent to the K-region (positions 9—10; Figure 7) of phenanthrene favored mutagenic activity. In addition to a methyl substituent at or adjacent to the K-region, the presence of an unsubstituted angular ring also

Table 6
DERIVATIVES OF ANTHRACENE ON MOUSE SKIN

Oxygenated anthracene derivatives[a]	Promotion: dose/ frequency	Initiator; mouse strain	Days to first papilloma	Comments[b]	Ref.
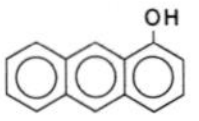	80 μg/3 × weekly	20 μg DMBA; ICR/Ha	486	5% TBA; 0.05 T/A	93
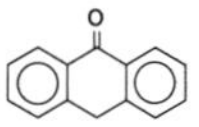	70 μg/3 × weekly	20 μg DMBA; ICR/Ha	—	0% TBA	94
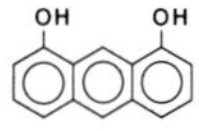	75 μg/3 × weekly	20 μg DMBA; ICR/Ha	—	0% TBA	93
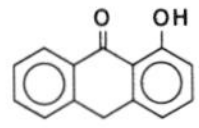	80 μg/3 × weekly	20 μg DMBA; ICR/Ha	154	33% TBA; 0.4 T/A	93
	80 μg/3 × weekly	20 μg DMBA; ICR/Ha	—	0% TBA	93
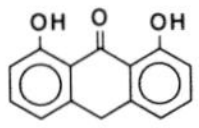	80 μg/3 × weekly	20 μg DMBA; ICR/Ha	68	40% TBA; 0.7 T/A	93
	80 μg/3 × weekly	20 μg DMBA; ICR/Ha	59	90% TBA; 4.7 T/A	94
	83 μg/5 × weekly	125 μg DMBA; C57 mice	252[c]	46% TBA	95
	83 μg/5 × weekly	125 μg DMBA; ICR mice	98[c]	97% TBA	95
	250 μg/5 × weekly	120 mg urethane; ICR mice	70	40% TBA	96
	250 μg/5 × weekly	120 mg urethane; CFW mice	63	75% TBA	96
	125 μg/5 × weekly	126 μg DMBA; AB mice	77	50% TBA	97
	10 μg/3 × weekly	25 μg DMBA; AB mice	—	0% TBA	98
	20 μg/3 × weekly	25 μg DMBA; AB mice	N.S.[d]	66% TBA; 2.14 T/A	98
	1.7 mg/3 × weekly	25 μg DMBA; AB mice	N.S.[d]	93% TBA; 3.07 T/A	98
	80 μg/5 × weekly	75 μg DMBA; CD1	77	53% TBA; 1.03 T/A[e]	99
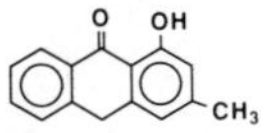	5 μg/3 × weekly	25 μg DMBA; Sencar mice	—	0% TBA	100
	50 μg/3 × weekly	25 μg DMBA; Sencar mice	N.S.[d]	88% TBA	100
	24 μg/3 × weekly	25 μg DMBA; Sencar mice	N.S.[d]	67% TBA	99
	53 μg/3 × weekly	25 μg DMBA; Sencar mice	N.S.[d]	88% TBA	99
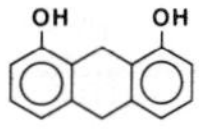	80 μg/3 × weekly	20 μg DMBA; ICR/Ha	—	0% TBA	93
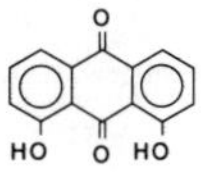	170 μg/3 × weekly	20 μg DMBA; ICR/Ha	—	0% TBA	94

Table 6 (continued)
DERIVATIVES OF ANTHRACENE ON MOUSE SKIN

Oxygenated anthracene derivatives[a]	Promotion: dose/ frequency	Initiator; mouse strain	Days to first papilloma	Comments[b]	Ref.
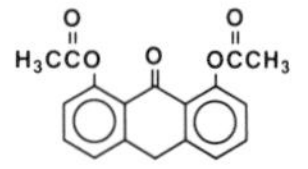 H_3CCO O $OCCH_3$	124 μg/3 × weekly	20 μg DMBA; ICR/Ha	322	7% TBA; 0.01 T/A	93
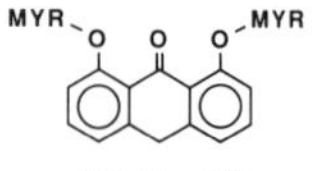 MYR O O O MYR	288 μg/3 × weekly	20 μg DMBA; ICR/Ha	—	0% TBA	93
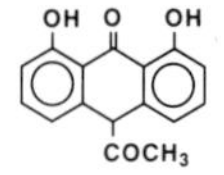 OH O OH; $COCH_3$	15 μg/3 × weekly	20 μg DMBA; ICR/Ha	89	20% TBA; 0.33 T/A	93
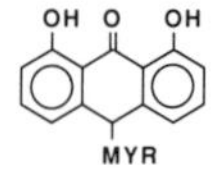 OH O OH; MYR	300 μg/3 × weekly	20 μg DMBA; ICR/Ha	55	60% TBA; 0.75 T/A	93
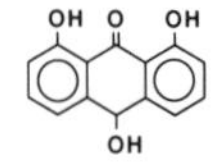 OH O OH; OH	85 μg/3 × weekly	20 μg DMBA; ICR/Ha	—	0% TBA	93
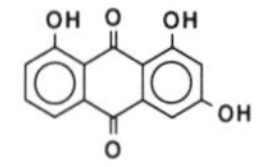 OH O OH; OH; O	0.5%/2 × weekly	DMBA; 1% Mineral oil	—	0% TBA	100
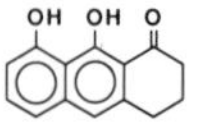 OH OH O	80 μg/3 × weekly	20 μg DMBA; ICR/Ha	—	0.02% TBA	93

[a] Myristyl groups are abbreviated MYR.
[b] Percent tumor-bearing animals is listed as % TBA. The average number of tumors of all animals in the experimental group is given as T/A in these tables.
[c] Days required for tumors in 20% of the mice.
[d] N.S., not stated.
[e] Sacrificed at 35 weeks.

appears to favor mutagenic activity. These two factors are consistent with almost all of the results obtained for those methylated phenanthrenes assayed in *S. typhimurium* TA100 as outlined in Table 7. Two exceptions to these generalizations are 1,9- and 4,9-dimethylphenanthrene. As described in Chapter 4, tumorigenic activity among several methylated PAH can be inhibited by *peri*-substitution adjacent to the angular ring involved in their ultimate metabolic activation. In these studies on the structure-mutagenicity relationships of methylated phenanthrenes, the presence of a methyl substituent adjacent to the available unsubstituted angular ring of both 1,9- and 4,9-dimethylphenanthrene correlated with their lack of mutagenic activity. Both 1,10- and 4,10-dimethylphenanthrene, which have a free *peri*-position adjacent to their unsubstituted angular ring, are active as mutagens.

It is known that phenanthrene is extensively metabolized to its K-region dihydrodiol, 9,10-dihydroxy-9,10-dihydrophenanthrene. In the vast majority of instances, phenanthrene has not been found to be mutagenic in *S. typhimurium*. Under exceptional experimental conditions, however, a positive mutagenic response was observed.[109] It is known that the 9,10-oxide of phenanthrene is weakly mutagenic.[110-112] This oxide could account for the low level of activity observed for phenanthrene in the presence of epoxide hydratase inhibitors. Under the assay conditions routinely employed to evaluate the methylated analogues of phenanthrene listed in Table 7, it is unlikely that this oxide contributes to the observed mutagenic activity. Recently, the mutagenic activity of the 9,10-oxide of phenanthrene was

Table 7
MUTAGENIC ACTIVITY OF PHENANTHRENE AND PHENANTHRENE DERIVATIVES IN *S. TYPHIMURIUM*

	TA98[a]	Ref.	TA100[a]	Ref.	TM677	Ref.
	(−)	14, 45, 51, 59	(−)	14,50,51,59	(−)	33, 58
Methyl isomers						
1	(+)	14, 107	(+)	14, 106, 107	(+)	33, 58
2	(−)	14, 107	(−)	14, 106, 107	(+)	33, 58
3	(−)	14, 107	(−)	14, 106, 107	(−)	58
4	(−)	14, 106	(−)	14, 106, 107		
9	(−)	14, 107	(+)	14, 106, 107	(+)	58
1,4	(+)	14, 107	(+)	106, 107		
1,9	(−)	107	(−)	107		
1,10	(+)	108	(+)	108		
2,7	(−)	14	(−)	14, 106, 107		
3,6	(−)	14	(−)	14, 106, 107		
4,5	(−)	14	(−)	14, 107		
4,9	(−)	107	(−)	107		
4,10	(−)	107	(+)	107		
1,3,4	(+)	108	(+)	108		
2,4,5,7	(−)	107	(−)	107		
3,4,5,6	(−)	107	(−)	107		
Alkyl derivatives						
2-Ethyl	(−)	107	(−)	107		
3-Ethyl	(−)	107	(−)	107		
9-Ethyl	(−)	107	(−)	107		
9-Propyl	(−)	107	(−)	107		
9-*i*-Propyl	(−)	107	(−)	107		
Halogenated derivatives						
9-Fluoro	(−)	107	(+)	107		
9-Chloro	(−)	107	(+)	107		
9-Bromo	(−)	107	(+)	107		

[a] The data listed provide a qualitative assessment of mutagenic activity in *S. typhimurium* TA98 and TA100. In all instances compounds were assayed at doses of 5—200 μg per plate. A (+) response is indicated for any compound which produced a twofold increase in His^+ revertants/plate.

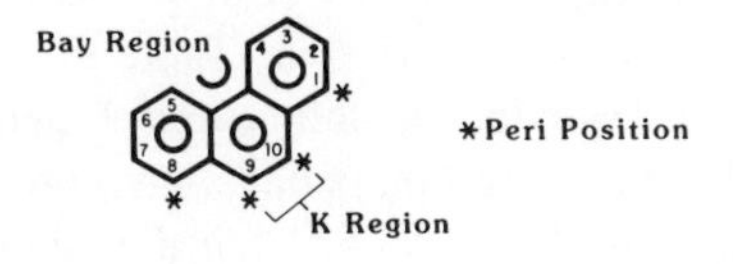

FIGURE 7. Structural features associated with the phenanthrene nucleus.

compared with that of 1,2- and 3,4-epoxide of 1,2,3,4-tetrahydrophenanthrene. Relative to these epoxides, the 9,10-oxide of phenanthrene was much less active as a mutagen in both *S. typhimurium* TA98 and TA100.[113] Both diastereomers of the 1,2-diol-3,4-epoxide of phenanthrene were also significantly more mutagenic than the 9,10-oxide. The observation that methyl substituents at or adjacent to the K-region of phenanthrene favors mutagenic activity, in fact, indicates that inhibition of the formation of a K-region epoxide is associated with mutagenic activity. Metabolism studies (see Section V.C) have supported this hypothesis. It is known that the presence of a halogen substituent frequently inhibits oxidative metabolism on the carbon to which it is attached. Additional evidence that inhibition of the metabolism of certain phenanthrene derivatives to K-region dihydrodiols was associated with their mutagenic activity was provided by assays on the mutagenic activity of halogenated phenanthrenes substituted at the 9 position. 9-Fluorophenanthrene and 9-chlorophenanthrene were both mutagenic in *S. typhimurium* TA100 in the presence of rat-liver homogenate. While 9-bromophenanthrene is mutagenic, this particular halogenated derivatives was less active than either the fluoro or chloro derivative. Studies on the mutagenic activity of a series of alkylphenanthrenes indicated that increased chain length of an alkyl substituent beyond that of a methyl substituent at the 9 position resulted in a loss of mutagenic activity. In view of those results and those with 9-bromophenanthrene, increased steric bulk at the 9 position of phenanthrene beyond that of a methyl or chloro substituent appears to be associated with a loss of mutagenic potency.

The genotoxic activity of phenanthrene and methylated derivatives of phenanthrene have been evaluated in bioassay systems other than *S. typhimurium*. In DNA-repair-deficient and DNA-repair-proficient *Escherischia coli,* phenanthrene did not exhibit any DNA-modifying activity.[59] Phenanthrene was also shown not to transform hamster fibroblast cells.[114] Using human lymphoblasts, both 1- and 9-methylphenanthrene were mutagenic.[34] While phenanthrene and 1-methylphenanthrene were inactive in the rat hepatocyte primary culture DNA repair test,[30,115,116] 1,4-dimethylphenanthrene was genotoxic.[115,116] In contrast to benzo[a]pyrene, phenanthrene and the K-region oxide of phenanthrene did not inhibit infectious phage.[117] The K-region oxide of phenanthrene has also been shown to transform hamster embryo cells[118] and cells derived from mouse prostate.[119] Studies on the binding of the 9,10-oxide of phenanthrene to DNA in vitro, however, have shown that almost negligible binding occurs relative to the K-region oxides of benzo[a]pyrene and 7,12-dimethylbenzo[a]anthracene.[120]

B. Tumorigenic Activity

Several bioassays have been performed to evaluate the tumorigenic activity of phenanthrene. A listing of representative bioassays performed with phenanthrene and the results of these assays are outlined in Table 8. As indicated by these data phenanthrene did not demonstrate significant tumorigenic activity under these various bioassay conditions. The tumorigenic activity of each of three possible isomeric dihydrodiols of phenanthrene was also evaluated on mouse skin[113] and in newborn mice.[124] None of these dihydrodiols had significant tumorigenic activity. While 1,2-dihydrophenanthrene and both diastereomers of the 1,2-diol-3,4-epoxide of phenanthrene were inactive as tumorigens in newborn mice, the 3,4-epoxide of 1,2,3,4-tetrahydrophenanthrene had significant tumorigenic activity.

The available data on the tumorigenic activity of monomethylated phenanthrenes indicate that these analogues are not active in mouse skin (Table 9). Thus, despite the well-documented mutagenic activity of 1-methylphenanthrene and 9-methylphenanthrene in *S. typhimurium* and human lymphoblasts, these compounds have not been shown, to date, to be tumorigenic in laboratory animals.[14,33,44] Among the dimethylated derivatives of phenanthrene, both 1,4- and 4,10-dimethylphenanthrene are active as tumor initiators on mouse skin (Table 9). Both 1,4- and 4,10-dimethylphenanthrene were active as mutagens in *S. typhimurium*. The structural features which favored mutagenic activity were a methyl substituent at or adjacent to the K-region, the presence of at least one unsubstituted angular ring, and an unsubstituted *peri*-position adjacent to this angular ring (Figure 7). In addition to conforming to these structural requirements, both of these tumorigenic phenanthrene derivatives had a methyl substituent in the bay region. Among the polymethylated derivatives of phenanthrene which have been evaluated for tumorigenic activity, 1,2,4- and 1,3,4-trimethylphenanthrene and 1,2,3,4-tetramethylphenanthrene were active. As can be noted from their structures, each of these analogues conform to the structural parameters which favor mutagenic activity with the one additional requirement of having a bay-region methyl substituent.

Table 8
REPRESENTATIVE BIOASSAYS ON THE TUMORIGENIC ACTIVITY OF PHENANTHRENE

Species/strain	No. of animals	Est. total dose/animal	Route and site of administration	Comments[a]	Ref.
Mice; —	100	1.0% sol.	Skin; topical	No tumors	26, 27
Mice; —	100	90 mg[b]	Skin; topical	Promotion with croton oil; 1% TBA	22
Mice; "S" Strain	20	540 mg	Skin; topical	As a tumor initiator; croton oil used as promoter 20% TBA; 0.6 T/A	64
Mice; —	20	300 μg	Skin; topical	Promotion with croton oil; 3 skin papillomas	121
Mice; CD1	30	178 μg	Skin; topical	As a tumor initiator with TPA; 7% TBA; 0.28 T/A	113
Mice; Newborn	57	40 μg	s.c.	2 Skin papillomas; 3 lung adenomas; 4 hepatocarcinoma	122, 123
Mice; Swiss Webster Newborn	100	250 μg	i.p.	17% Pulmonary tumors; compared to controls, little or no tumorigenic activity observed	124
Mice; C57B1	27	5.0 mg	s.c.	No tumors	68
Rats; Sprague-Dawley	25	6 mg	i.p.	No tumors	125
Rats; Sprague-Dawley	10	200 mg	Intragastric	No mammary tumor induction	126

[a] Percent tumor-bearing animals is listed as % TBA. The average number of tumors of all animals in the experimental group is given as T/A in these tables.

[b] Calculation of estimated total dose assumes that approximately 100 μℓ of solution was painted on the skin of mice. These values are only crude approximations in view of the limited experimental detail available.

The structure-tumorigenic activity relationships observed for methylated phenanthrenes parallels those structural requirements which have been shown to favor the tumorigenic activity of higher molecular weight methylated polycyclic aromatic hydrocarbons which contain a bay region.[83,132] These structural requirements favoring tumorigenic activity and the possible molecular basis for the observed structure-activity of methylated four- and five-ring PAH are discussed in detail in Chapter 4. The major difference in the structure-activity relationships, however, between methylated analogues of phenanthrene and chrysene, for example, is that a methyl substituent is required at or adjacent to the K-region of phenanthrene for activity within this group of PAH. As will be discussed (see Section V.C), the role of these methyl substituents at or adjacent to the K-region is to inhibit detoxification. Thus, inhibition of detoxification pathways among certain PAH by appropriate methyl group substitution may be as significant as or more significant than altering the intrinsic biological activity of its ultimate activated form.

The molecular basis for the tumorigenic activity of methylated phenanthrenes has not been determined. Comparative studies on the types of metabolites formed from phenanthrene, 1- and 9-methylphenanthrene, and 1,4-dimethylphenanthrene have been performed. The results of these investigations have provided insight into the possible molecular basis for the differences in their biological activity.

C. Metabolism Studies

One of the major metabolites of phenanthrene identified from both in vivo and in vitro studies is 9,10-dihydro-9,10-dihydroxyphenanthrene.[133,134] In studies using rat-liver preparations, the 9,10-dihydrodiol was the dominant metabolite with lesser amounts of both the 1,2- and 3,4-dihydrodiols also being formed. These results are in marked contrast to similar metabolism studies performed with both 1- and 9-methylphenanthrene.[14,106] Using a similar

Table 9
BIOASSAYS ON TUMORIGENIC ACTIVITY OF METHYLATED PHENANTHRENES

Compound	Species/ strain	Est. total dose/animal	Route and site of administration	Comments[a]	Ref.
1-Methylphenanthrene	CD1 Mice	1.0 mg	Skin; topical	As a tumor initiator; 0% TBA	106
2-Methylphenanthrene	CD1 Mice	1.0 mg	Skin; topical	As a tumor initiator; 0% TBA	106
3-Methylphenanthrene	CD1 Mice	1.0 mg	Skin; topical	As a tumor initiator; 0% TBA	106
4-Methylphenanthrene	CD1 Mice	1.0 mg	Skin; topical	As a tumor initiator; 0% TBA	106
9-Methylphenanthrene	CD1 Mice	1.0 mg	Skin; topical	As a tumor initiator; 0% TBA	106
1,4-Dimethylphenanthrene	CD1 Mice	1.0 mg	Skin; topical	As a tumor initiator; 95% TBA; 8.20 T/A	106, 107
	CD1 Mice	0.3 mg	Skin; topical	80% TBA; 3.20 T/A	106, 107
1,9-Dimethylphenanthrene	CD1 Mice	4.0 mg	Skin; topical	No tumors	127, 128
	CD1 Mice	1.0 mg	Skin; topical	As a tumor initiator; 0% TBA	107
2,7-Dimethylphenanthrene	CD1 Mice	1.0 mg	Skin; topical	As a tumor initiator; 5% TBA; 0.05 T/A	107
3,6-Dimethylphenanthrene	CD1 Mice	1.0 mg	Skin; topical	As a tumor initiator; 0% TBA	107
3,9-Dimethylphenanthrene	Mice	40 mg[b]	Skin; topical	0% TBA	130
4,5-Dimethylphenanthrene	CD1 Mice	1.0 mg	Skin; topical	As a tumor initiator; 5% TBA; 0.05 T/A	107
4,9-Dimethylphenanthrene	CD1 Mice	1.0 mg	Skin; topical	As a tumor initiator; 10% TBA; 0.10 T/A	107
4,10-Dimethylphenanthrene	CD1 Mice	1.0 mg	Skin; topical	55% TBA; 1.45 T/A	107
	CD1 Mice	0.3 mg	Skin; topical	35% TBA; 1.57 T/A	107
1-Methyl-3-isopropyl-phenanthrene	Mice	2—9 mg	s.c.	0% TBA	129
1-Methyl-7-isopropyl-phenanthrene	Mice	(1.0% oil)[c]	Skin; topical	No tumors	26, 27
1,3,4-Trimethylphenanthrene	CD1 Mice	1.0 mg	Skin; topical	As a tumor initiator; 84% TBA; 3.4 T/A	131
	CD1 Mice	40 mg[b]	Skin; topical	20% TBA; 0.2 T/A	130
2,6,9-Trimethylphenanthrene	Mice	40 mg[b]	Skin; topical	0% TBA	130
3,6,9-Trimethylphenanthrene	Mice	40 mg[b]	Skin; topical	0% TBA	130
1,2,3,4-Tetramethylphenanthrene	Mice	(0.3% oil)[b]	Skin; topical	40% TBA; 0.9 T/A	85
	Mice	5.0 mg	s.c.	0% TBA	85
2,4,5,7-Tetramethylphenanthrene	CD1 Mice	1.0 mg	Skin; topical	As a tumor initiator; 0% TBA	107
3,4,5,6-Tetramethylphenanthrene	CD1 Mice	1.0 mg	Skin; topical	As a tumor initiator; 0% TBA	107

[a] Percent tumor-bearing animals is listed as % TBA. The average number of tumors of all animals in the experimental group is given as T/A in these tables.

[b] Calculation of estimated total dose assumes that approximately 100 μℓ of solution was painted on the skin of mice. These values are only crude approximations in view of the limited experimental detail available.

[c] No further detail available.

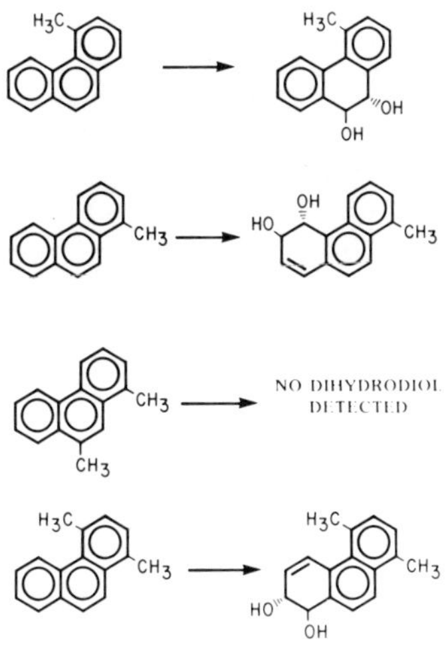

FIGURE 8. Major dihydrodiols formed as metabolites of specific methylated phenanthrenes.

incubation mixture as employed to assay the activity of these mutagenic methylphenanthrenes, the 9,10-dihydrodiol was not detected among their metabolites. In the case of both 1-methylphenanthrene and 9-methylphenanthrene, the major dihydrodiols which were detected were the 3,4- and 5,6-dihydrodiols. Assays on the in vitro metabolites of both 1- and 9-methylphenanthrene indicated that these bay-region dihydrodiols were in each case their major proximate mutagens.

Metabolism studies performed with 3- and 4-methylphenanthrene revealed that the diol formed in both instances was the K-region dihydrodiol. Since phenanthrene as well as 3- and 4-methylphenanthrene are inactive as mutagens, these data suggest that inhibition of 9,10-dihydrodiol formation is associated with the mutagenic activity of both 1- and 9-methylphenanthrene. These results also suggest that either simple oxides in the bay region or further activation of the bay-region dihydrodiols of either 1- or 9-methylphenanthrene is ultimately associated with their activation to mutagens.

Studies were also performed on the types of dihydrodiols formed from 3,6- and 4,9-dimethylphenanthrene (which are inactive as mutagens or tumor-initiations) as compared to 1,4- and 4,10-dimethylphenanthrene (which are active as mutagens and tumor-initiators). The only dihydrodiols detected among the metabolites of 1,4- and 4,10-dimethylphenanthrene were their respective 7,8-dihydrodiols. These specific dihydrodiols can serve in each instance as the requisite precursor for the formation of a bay-region dihydrodiol epoxide. In contrast to these results, the major metabolites of 3,6-dimethylphenanthrene were 9-hydroxy-3,6-dimethylphenanthrene and 9,10-dihydro-9,10-dihydroxy-3,6-dimethylphenanthrene. Among the metabolites of 4,9-dimethylphenanthrene, no dihydrodiols were detected among the metabolites isolated from incubations with rat-liver homogenate.

These data on the metabolism of dimethylphenanthrenes suggest that dihydrodiol precursors to bay-region diol-epoxides are associated with the tumorigenic activity of methylated derivatives of phenanthrene (Figure 8). Inhibition of 9,10-dihydrodiol formation by the presence of a methyl substituent either on or adjacent to the K-region is also essential for tumorigenic activity. Studies on the structure-activity of polymethylated phenanthrenes indicate that for tumorigenic activity it is required that there be a bay-region methyl group. Whether this bay-region methyl group alters the intrinsic activity of the diol-epoxide or is responsible for directing initial oxidative metabolism away from the bay region (inhibiting bay-region diol formation) has not been determined. Both diastereomers of 1,2,3,4-tetrahydro-1,2-dihydroxy-3,4-epoxyphenanthrene have been shown to be inactive as tumorigenic agents.[113]

Metabolism studies suggest that the tumorigenic methylated phenanthrenes are being ultimately activated to bay-region dihydrodiol epoxides. The unique biological potency of bay-region diol-epoxides which have a methyl group in the bay region, as demonstrated in several studies, represents the most likely explanation for this additional structural requirement for tumorigenicity among methylated phenanthrenes.

ACKNOWLEDGMENTS

The authors thank Ms. Debbie Conroy for her dedication and efforts in the preparation of this chapter. The authors also thank the numerous colleagues at the American Health Foundation, especially Dr. Hoffmann, who have contributed to our studies on tricyclic PAH. The support of the Department of Energy by the Office of Health and Environmental Research under Contract DE-AC02-8CER60080 in the investigation of the structure-activity of tricyclic PAH is gratefully acknowledged.

REFERENCES

1. **Gelboin, H. V. and Ts'o, P. O. P.,** *Polynuclear Aromatic Hydrocarbons,* Vol. 1, Academic Press, New York, 1978; Vol. 2, Academic Press, New York 1978; Vol. 3, Academic Press, New York, 1981.
2. **Dipple, A., Moschel, R. C., and Bigger, C. A. H.,** Polynuclear aromatic carcinogens, in *Chemical Carcinogens,* Vol. 1, 2nd ed., Searle, C. E., Ed., American Chemical Society, Washington, D.C., 1984, chap. 2.
3. **Arcos, J. C. and Argus, M. F.,** *Chemical Induction of Cancer,* Vol. 2A, Academic Press, New York, 1974.
4. **Grimmer, G.,** *Environmental Carcinogens: Polycyclic Aromatic Hydrocarbons,* CRC Press, Boca Raton, Florida, 1983.
5. International Agency for Research on Cancer, IARC Monographs on the Evaluation of Carcinogenic Risk of Chemicals to Man, Vol. 3, Certain Polycyclic Aromatic Hydrocarbons and Heterocyclic Compounds, IARC, Lyon, France, 1973.
6. International Agency for Research on Cancer, IARC Monographs on the Evaluation of Carcinogenic Risk of Chemical to Humans, Vol. 32, Part 1, Chemical Environmental and Experimental Data, IARC, Lyon France, 1983.
7. **Guerin, M. R., Epler, J. L., Griest, W. H., Clark, B. R., and Rao, T. K.,** Polycyclic aromatic hydrocarbons from fossil fuel conversion processes, *Carcinogenesis,* Vol. 3, *Polynuclear Aromatic Hydrocarbons,* Jones, P. W. and Freudenthal, R. I., Eds., Raven Press, New York, 1978, 21.
8. **Kubota, H., Griest, W. H., and Guerin, M. R.,** Determination of carcinogens in tobacco smoke and coal-derived samples. Trace polynuclear aromatic hydrocarbons, in *Trace Substances in Environmental Health,* Vol. 9, Hemphill, D. D., Ed., University of Missouri, Columbia, Missouri, 1975, 281.
9. **Severson, R. F., Snook, M. E., Akin, F. J., and Chortyk, O. T.,** Correlation of biological activity with polynuclear aromatic hydrocarbon content of tobacco smoke fractions, *Carcinogenesis,* Vol. 3, *Polynuclear Aromatic Hydrocarbons,* Jones, P. W. and Freudenthal, R. I., Eds., Raven Press, New York, 1978, 115.
10. **Howard, A. G. and Mills, G. A.,** Identification of polynuclear aromatic hydrocarbons in diesel particulate emissions, *Int. J. Environ. Anal. Chem.,* 14, 43, 1983.

11a. **Hartwell, J. L., Ed.,** Survey of Compounds Which Have Been Tested for Carcinogenicity, Public Health Service, Washington, D.C., 1951.

11b. **Shubik, P. and Hartwell, J. L., Ed.,** Survey of Compounds Which Have Been Tested for Carcinogenicity, Suppl. 1, Public Health Services, Washington, D.C., 1957.

11c. **Shubik, P. and Hartwell, J. L., Ed.,** Survey of Compounds Which Have Been Tested for Carcinogenicity, Suppl. 2, Public Health Service, Washington, D.C., 1969.

11d. **Thompson, J. I., Ed.,** Survey of Compounds Which Have Been Tested for Carcinogenicity, 1961—1967, Public Health Service, Washington, D.C., 1973.

11e. **Thompson, J. I., Ed.,** Survey of Compounds Which Have Been Tested for Carcinogenicity, 1969, Public Health Service, Washington, D.C., 1972.

11f. **Thompson, J. I., Ed.,** Survey of Compounds Which Have Been Tested for Carcinogenicity, 1970—1971, Public Health Service, Washington, D.C., 1974.
11g. **Thompson, J. I., Ed.,** Survey of Compounds Which Have Been Tested for Carcinogenicity, 1972—1973, National Cancer Institute, Washington, D.C., 1975.
11h. **Thompson, J. I., Ed.,** Survey of Compounds Which Have Been Tested for Carcinogenicity, National Cancer Institute, 1974—1975, Washington, D.C., 1983.
11i. **Thompson, J. I., Ed.,** Survey of Compounds Which Have Been Tested for Carcinogenicity, 1976—1977, National Cancer Institute, Washington, D.C., 1984.
11j. **Thompson, J. I., Ed.,** Survey of Compounds Which Have Been Tested for Carcinogenicity, Vol. 1978, National Cancer Institute, Washington, D.C., 1980.
11k. **Thompson, J. I., Ed.,** Survey of Compounds Which Have Been Tested for Carcinogenicity, 1979—1980, National Cancer Institute, Washington, D.C., 1984.
11l. **Thompson, J. I., Ed.,** Survey of Compounds Which Have Been Tested for Carcinogenicity: 1980 Cumulative Indexes, National Cancer Institute, Washington, D.C., 1984.
12. **Peake, E. and Parker, K.,** Polynuclear aromatic hydrocarbons and the mutagenicity of used crankcase oils, in *Polynuclear Aromatic Hydrocarbons: Chemistry and Biological Effects,* Bjørseth, A. and Dennis, A. J., Eds., Battelle Press, Columbus, Ohio, 1979, 1025.
13. **Griest, W. H., Tomkins, B. A., Epler, J. L., and Rao, T. K.,** Characterization of multialkylated polycyclic aromatic hydrocarbons in energy-related materials, in *Polynuclear Aromatic Hydrocarbons,* Jones, P. W. and Leber, P., Eds., Ann Arbor Science, Ann Arbor, Mich., 1979, 395.
14. **LaVoie, E. J., Tulley, L., Bedenko, V., and Hoffman, D.,** Mutagenicity, tumor-initiating activity, and metabolism of tricyclic polynuclear aromatic hydrocarbons, in *Polynuclear Aromatic Hydrocarbons: Chemistry and Biological Effects,* Bjørseth, A. and Dennis, A. J., Eds., Battelle Press, Columbus, Ohio, 1980, 1041.
15. **LaVoie, E. J., Tulley-Freiler, L., Bedenko, V., Girach, Z., and Hoffman, D.,** Comparative studies on the tumor-initiating activity and metabolism of methylfluorene and methylbenzofluorenes, in *Polynuclear Aromatic Hydrocarbons: Chemical Analysis and Biological Fate,* Cooke, M. C. and Dennis, A. J., Eds., Battelle Press, Columbus, Ohio, 1981, 417.
16. **LaVoie, E. J., Coleman, D. T., Geddie, N. G., and Rice, J. E.,** Studies on the mutagenicity and tumor-initiating activity of methylated fluorene derivatives, *Chem. Biol. Interact.,* 52, 301, 1985.
17. **LaVoie, E. J., Tulley, L., Bedenko, V., and Hoffmann, D.,** Mutagenicity of methylated fluorenes and benzofluorenes, *Mutat. Res.,* 91, 167, 1981.
18. **Cavalieri, E., Rogan, R., and Rogan, E.,** Hydroxylation and conjugation at the benzylic carbon atom: a possible mechanism of carcinogenic activation for some methyl-substituted aromatic hydrocarbons, in *Polynuclear Aromatic Hydrocarbons,* Jones, P. W. and Leber, P., Eds., Ann Arbor Science, Ann Arbor, Mich., 1979, 517.
19. **Flesher, J. W. and Sydnor, K. L.,** Carcinogenicity of derivatives of 7,12-dimethylbenz[a]anthracene, *Cancer Res.,* 31, 1951, 1971.
20. **Sprinzak, Y.,** Reactions of active methylene compounds in pyridine solution. The ionic autooxidation of fluorene and its derivatives, *J. Chem. Soc.,* 77, 5449, 1955.
21. **Riegel, B., Wartman, W. B., Hill, W. T., Reeb, B. B., Shubik, P., and Stanger, D. W.,** Delay of methylcholanthrene skin carcinogenesis in mice by 1,2,5,6-dibenzofluorene, *Cancer Res.,* 11, 301, 1951.
22. **Graffi, A., Vlamynck, E., Hoffmann, F., and Schulz, I.,** Untersuchungen über die geshwulstaustauslösende Wirkung verschiedener chemischer Stoffe in der Kombination mit Crotonöl., *Arch. Geschwulstforsch.,* 5, 110, 1953.
23. **Shubik, P. and Porta, G. D.,** Carcinogenesis and acute intoxication with large doses of polycyclic aromatic hydrocarbons, *Arch. Pathol.,* 64, 691, 1957.
24. **Morris, H. P., Velat, C. A., Wagner, B. P., Dahlgard, M., and Ray, F. E.,** Studies of carcinogenicity in the rat of derivatives of aromatic amines related to N-2-fluorenylacetamide, *J. Natl. Cancer Inst.,* 24, 149, 1960.
25. **Wilson, R. H., De Eds, F., and Cox, A. J., Jr.,** The carcinogenic activity of 2-acetaminofluorene. IV. Action of related compounds, *Cancer Res.,* 7, 453, 1947.
26. **Kennaway, E. L.,** On the cancer-producing factor in tar, *Br. Med. J.,* 1, 564, 1924.
27. **Kennaway, E. L.,** On the cancer-producing tars and tar-fractions, *J. Ind. Hyg.,* 5, 462, 1924.
28. **Griswald, D. P., Jr., Casey, A. E., Weisburger, E. K., Weisburger, J. H., and Schabel, F. M., Jr.,** *Cancer Res.,* 26, 619, 1966.
29. **Cook, J. W.,** The production of cancer by pure hydrocarbons, II, *Proc. R. Soc., London Ser. B,* 111, 485, 1932.
30. **Probst, G. S., McMahon, R. E., Hill, L. E., Thompson, C. Z., Epp, J. K., and Neul, S. B.,** Chemically-induced unscheduled DNA synthesis in primary rat hepatocyte cultures: a comparison with bacterial mutagenicity using 210 compounds, *Environ. Mutagen.,* 3, 11, 1981.

31. **Rice, J. E., Rivenson, A., Braley, J., and LaVoie, E. J.,** Methylated derivatives of pyrene and fluorene: evaluation of genotoxicity in the hepatocyte/DNA repair test and tumorigenic activity in newborn mice, *J. Toxicol. Environ. Health,* 21, 525, 1987.
32. **Bjørseth, A.,** Analysis of polycyclic aromatic hydrocarbons in environmental samples by glass capillary gas chromatography, *Carcinogenesis,* Vol. 3, *Polynuclear Aromatic Hydrocarbons,* Jones, P. W. and Freudenthal, R. I., Eds., Raven Press, New York, 1978, 75.
33. **Kaden, D. A., Hites, R. A., and Thilly, W. G.,** Mutagenicity of soot and associated polycyclic aromatic hydrocarbons, *Cancer Res.,* 39, 4152, 1979.
34. **Thilly, W. G., Longwell, J., and Andon, B. M.,** General approach to the biological analysis of complex mixtures, *Environ. Health. Perspect.,* 48, 129, 1983.
35. **Hecht, S. and Amin, S.,** Unpublished data.
36. **Gibson, T. L., Smart, V. B., and Smith, L. L.,** Non-enzymatic activation of polycyclic aromatic hydrocarbons as mutagens, *Mutat. Res.,* 49, 153, 1978.
37. **Harvey, R. G. and Halomen, M.,** Interaction between carcinogenic hydrocarbons and nucleosides, *Cancer Res.,* 28, 2183, 1968.
38. **Takemura, N., Hashida, C., and Shimizu, H.,** Carcinogenicity and mutagenicity of 5-nitroacenaphthene and tetrachlorobenzidine, Proc. 3rd Int. Symp. Detection and Prevention of Cancer, *Prevention and Detection of Cancer,* Vol. 2, Nieburgs, H. E., Ed., Marcel Dekker, New York, 1977, 1329.
39. **Steinbäck, F.,** Local and systemic effects of commonly used cutaneous agents: lifetime studies of 16 compounds in mice and rabbits, *Acta. Pharmacol. Toxicol.,* 41, 417, 1977.
40. **Yahagi, T., Shimizu, H., Nagao, M., Takimura, N., and Sugimura, T.,** Mutagenicity of S-nitronaphthene in Salmonella, *Gann,* 66, 581, 1975.
41. **Takimura, N., Hashida, C., and Terasana, M.,** Carcinogenic action of 5-nitroacenaphthene, *Br. J. Cancer,* 30, 481, 1974.
42. **El-Bayoumy, K. and Hecht, S. S.,** Identification of mutagenic metabolite formed by C-hydroxylation and nitroreduction of 5-nitroacenaphthene in rat liver, *Cancer Res.,* 42, 1243, 1982.
43. **LaVoie, E. J., Coleman, D. T., Tonne, R. L., and Hoffmann, D.,** Mutagenicity, tumor-initiating activity and metabolism of methylated anthracenes, in *Polynuclear Aromatic Hydrocarbons,* Cooke, M. E. and Dennis, A. J., Eds., Battelle Press, Columbus, Ohio, 1983, 785.
44. **LaVoie, E. J., Coleman, D. T., Rice, J. E., Geddien, N. G., and Hoffmann, D.,** Tumor-initiating activity, mutagenicity, and metabolism of methylated anthracenes, *Carcinogenesis,* 6, 1483, 1985.
45. **Hermann, M.,** Synergistic effects of individual polycyclic aromatic hydrocarbons on the mutagenicity of their mixtures, *Mutat. Res.,* 90, 399, 1981.
46. **Falck, K., Partanen, P., Sorsi, M., Guovaniemi, O., and Vainio, H.,** Mutascreen, an automated bacterial mutagenicity assay, *Mutat. Res.,* 150, 119, 1985.
47. **Tikkanen, L., Mtasushima, T., and Natori, S.,** Mutagenicity of anthraquinones in the Salmonella preincubation test, *Mutat. Res.,* 116, 297, 1983.
48. **Herbold, B. A., Arni, P., Engelhardt, G., Geriche, D., and Longstaff, E.,** Comparative Ames-test study with dyes, *Mutat. Res.,* 97, 429, 1982.
49. **Brown, J. P., Dietrich, P. S., and Brown, R. J.,** Frameshift mutagenicity of certain naturally occurring phenolic compounds in the Salmonella/microsome test: activation of anthraquinone and flavonol glycosides by gut bacterial enzymes, *Biochem. Soc. Trans.,* 5, 1489, 1977.
50. **Sakai, M., Yoshida, D., and Mizusaki, S.,** Mutagenicity of polycyclic aromatic hydrocarbons and quinones on *Salmonella typhimurium* TA97, *Mutat. Res.,* 156, 61, 1985.
51. **Simon, V. F.,** *In vitro* mutagenicity assays of chemical carcinogens and related compounds with *Salmonella typhimurium, J. Natl. Cancer Inst.,* 62, 893, 1979.
52. **de Serres, F. S. and Ashby, J.,** Evaluation of short-term tests for carcinogens, *Progress in Mutation Research,* Vol. 1, Elsevier/North-Holland, New York, 1981.
53. **Simon, V. F.,** *In vitro* assays for recombinogenic activity of chemical carcinogens and related compounds with *Saccharomyces cerevisciae* D_3, *J. Natl. Cancer Inst.,* 62, 901, 1979.
54. **DeFlora, S., Zjanacchi, P., Camoirano, A., Bennicelli, C., and Bradolati, G. S.,** Genotoxic activity and potency of 135 compounds in the Ames reversion test and in a bacterial DNA-repair test, *Mutat. Res.,* 133, 161, 1984.
55. **Vogel, E. W., Zjijlstra, J. A., and Blijleven, W. G. H.,** Mutagenic activity of selected aromatic amines and polycyclic hydrocarbons in Drosophila melanogaster, *Mutat. Res.,* 107, 53, 1983.
56. **Zijlstra, J. A. and Vogel, E. W.,** Mutagenicity of 7,12-dimethylbenz[a]anthracene and some other aromatic mutagenic in Drosophilis melanogaster, *Mutat. Res.,* 125, 243, 1984.
57. **Quillardet, P., de Bellecombe, C., and Hofnung, M.,** The SOS Chromotest, a colorimetric bacterial assay for genotoxins: validation study with 83 compounds, *Mutat. Res.,* 147, 77, 1985.
58. **Barfknecht, T. R., Andon, B. M., Thilly, W. G., and Hites, R. A.,** Soot and mutation in bacteria and human cells, in *Polynuclear Aromatic Hydrocarbons: Chemical Analysis and Biological Fate,* Cooke, M. and Dennis, A. J., Eds., Battelle Press, Columbus, Ohio, 1981, 231.

59. **Rosenkrantz, H. S. and Poirer, L. A.,** Evaluation of the mutagenicity and DNA-modifying activity of carcinogens and noncarcinogens in microbial systems, *J. Natl. Cancer Inst.,* 62, 873, 1979.
60. **Connell, J. R. and Duncan, S. J.,** The effect of non-phorbol promoters as compared with phorbol myristate acetate on sister chromatid exchange induction in cultured chinese hamster cells, *Cancer Lett.,* 11, 351, 1981.
61. **Poirier, M. C., DeCicco, B. T., and Liebermann, M. W.,** Nonspecific inhibition of DNA repair synthesis by tumor promoters in human diploid fibroblasts damaged with N-acetoxy-2-acetylaminofluorene, *Cancer Res.,* 35, 1392, 1975.
62. **Parry, J. M., Parry, E. M., and Barrett, J. C.,** Tumor promoters induce mitotic aneuploidy in yeast, *Nature,* 294, 263, 1981.
63. **Silinskas, K. C., Kateley, S. A., Tower, J. E., Maher, V. M., and McCormick, J. J.,** Induction of anchorage-independent growth in human fibroblasts by propane sultone, *Cancer Res.,* 41, 1620, 1981.
64. **Solomon, M. H. and Roe, F. J. C.,** Further tests for tumor-initiating activity: N,N-di-(2-chloroethyl)-p-aminophenylbutyric and (CB/348) as an initiator of skin tumor formation in the mouse, *Br. J. Cancer,* 10, 363, 1956.
65. **Wynder, E. L. and Hoffmann, D.,** A study of tobacco carcinogenesis. VII. The role of higher polycyclic hydrocarbons, *Cancer,* 12, 1079, 1959.
66. **Scribner, J. D.,** Tumor initiation by apparently noncarcinogenic polycyclic aromatic hydrocarbons, *J. Natl. Cancer Inst.,* 50, 1717, 1973.
67. **Bergolts, B. M. and Ilima, A. A.,** Fate of some carcinogenic and noncarcinogenic hydrocarbons in living organisms, *Biokhimija,* 16, 262, 1951.
68. **Steiner, P. E.,** Carcinogenicity of multiple chemicals simultaneously administered, *Cancer Res.,* 15, 632, 1955.
69. **Schmähl, D.,** Prüfung von Naphthalin und Anthracen auf cancerogene Wirkung an Ratten, *Z. Krebsforsch.,* 60, 697, 1955.
70. **Stanton, M. F., Miller, E., Wrench, C., and Blackwell, R.,** Experimental induction of epidermoid carcinoma in the lungs of rats by cigarette smoke condensate, *J. Natl. Cancer Inst.,* 49, 867, 1972.
71. **Russell, H.,** An unsuccessful attempt to induce gliomata in rabbits with cholanthrene, *J. Path. and Bact.,* 59, 481, 1947.
72. **Kennaway, E. L. and Hieger, I.,** Carcinogenic substances and their fluorescence spectra, *Br. Med. J.,* 1, 1044, 1930.
73. **Velluz, L.,** Cancerologie — Étudie de comparaison, dans la série polycyclique, entré l'oxydabilite reversible et le pouvoir carcinogénétique, *C. R. Acad. Sci.,* 206, 1514, 1938.
74. **Hadler, H. I., Darchun, V., and Lee, K.,** Initiation and promotion activity of certain polynuclear hydrocarbons, *J. Natl. Cancer Inst.,* 23, 1383, 1959.
75. **Shear, M. J. and Lieter, J.,** Studies in carcinogenesis. XIV. 3-Substituted and 10-substituted derivatives of 1,2-benzanthrene, *J. Natl. Cancer Inst.,* 1, 303, 1940.
76. **Pietra, G., Rappaport, H., and Shubik, P.,** The effects of carcinogenic chemicals in newborn mice, *Cancer,* 14, 308, 1961.
77. **Kennaway, E. L., Kennaway, N. M., and Warren, F. L.,** The examination of dimethylanthracenes for carcinogenic properties, *Cancer Res.,* 2, 157, 1942.
78. **Cook, J. W. and Kennaway, E. L.,** Chemical compounds as carcinogenic agents, *Am. J. Cancer,* 39, 381, 1940.
79. **Purchase, T. F. H., Longstaff, E., Ashby, J., Syles, J. A., Anderson, D., Lefevre, P. A., and Westwood, F. R.,** An evaluation of 6 short-term tests for detecting organic chemical carcinogens, *Br. J. Cancer,* 37, 873, 1978.
80. **Peck, R. M. and Peck, E. B.,** Relationship between carcinogenesis in vivo and alkylation and solvolysis in vitro, *Cancer Res.,* 40, 782, 1980.
81. **Lijinsky, W. and Saffioti, U.,** Relationships between structure and skin tumorigenic activity among hydrogenated derivatives of several polycyclic aromatic hydrocarbons, *Ann. Ital. Dermatol. Clin. Sper.,* 19, 34, 1965.
82. **Wislocki, P. G., Fiorentine, K. M., Fu, P. P., Yang, S. K., and Lu, A. Y. H.,** Tumor-initiating ability of twelve monomethylbenz[a]anthracenes, *Carcinogenesis,* 3, 215, 1985.
83. **DiGiovanni, J., Diamond, L., and Harvey, R. G.,** Enhancement of the skin tumor-initiating activity of polycyclic aromatic hydrocarbons by methyl-substitution at non-benzo 'bay region' positions, *Carcinogenesis,* 4, 403, 1983.
84. **Bradbury, J. T., Bachmann, W. E., and Lewisohn, M. G.,** The production of cancer by some new chemical compounds, *Cancer Res.,* 1, 685, 1941.
85. **Badger, G. M., Cook, J. W., Hewett, C. L., Kennaway, E. L., Kennaway, N. M., and Martin, R. H.,** The production of cancer by pure hydrocarbons, VI, *Proc. R. Soc. London, Ser. B.,* 131, 170, 1942.

86. **Boyland, E. and Levi, A. A.,** Production of dihydroxydihydroanthracene from anthracene, *Biochem. J.*, 29, 2679, 1935.
87. **Boyland, E. and Levi, A.,** Product of dihydroxydihydroanthraceneglucuronic acid from anthracene, *Biochem. J.*, 30, 728, 1936.
88. **Sims, P.,** Metabolism of polycyclic compounds. The metabolism of anthracene and some related compounds in rats, *Biochem. J.*, 92, 621, 1964.
89. **Von Teingeln, L. S., Yang, D. T. C., Huang, L. W., Stennis, R. V., Lai, C. C., and Fu, P. P.,** Stereoselective metabolism of methylated anthracenes: absolute configuration of *trans*-dihydrodiol metabolites, in *Polynuclear Aromatic Hydrocarbons: Chemistry, Characterization and Carcinogenesis,* 9th Int. Symp. on Polynuclear Aromatic Hydrocarbons, Cooke, M. and Dennis, A. J., Eds., Battelle Press, Columbus, Ohio, 933, 1986.
90. **Lamparczyk, H. S., Farmer, P. B., Cary, P. D., Grover, P. L., and Sims, P.,** The metabolism of 9,10-dimethylanthracene by rat liver microsomal preparations, *Carcinogenesis,* 5, 1045, 1984.
91. **Miller, E. C. and Miller, J. A.,** Low carcinogenicity of the K-region epoxides of 7-methylbenz[a]anthracene and benz[a]anthracene in the mouse and rat, *Proc. Soc. Exp. Biol. Med.*, 124, 915, 1967.
92. **Boyland, E. and Sims, P.,** The carcinogenic activities in mice of compounds related to benz[a]anthracene, *Int. J. Cancer,* 2, 500, 1967.
93. **Van Duuren, B. L., Segal, A., Tseng, S.-S., Rusch, G. M., Loewengart, G., Mate, U., Roth, D., Smith, A., Melchionne, S., and Seidman, I.,** Structure and tumor-promoting activity of analogues of anthralin (1,8-dihydroxy-9-anthrone), *J. Med. Chem.*, 21, 26, 1978.
94. **Segel, A., Katz, C., and Van Duuren, B. L.,** Structure and tumor-promoting activity of anthralin (1,8-dihydroxy-9-anthrone) and related compounds, *J. Med. Chem.*, 14, 1152, 1971.
95. **Bock, F. G. and Burns, R.,** Tumor-promoting properties of anthralin (1,8,9-Anthratriol), *J. Natl. Cancer Inst.*, 30, 393, 1963.
96. **Yasuhura, K.,** Skin papilloma production by anthralin painting after urethan initiation in mice, *Gann,* 59, 187, 1968.
97. **Langbein, W.,** Beitrag zur kokanzerogenen Wirkung von Cignolin beim Tier, *Radiobiol. Radiother.*, 13, 233, 1972.
98. **DiGiovanni, J., Decina, P. C., Prichett, W. P., Cantor, J., Aalfs, K. K., and Coombs, M. M.,** Mechanism of mouse skin tumor promotion by chrysarobin, *Cancer Res.*, 45, 2584, 1985.
99. **LaVoie, E. J., Bedenko, V., and Hoffmann, D.,** Unpublished results.
100. **Saffiotti, V. and Shubik, P.,** Studies on promoting action in skin carcinogenesis, *Natl. Cancer Inst. Monogr.*, 10, 489, 1963.
101. **DiGiovanni, J. and Boutwell, R. K.,** Tumor promoting activity of 1,8-dihydroxy-3-methyl-9-anthrone (chrysarobin) in female SENCAR mice, *Carcinogenesis,* 4, 281, 1983.
102. **Van Duuren, B. L. and Goldschmidt, B. M.,** Cocarcinogenic and tumor-promoting agents in tobacco carcinogenesis, *J. Natl. Cancer Inst.*, 56, 1237, 1976.
103. **Snook, M. E., Severson, R. F., Higman, H. C., Arrendale, R. F., and Chortyk, O. T.,** Polynuclear aromatic hydrocarbons of tobacco smoke: isolation and identification, *Beitr. Tabakforsch.*, 8, 250, 1976.
104. **Yu, M.-L. and Hites, R. A.,** Identification of organic compounds in diesel engine soot, *Anal. Chem.*, 53, 951, 1981.
105. **Lee, M. L., Novotny, M., and Bartte, K. D.,** Gas chromatography/mass spectrometric and nuclear magnetic resonance determination of polynuclear aromatic hydrocarbons in airborne particulates, *Anal. Chem.*, 48, 1566, 1976.
106. **LaVoie, E. J., Tulley-Freiler, L., Bedenko, V., and Hoffmann, D.,** Mutagenicity, tumor-initiating activity and metabolism of methylphenanthrenes, *Cancer Res.*, 41, 3441, 1981.
107. **LaVoie, E. J., Bedenko, V., Tulley-Freiler, L., and Hoffmann, D.,** Tumor-initiating activity and metabolism of polymethylated phenanthrenes, *Cancer Res.*, 42, 4045, 1982.
108. **LaVoie, E. J., Rice, J. E., and Geddie, N.,** Unpublished results.
109. **Bücker, M., Glatt, H. R., Platt, K. L., Avnir, D., Ittah, Y., Blum, J., and Oesch, F.,** Mutagenicity of phenanthrene and phenanthrene K-region derivatives, *Mutat. Res.*, 66, 337, 1979.
110. **Cookson, M. J., Grover, P. L., and Sims, P.,** Mutagenicity of 'K-region' epoxides of polycyclic hydrocarbons towards bacteriophage T_2, *Proc. Biochem. Soc.*, 125, 100, 1971.
111. **Ames, B. N., Sims, P., and Grover, P. L.,** Epoxides of carcinogenic polycyclic hydrocarbons are frameshift mutagens, *Science,* 176, 46, 1972.
112. **Miyata, N., Shudo, K., Kitahara, Y., Huang, G. F., and Okamoto, T.,** Mutagenicity of K-region epoxides of polycyclic aromatic hydrocarbons: structure-activity relationship, *Mutat. Res.*, 37, 187, 1976.
113. **Thomas, P. E., Mah, H. D., Karie, J. M., Yagi, H., Jerina, D. M., and Conney, A. H.,** Mutagenicity and tumorigenicity of phenanthrene and chrysene epoxides and diol epoxides, *Cancer Res.*, 39, 3069, 1979.
114. **Parodi, S., Taninger, M., Russo, P., Pala, M., Vecchio, D., Fassina, G., and Santi, L.,** Quantitative productivity of its transformation *in vitro* assay compound with the Ames test, *J. Toxicol. Environ. Health,* 12, 483, 1983.

115. **Tong, C., Laspia, M. F., Telang, S., and Williams, G. M.,** The use of adult rat liver cultures in the detection of the genotoxicity of various polycyclic aromatic hydrocarbons, *Environ. Mutagen.*, 3, 477, 1981.
116. **Tong, C., Brat, V., Telang, S., Laspia, M. F., Fazio, M., Russ, B., and Williams, G.,** Effects of genotoxic polycyclic aromatic hydrocarbons in rat liver culture systems, in *Polynuclear Aromatic Hydrocarbons: Formation, Metabolism, and Measurement,* Cooke, M. and Dennis, A. J., Eds., Battelle Press, Columbus, Ohio, 1983, 1189.
117. **Hsu, W.-T., Harvey, R. G., Lin, E. J. S., and Weiss, A.,** A bacteriophage system for screening and study of biologically active polycyclic aromatic hydrocarbons and related compounds, *Proc. Natl. Acad. Sci.,* 74, 1378, 1977.
118. **Huberman, E., Kuroki, T., Marquardt, H., Selkirk, J. K., Heidelberger, C., Grover, P. L., and Sims, P.,** Transformation of hamster embryo cells by epoxides and other derivatives of polycyclic hydrocarbons, *Cancer Res.*, 32, 1391, 1972.
119. **Marquardt, H., Kuroki, T., Huberman, E., Selkirk, J. K., Heidelberger, C., Grover, P. L., and Sims, P.,** Malignant transformation of cells derived from mouse prostate by epoxides and other derivatives of polycyclic hydrocarbons, *Cancer Res.*, 32, 716, 1972.
120. **Blobstein, S. H., Weinstein, I. B., Dansette, P., Yagi, H., and Jerina, D. M.,** Bonding of K- and non-K-region arene oxides and phenols of polycyclic hydrocarbons to polyguanylic acid, *Cancer Res.*, 36, 1293, 1976.
121. **Roe, F. J. C., Ball, J. K., Pierce, W. E. H., and Walters, M. A.,** *Br. Emp. Cancer Campaign,* 40, 41, 1962.
122. **Roe, F. J. C. and Grant, G. A.,** Tests of pyrene and phenanthrene for incomplete and anticarcinogenic activity, *Br. Emp. Cancer Campaign,* 41 (Part 2), 59, 1964.
123. **Grant, G. and Roe, F. J. C.,** The effect of phenanthrene on tumor induction by 3,4-benzopyrene administered to newly born mice, *Br. J. Cancer,* 17, 261, 1963.
124. **Buening, M. K., Levin, W., Karle, J. M., Yagi, H., Jerina, D. M., and Conney, A. H.,** Tumorigenicity of bay-region epoxides and other derivatives of chrysene and phenanthrene in newborn mice, *Cancer Res.*, 39, 5063, 1979.
125. **Marron, T. V.,** Return of hepatic vitamin A in rats after depletion of methylcholanthrene, *Proc. Soc. Exp. Biol. Med.*, 48, 219, 1941.
126. **Huggins, C. and Yang, N. C.,** Induction and extinction of mammary cancer, *Science,* 137, 257, 1962.
127. **Barry, G., Cook, J. W., Haslewood, G. A. D., Hewett, C. L., Hieger, I., and Kennaway, E. L.,** The production of cancer by pure hydrocarbons, *Proc. R. Soc. London, Ser. B.*, 117, 318, 1935.
128. **Bachmann, W. E., Cook, J. W., Dansi, A., deWorms, C. G. M., Haslewood, G. A. D., Hewett, C. L., and Robinson, A. M.,** The production of cancer by pure hydrocarbons, IV, *Proc. R. Soc. London, Ser. B.*, 123, 343, 1937.
129. **Berenblum, I.,** Unpublished results.
130. **Dannenberg, H.,** Beitrag zur krebserzeugenden Wirkung aromatischer Kohlenwassertoffe und verwandter Verbindungen, *Z. Krebsforsch.*, 63, 102, 1959.
131. **LaVoie, E. J.,** Unpublished results.
132. **Hecht, S., Amin, S., Rivenson, A., and Hoffmann, D.,** Tumor-initiating activity of 5,11-dimethylchrysene and the structural requirements favoring carcinogenicity of methylated polynuclear aromatic hydrocarbons, *Cancer Lett.*, 8, 65, 1979.
133. **Boyland, E. and Sims, P.,** Metabolism of polycyclic compounds, The metabolism of phenanthrene in rabbits and rats: dihydrodihydroxy compounds and related glucosiduronic acids, *Biochem. J.*, 84, 571, 1962.
134. **Sims, P.,** Qualitative and quantitative studies on the metabolism of a series of aromatic hydrocarbons by rat liver preparations, *Biochem. Pharmacol.*, 17, 795, 1970.

Chapter 7

STRUCTURE-ACTIVITY RELATIONSHIPS IN THE METABOLISM AND BIOLOGICAL ACTIVITY OF CYCLOPENTA-FUSED POLYCYCLIC AROMATIC SYSTEMS

A. Gold, R. Sangaiah, and S. Nesnow

TABLE OF CONTENTS

I. INTRODUCTION

The cyclopentapolycyclic aromatic hydrocarbons (cyclopenta-PAH) to be discussed in this chapter are characterized by a cyclopenta ring fused to the periphery of an even-alternant aromatic nucleus in such a manner that two carbon atoms of the five-membered ring form an etheno bridge connecting two peripheral carbon atoms of the parent compound. The smallest cyclopenta-PAH is acenaphthylene, which contains three rings. Acenaphthylene has been well characterized[1] and is not a bacterial mutagen,[2] so despite its presence in the environment, it will not be further discussed. The four-ring homologues in the series are acephenanthrylene (1) and aceanthrylene (2). All possible five-ring homologues except for cyclopenta[cd]pyrene (3) can be generated by fusion of a benzo ring to parent compounds 1 and 2 and are named accordingly, as shown in Figure 1, which gives the structures of all compounds treated in the text.

Since the characterization of cyclopenta[cd]pyrene (CPP) as an important component of kerosene and gasoline soots[1,3,4] and its identification as a potent mutagen,[5] there has been considerable interest in the peripherally fused cyclopenta-PAH as potential environmental carcinogens. CPP was shown to be capable of morphologically transforming C3H10T1/2[6] cells and also to be an initiator in CD-1 mice.[7-9] CPP is distinguished from most other known carcinogenic PAH by lack of the bay-region feature, which had been regarded as a prerequisite for biological activity via metabolism to a diol-epoxide.[10] Hence, the initial work on CPP indicated the need for a more basic understanding of the relationships between molecular structure and biological activity. A number of cyclopenta-PAH in addition to CPP have been more recently identified in pyrogenic samples. Acephenanthrylene (1) has been identified in extracts of the smoke of kerosene, wood, and tobacco and also in carbon black,[11] benz[e]- and benz[j]aceanthrylene in coal emission condensate,[12] and benz[l]aceanthrylene in extracts of wood smoke. The systematic synthesis, bioassay, and metabolism studies of these novel PAH are therefore interesting both from a theoretical point of view for investigation of structure-activity relationships using the cyclopenta-PAH as probes, and from a practical point of view to assess the impact they might have on human health and to provide pure compounds as standards for use in analysis of environmental samples.

II. SYNTHESIS

Synthesis has been accomplished either by Friedel-Crafts cyclization of an appropriate arylacetic acid precursor to form the five-membered ring or by building the desired carbon skeleton from readily available precursor(s) already containing the cyclopenta feature. In general, cyclizations of arylacetic acids have been found not to proceed unless the site of the desired acylium ion attack is highly activated towards electrophilic substitution. Furthermore, extensive decarbonylation may occur if the benzyl carbonium ion so formed is highly stabilized, yielding alkylated PAH as undesirable, potentially biologically active impurities. A description of synthetic routes is given below. All of the cyclopenta-PAH have electron impact mass spectra typical of PAH, consisting of a base peak at the mass-to-charge ratio of the molecular ion and a second peak at half the molecular weight corresponding to a doubly charged molecular ion. Accurate mass measurements on all molecular ions corresponded to the required elemental compositions.

A. Acephenanthrylene

(Yellow-Orange Needles from Methanol, mp 143—144°)

Synthesis in very low yield was initially reported by photochemical cyclization of 1-(*o*-iodobenzylidene)indane,[13] however, a more convenient route[11] proceeds through fusion of

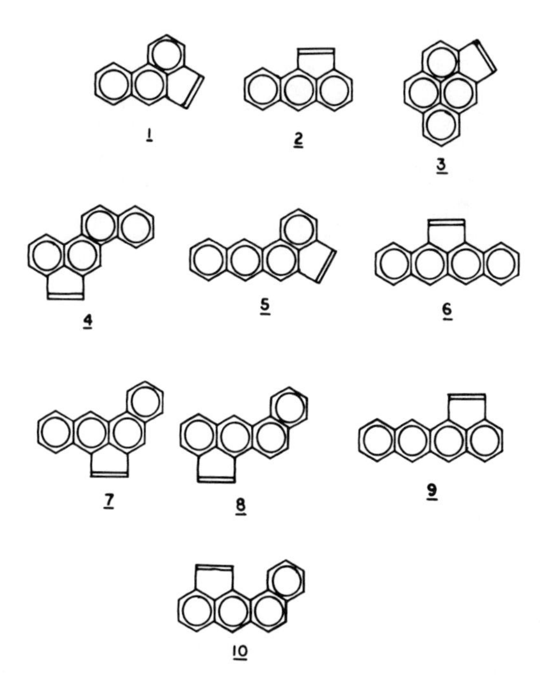

FIGURE 1. Structures of cyclopenta PAH discussed in this review: 1, acephenanthrylene; 2, aceanthrylene; 3, cyclopenta[cd]pyrene; 4, benz[j]acephenanthrylene; 5, benz[k]acephenanthrylene; 6, benz[d]aceanthrylene; 7, benz[e]aceanthrylene; 8, benz[j]aceanthrylene; 9, benz[k]aceanthrylene; 10, benz[l]aceanthrylene.

a six-membered ring at the C_4–C_5 bond of acenaphthene followed by dehydrogenation to acephenanthrylene by treatment with DDQ.

Sn, HCl → CO_2H → H_2SO_4 →

11 → , → 1

(Scheme 1)

B. Benz[j]acephenanthrylene

(Orange Needles from Benzene, mp 170—171°)

Benz[j]acephenanthrylene has been synthesized in a multistep procedure (Scheme 2) from the 7-oxo intermediate (11) of the acephenanthrylene synthesis[11] shown in Scheme 1. A four-carbon unit was added in two steps to form a γ-substituted butyric acid which cyclized smoothly in hydrofluoric acid to the benz[j]acephenanthrylene skeleton. A reduction-dehydration-aromatization sequence led to benz[j]acephenanthylene in 26% overall yield.

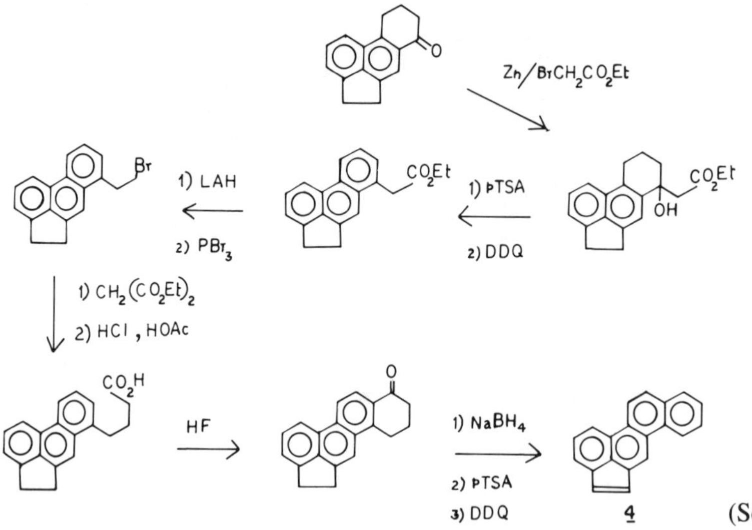

(Scheme 2)

C. Benz[k]acephenanthrylene

(Yellow-Orange Plates from Hexane, mp 233—234°)

The benz[k]acephenanthrylene skeleton was readily constructed by cyclodehydration of 2-(5-acenaphthoyl)benzoic acid, and the target PAH obtained[14] in 35% overall yield following reduction of the quinone and dehydrogenation with DDQ according to Scheme 3.

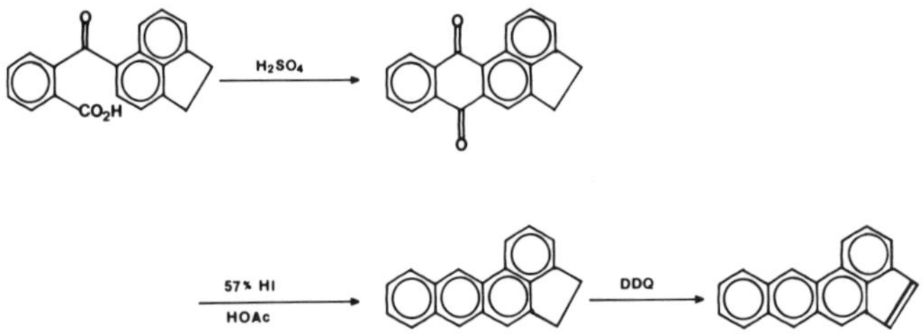

(Scheme 3)

D. Aceanthrylene

(Red Needles from Hexane, mp 94—95°)

The three schemes[15-17] recently reported all have in common the approach of adding the five-membered ring to an anthracene skeleton. The cyclopenta-fused ring has been formed directly by the Friedel-Crafts condensation of anthracene with either chloroacetyl chloride[16] or oxalyl chloride[15] and in a stepwise strategy via (1-anthryl)acetic acid.[17] Condensation of anthracene with chloroacetyl chloride (Scheme 4) involves the fewest number of steps:

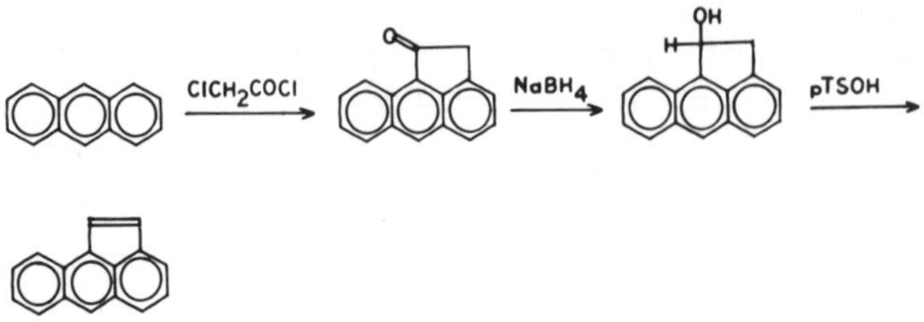

(Scheme 4)

E. Benz[d]aceanthrylene

(mp > 300°)

After unsuccessful attempts at addition of oxalyl chloride across the *meso* carbons (C_5, C_6) of naphthacene, 1,2,3,4-tetrahydronaphthacene was synthesized (Scheme 5) for use as substrate. The quinones derived from acid treatment[18] of 2(tetrahydronaphthoyl)benzoic acid[19] were separated[20] by fractional crystallization and, in accord with expectations, the Friedel-Crafts reaction on tetrahydronaphthacene yielded the desired benz[d]aceanthrylene skeleton. The resulting diketone was reduced and dehydrated by published methods[15,16] to the 5-oxo derivative. Benz[d]aceanthrylene was obtained by Wolff-Kishner reduction of the 5-oxo derivative followed by dehydrogenation with three equivalents of DDQ (Scheme 5).

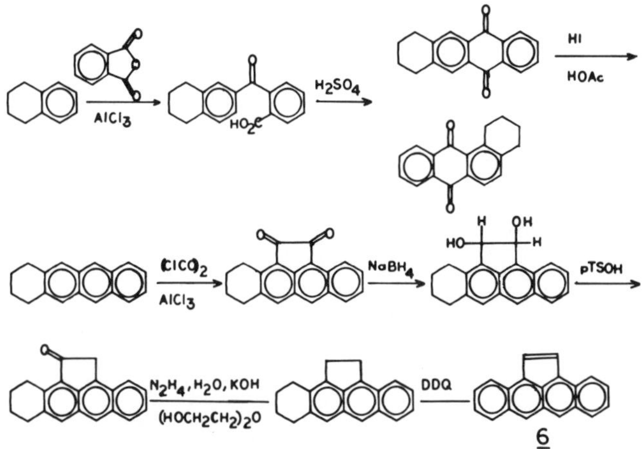

AlCl3 H2SO4 HI HOAc ClCO2 AlCl3 NaBH4 pTSOH N2H4·H2O, KOH (HOCH2CH2)2O DDQ 6

(Scheme 5)

F. Benz[e]aceanthrylene

(Orange Needles from Methanol, mp 138°)

The benz[e]aceanthrylene skeleton was obtained[14] by cyclodehydration of (7-benzanthryl)acetic acid in hydrogen fluoride. Ring closure occurred to both neighboring peri-positions (1:3, C_6:C_8) and the ketones were separated by normal phase preparative HPLC. Benz[e]aceanthrylene was obtained in a straight-forward manner by reduction and dehydration of the ketone product from cyclization to C_6 (Scheme 6).

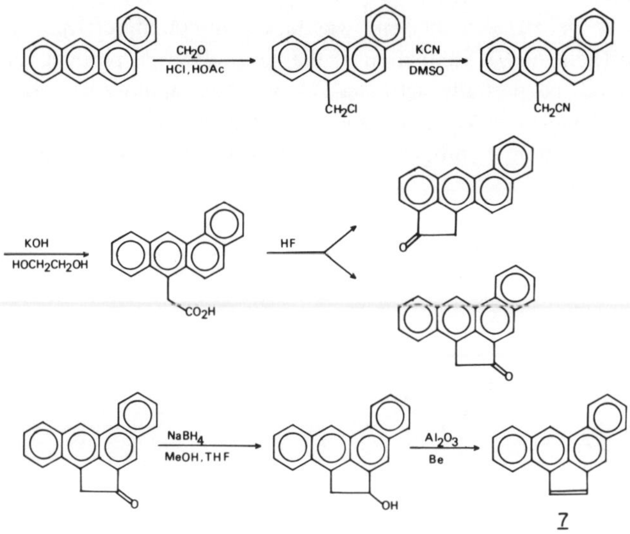

CH2O HCl, HOAc KCN DMSO KOH HOCH2CH2OH HF NaBH4 MeOH, THF Al2O3 Be 7

(Scheme 6)

G. Benz[j]aceanthrylene

(Orange Plates from Hexane, mp 170—171°)

Benz[j]aceanthrylene was readily synthesized[14] by reduction and dehydration of the ketone product from cyclodehydration of (7-benzanthryl)acetic acid to C_8 described above (Scheme 7).

NaBH4 MeOH.THF → Al2O3 Be → 8

(Scheme 7)

H. Benz[k]aceanthrylene

(Violet Needles from Methanol, mp > 300°)

The benz[k]aceanthrylene skeleton was derived from Friedel-Crafts acylation of benzene with acenaphten-2,3-dicarboxylic acid anhydride and treatment of the acylation product in a fused salt mixture to yield the 7,12-quinone.[21] Reduction[19] to the partially saturated hydrocarbon and dehydrogenation by DDQ (two equivalents) gave benz[k]aceanthrylene in 8% overall yield (Scheme 8).

+ AlCl3 → CO2H + CO2H ↓ AlCl3 + NaCl ← 1) Zn/NaOH 2) DDQ 9

(Scheme 8)

I. Benz[l]aceanthrylene

(Orange Prisms from Methanol, mp 157—158°)

The approach to benz[l]aceanthrylene was based on commercially available 9,10-dihydrophenanthrene (Scheme 9). Reduction of the succinoylation product followed by cyclodehydration yielded the partially saturated 11-oxo benzanthracene derivative which was transformed to (11-benzanthryl)acetic acid. The benz[l]aceanthrylene skeleton was then formed[14] by cyclization in hydrogen fluoride and a reduction dehydration sequence led smoothly to the PAH.

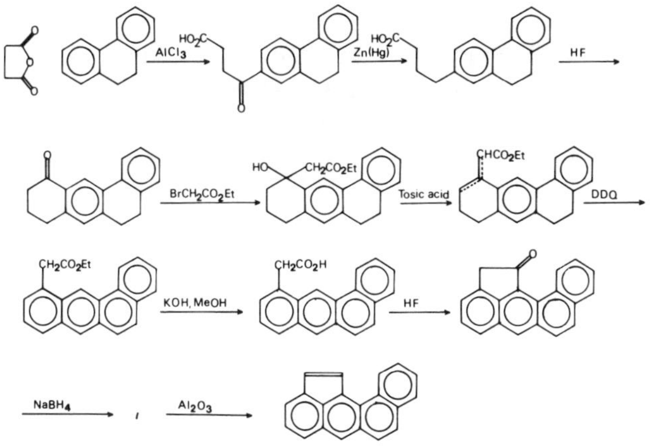

(Scheme 9)

J. Cyclopenta[cd]pyrene

(Orange Plates from Methanol, mp 174—176°)

The high level of interest in cyclopenta[cd]pyrene is evident in the publication of seven synthetic routes[22-28] to the PAH. Six of the schemes lead ultimately to the cyclization of (4-pyrenyl)acetic acid and transformation of the resulting 3-oxo precursor by reduction and dehydration. The most efficient of these schemes[25] gives an overall yield of 38% from pyrene. For three of the routes[22,23,28] in which the (4-pyrenyl)acetic acid is obtained by oxidation of 4-acetylpyrene, use of thallium trinitrate ($Tl(NO_3)_3$) as oxidant[28] will enhance the reported overall yields. The highest yield[27] (64%) has been reported from direct alkylation of the pyrene dianion to [4-(4,5-dihydropyrenyl)]acetic acid, cyclization of the partially saturated acid to the corresponding dihydro ketone, Wolff-Kishner reduction and aromatization by DDQ (Scheme 10).

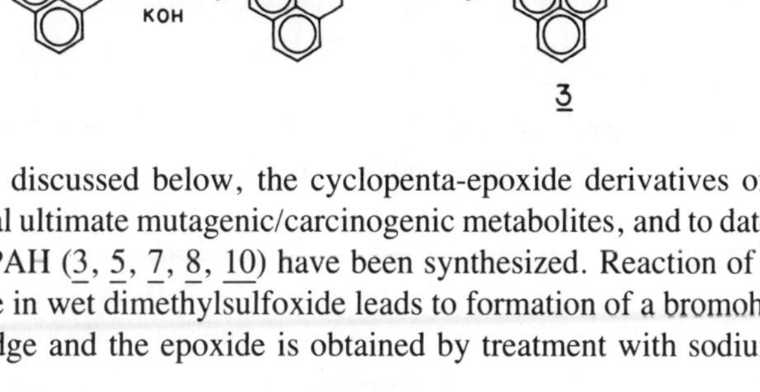

(Scheme 10)

For reasons to be discussed below, the cyclopenta-epoxide derivatives of the PAH are of interest as potential ultimate mutagenic/carcinogenic metabolites, and to date, the cyclopenta-epoxides of five PAH (3, 5, 7, 8, 10) have been synthesized. Reaction of the PAH with N-bromosuccinimide in wet dimethylsulfoxide leads to formation of a bromohydrin exclusively on the etheno bridge and the epoxide is obtained by treatment with sodium methoxide.

III. BIOASSAY

A. Bacterial Mutagenicity

All the cyclopenta compounds discussed in this text have been bioassayed in *Salmonella typhimurium* strain TA98 and, with two possible exceptions, are active mutagens in the

Table 1
ACTIVITY OF CYCLOPENTA PAH IN *S. TYPHIMURIUM* STRAIN TA98

PAH	Rev/nmole +S9	Rev/nmole −S9	Ref.
Acephenanthrylene[a]	0.8	0	31
Benz[j]acephenanthrylene[b,c]	8.6	0	UR
Benz[k]acephenanthrylene[d]	5.3	0	30
Aceanthrylene[a]	20	0	31, 32
Benz[d]aceanthrylene[b,c]	29	0	UR
Benz[e]aceanthrylene[d]	17	0	30
Benz[j]aceanthrylene[d]	13	0	30
Benz[k]aceanthrylene[b,c,e]	9	9	UR
Benz[l]aceanthrylene[d]	10	0	30
Cyclopenta[cd]pyrene[f]	80	0	5

Note: UR, Unpublished results.

[a] 2-Anthramine, 0.5 μg/plate + 0.3 mg/plate Aroclor® 1254-induced S9, 2100 rev/plate positive control in TA98.
[b] Benzo[a]pyrene + 1.27 mg/plate Aroclor® 1254-induced S9, 3.2 rev/nmole from reference curve in TA98.
[c] S9 concentrations not optimized for cyclopenta PAH or benzo[a]pyrene.
[d] Benzo[a]pyrene, 30 μg/plate + 3 mg/plate Aroclor® 1254-induced S9, 5.5 rev/nmole positive control in TA98.
[e] The high level of direct-acting mutagenicity suggests the possibility of endo-peroxide formation (see text).
[f] Benzo[a]pyrene + 50 μℓ/plate Aroclor® 1254-induced S9, 121 rev/nmole from reference curve in TA100.

presence of exogenous metabolic activation. Acephenanthrylene displays only marginal activity, while the mutagenicity of benz[k]aceanthrylene may be masked by formation of an endoperoxide under assay conditions. The naphthacene nucleus readily forms endoperoxides at the highly nucleophilic meso positions[29] and the potent direct-acting mutagenicity observed for benz[k]aceanthrylene may indicate generation of the analogous endoperoxide in the assay medium. The activities of all cyclopenta-PAH in TA98, along with activities of positive controls or reference standards, are given in Table 1. For assays in which benzo[a]pyrene was a positive control or concurrently tested as a reference standard, activities of the cyclopenta-PAH (except for acephenanthrylene) are comparable to or greater than benzo[a]pyrene.

Activities of benzo[a]pyrene and the series of cyclopenta-PAH derived from benz[a]anthracene (5, 7, 8, 10; Table 1) have been directly compared in *S. typhimurium* TA98 using the optimal S9 concentration for each compound.[30] The order of activities is benz[e]aceanthrylene > benz[j]aceanthrylene > -benz[l]aceanthrylene which is about two-fold more active than either benz[k]acephenanthrylene or benzo[a]pyrene. The spectrum of activity of aceanthrylene, the e-, j-, and l-benzaceanthrylene isomers, and benz[k]acephenanthrylene over the five Ames tester strains is similar to that of benzo[a]pyrene. All are frame-shift mutagens giving the highest activity in strains TA98 and TA100, approximately tenfold less activity in TA1538, and marginal or no activity in TA1537. None of the cyclopenta-PAH tested revert the base-pair sensitive strain TA1535.

Cyclopenta[cd]pyrene is also a frame-shift mutagen and differs from the other cyclopenta-PAH in being as mutagenic in TA1537 and TA1538 as in TA98 and TA100, indicating that addition of the plasmid pKM101 does not produce the expected enhancement in mutagenicity.[5]

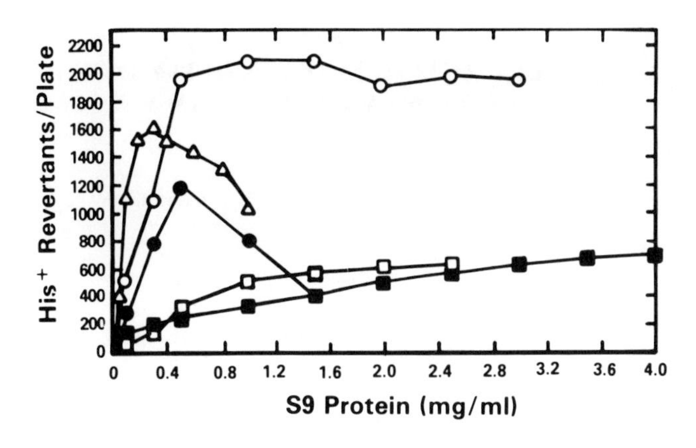

A

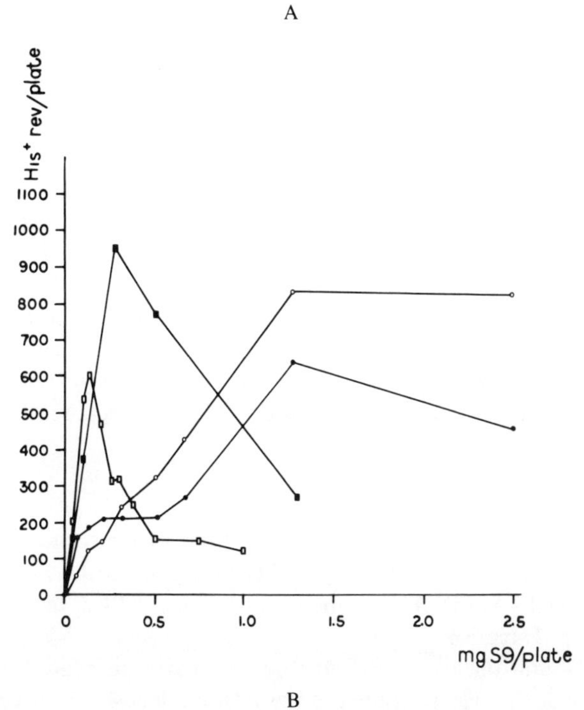

B

FIGURE 2. Dependence on S9 concentration of the mutagenic response of *S. typhimurium*. (A) 30 μg/plate PAH with TA98; open circles, benz[e]aceanthrylene; open triangles, benz[j]aceanthrylene; solid circles, benz[l]aceanthrylene; open squares, benz[k]acephenanthrylene; solid squares, benzo[a]pyrene; (B) 30 μ/plate PAH with TA98 except for cyclopenta[cd]pyrene at 1 μg/plate with TA100; solid squares, aceanthrylene; open circles, benz[j]acephenanthrylene; open squares, cyclopenta[cd]pyrene; solid circles, benzo[a]pyrene. The curves in panels A and B represent experiments run at different times.

The S9-concentration dependence of mutagenicity of cyclopenta[cd]pyrene, benz[j]acephenanthrylene, aceanthrylene, and the four isomeric cyclopenta-PAH derived from benzanthracene has been determined (Figure 2A,B). Cyclopenta[cd]pyrene,[5] aceanthrylene,[31,32] and the e-, j- and l-benzaceanthrylene[30] isomers all differ markedly from benzo[a]pyrene in showing optimal activity at S9 concentrations one fifth to one tenth the optimal S9 concentration for benzo[a]pyrene. The activity of benz[e]aceanthrylene remains at a plateau, while the mutagenicity of the j- and l-isomers and cyclopenta[cd]pyrene de-

Table 2
DIRECT-ACTING MUTAGENICITY OF CYCLOPENTA EPOXIDES IN *S. TYPHIMURIUM*

Epoxide	(Rev/nmole)
Benz[e]aceanthrylene-5,6-oxide[a]	250
Benz[j]aceanthrylene-1,2-oxide[a]	250
Benz[l]aceanthrylene-1,2,oxide[a]	150
Benz[k]acephenanthrylene-4,5-oxide[a]	18
Cyclopenta[cd]pyrene-3,4-oxide[b]	1440[5]

[a] Strain TA98, corrected for background.
[b] Strain TA100, corrected for background.

creases rapidly with increasing S9 concentration. Benz[k]acephenanthrylene displays peak activity at an intermediate level of S9 and the broad S9-dependence curve is similar in appearance to that of benzo[a]pyrene. The S9 dependence of benz[j]acephenanthrylene activity is indistinguishable from that of benzo[a]pyrene (Figure 2B).

The cyclopenta epoxides that have been synthesized are all direct-acting bacterial mutagens (Table 2). The oxides of cyclopenta[cd]pyrene[33] and the three benzaceanthrylene isomers are 10- to 20-fold more mutagenic than the parent PAH, while the 4,5-oxide of benz[k]acephenanthrylene is relatively less active, being only threefold more mutagenic than benz[k]acephenanthrylene. Benz[l]aceanthrylene-1,2-oxide has been directly compared to the *syn* and *anti*-benzo[a]pyrene diol-epoxides in TA98[30] and the order of mutagenicity was reported to be benz[l]aceanthrylene < *anti* benzo[a]pyrene diol-epoxide < *syn* benzo[a]pyrene diol-epoxide. On a molar basis, the *syn*-diol-epoxide was 11-fold more active than cyclopenta epoxide.

B. Mammalian Cell Mutagenicity

Cyclopenta-PAH and some of their oxides have been evaluated in two mammalian mutagenesis bioassay systems: Chinese hamster V79 lung cells at the HGPRT locus (6-thioguanine resistance) and L5178Y $TK^{+/-}$ mouse lymphoma cells at the TK locus (trifluorothymidine resistance; Table 3).

Induced gene mutation by the four metabolically activated cyclopenta-fused benz[a]anthracene isomers was measured at the HGPRT locus in V79 cells and compared to that of benzo[a]pyrene as reference.[30] With an Aroclor® 1254-induced rat-liver activating system, all five PAH showed a similar S9-concentration dependence of mutagenicity, reaching a plateau at 0.5 mg protein per mℓ. At the 10 μg/mℓ concentration, the e, j, and l isomers were equivalent in activity, with benz[k]acephenanthrylene being eight times less active. The e, j, and l isomers were four times as active as benzo[a]pyrene. When irradiated Syrian hamster embryo cells were used as an activating system, benz[l]aceanthrylene produced approximately half the activity of benzo[a]pyrene and the other cyclopenta benz[a]anthracenes were inactive. Benz[l]aceanthrylene-1,2-oxide was a direct acting mutagen in V79 cells, about half as potent on a molar basis as *syn*-BPDE which produced half the activity of the *anti* isomer.

Using the L5178Y $TK^{+/-}$ system which measures both gene mutation and chromosomal effects, benz[l]aceanthrylene was slightly less active than benzo[a]pyrene. Both PAH were shown to produce chromosomal aberrations in these cells.[61]

Both cyclopenta[cd]pyrene and its 3,4-oxide derivative have been assayed in the L5178Y $TK^{+/-}$ system.[6] The PAH is mutagenic with exogenous activation by Aroclor® 1254-induced S9 giving a mutant frequency 6.5 times that of the 2-acetylaminofluorene positive control

Table 3
MUTAGENICITY OF CYCLOPENTA-PAH IN MAMMALIAN CELLS

Cyclopenta-PAH (Positive Control)	Cell line genetic endpoint	Mutants/10^6 survivors
Benz[e]aceanthrylene	V79/HGPRT	500(10)[a]
Benz[j]aceanthrylene	V79/HGPRT	500(10)[a]
Benz[k]acephenanthrylene	V79/HGPRT	60(10)[a]
Benz[l]aceanthrylene	V79/HGPRT	500(10)[a]
(Benzo[a]pyrene)	V79/HGPRT	125(10)[a]
Benz[l]aceanthrylene	L5178Y/TK$^{+/-}$	350(5)[b]
(Benzo[a]pyrene)	L5178Y/TK$^{+/-}$	400(3)[b]
Cyclopenta[cd]pyrene	L5178Y/TK$^{+/-}$	188(1.6)[c]
(2-Acetylaminofluorene)	L5178Y/TK$^{+/-}$	581(30)[c]

[a] Data taken from Reference 32. Values represent the mean of triplicate determinations using Aroclor®-1254 induced rat-liver microsomes at 1.5 mg/mℓ protein, an exposure time of 4 hr. Values in parentheses are the PAH concentration in μg/mℓ. The spontaneous mutation frequency was 10 to 15 6-thioguanine-resistant mutants/10^6 survivors.

[b] Data taken from Reference 61. Values represent the average of duplicate determinations using Aroclor®-1254 induced rat-liver S9 and an exposure time of 4 hr. The spontaneous mutation frequency was 90—100 trifluorothymidine-resistant mutants/10^6 survivors.

[c] Data from Reference 6. Values represent the results of duplicate determinations using an Aroclor®-1254 induced rat-liver S9 and an exposure time of 4 hr. The spontaneous mutation frequency was 79 trifluorothymidine-resistant mutants/10^6 survivors.

on a molar basis. The 3,4-oxide was ~500 times more potent than the ethyl methane sulfonate positive control in this assay.

Only one in vivo genetic toxicology study has been performed to date with the cyclopenta-PAH. Benz[l]aceanthrylene and benzo[a]pyrene were administered to C57BL/6 mice by gavage and sister chromatid exchange (SCE) measured in peripheral blood lymphocytes. Benz[l]aceanthrylene produced a sharper elevation in the SCE frequency than benzo[a]pyrene.[61]

C. Morphological Cell Transformation and Tumorigenicity

Cyclopenta[cd]pyrene has been the most thoroughly tested of the cyclopenta-PAH to date. Cyclopenta[cd]pyrene produced a dose-related response in the C3H10T1/2 morphological transformation assay, but was 20 times less active than benzo[a]pyrene on a molar basis, estimated according to the sum of type II and type III foci formed.[6] The 3,4-oxide also transformed C3H10T1/2 cells. In CD-1 mice, cyclopenta[cd]pyrene was approximately 20 times less potent than benzo[a]pyrene as a tumor initiator, and was significantly less active than chrysene but equivalent in activity to the weak tumor initiator benz[a]anthracene.[7] In a comparison of the tumorigenic potency of cyclopenta[cd]pyrene and benzo[a]pyrene by repeated application to the backs of Swiss mice, cyclopenta[cd]pyrene produced tumors in 20% of test animals over a period during which the incidence was 100% for an equivalent dose of benzo[a]pyrene. Eventually, 57% of the cyclopenta[cd]pyrene-treated animals developed tumors.[8]

Acephenanthrylene, aceanthrylene, and several of the benzo isomers of these two parent ring systems have been assayed in the C3H10T1/2 system.[34,35] Acephenanthrylene, aceanthrylene, and benz[k]acephenanthrylene were inactive but benz[e]-, benz[j]-, and benz[l]aceanthrylene transformed C3H10T1/2 cells (Table 4). Benz[j]aceanthrylene was the

Table 4
MORPHOLOGICAL TRANSFORMATION OF C3H10T1/2C18 CELLS BY CYCLOPENTA-FUSED PAH

Cyclopenta fused PAH	Concentration (μg/mℓ)	% Dishes with type II or type III foci
Aceanthrylene	16	0
Acephenanthrylene	16	0
Benz[e]aceanthrylene	10	58
Benz[l]aceanthrylene	10	85
Benz[j]aceanthrylene	1.0	94
Benz[k]acephenanthrylene	10	0
Benz[j]acephenanthrylene	5	25[a]
Benzo[a]pyrene	1.0	87[b]

[a] References 34 and 38.
[b] Mean of three replicate determinations.

most active of the three, equivalent in transforming efficiency to benzo[a]pyrene, while the other PAH were less effective by a factor of ten. In an initiation-promotion study comparing tumor-initiating activity of benz[e]aceanthrylene and benz[l]aceanthrylene to that of benzo[a]pyrene in SENCAR mice,[36] benz[l]aceanthrylene was approximately four times more potent than benzo[a]pyrene and benz[e]aceanthrylene, which were equivalent in activity.

IV. METABOLISM

Chromatographic profiles and major metabolites generated by rat-liver microsomes have been determined for all cyclopenta compounds reported in this review with the exception of the benz[d]- and benz[k]-isomers of aceanthrylene. HPLC traces of the rat-liver metabolism mixtures are given in Figure 3, and the metabolite structures in Figure 4. The metabolism of acephenanthrylene, benz[j]acephenanthrylene, and aceanthrylene has not been reported in the literature and identification of the metabolites of these PAH will therefore be discussed. Oxidation of the cyclopenta ring is an important if not predominant metabolic pathway for all cyclopenta-PAH except benz[l]aceanthrylene. The cyclopenta dihydrodiols are readily identified by UV-vis spectra, since saturation of the five-membered ring leaves a chromophore identical to the alternant PAH nucleus from which the corresponding cyclopenta compounds are derived. The ^{1}H NMR spectra are also distinctive because of the absence both of vinylic resonances and of the AX quartet of a bridging etheno group. Other metabolites are identified by analysis of ^{1}H NMR and UV-vis spectra. Except for benz[j]aceanthrylene, benz[l]aceanthrylene, and cyclopenta[cd]pyrene, which have been radio-labeled for quantitative metabolism studies, relative yields of metabolites have been estimated by signal/noise ratios of ^{1}H NMR spectra of samples accumulated from an equal number of semipreparative HPLC runs, made up to equal volumes, and generated from the same number of transients (4×10^4). The mass spectral fragmentation patterns and accurate mass determinations on the molecular ions of all metabolites are consistent with assigned structures.

A. Acephenanthrylenes

Metabolism of acephenanthrylene by Aroclor® 1254-induced rat liver microsomes produced a single major metabolite readily identified by its phenanthrene-like UV-vis spectrum (Figure 5) as the 4,5-dihydrodiol (cyclopenta dihydrodiol). Consistent with this structure, the ^{1}H NMR (Figure 6) shows an intact bay region and no etheno bridge AX quartet.

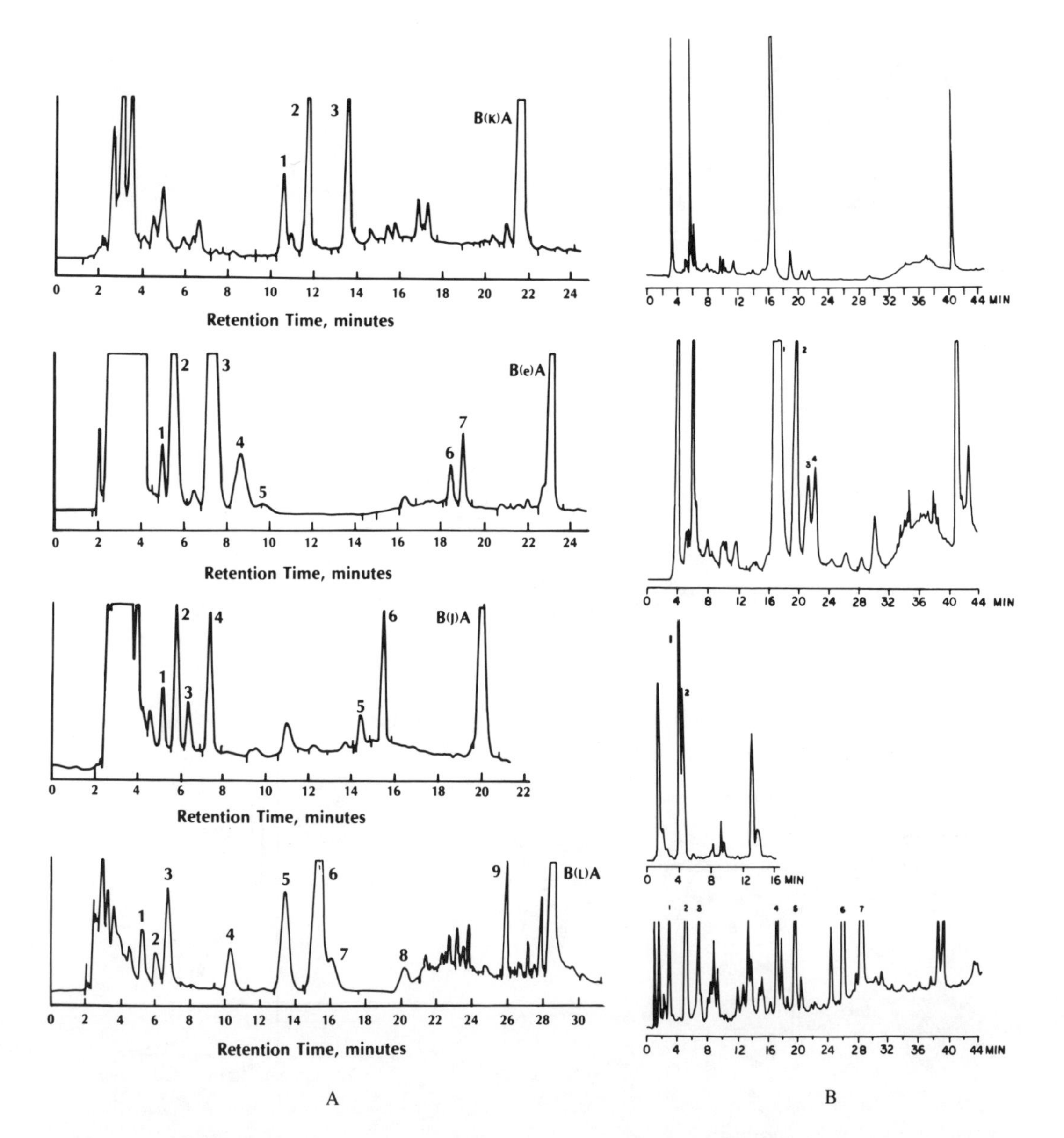

FIGURE 3. HPLC traces (254 nm) of Aroclor®-1254 induced rat-liver S9 product distributions from metabolism of the following PAH: panel A (top down), benz[k]acephenanthrylene; benz[e]aceanthrylene; benz[j]aceanthrylene, benz[l]aceanthrylene. Panel B (top down), acephenanthrylene; aceanthrylene; cyclopenta[cd]pyrene; benz[j]acephenanthrylene. Major metabolites are numbered and the structures of those identified are shown in Figure 4.

Metabolism of benz[j]acephenanthrylene under similar conditions generated a much more complex profile from which the two major metabolites have been identified. The compound corresponding to peak 6 had a chrysene-like UV-vis spectrum (Figure 7A). The ^1H NMR spectrum (Figure 8A), showed no etheno AX quartet, while four highly deshielded bay-region protons (a singlet for H_6 and three doublets) indicated that the chrysene aromatic nucleus remained intact, allowing the 4,5-dihydrodiol structure (cyclopenta-dihydrodiol) to be unambiguously assigned. The UV-vis spectrum of metabolite 7 (Figure 7B) was similar to that of acephenanthrylene, suggesting partial saturation of a terminal benzo ring. Supporting this conclusion is the appearance of only two bay-region doublets in the ^1H NMR

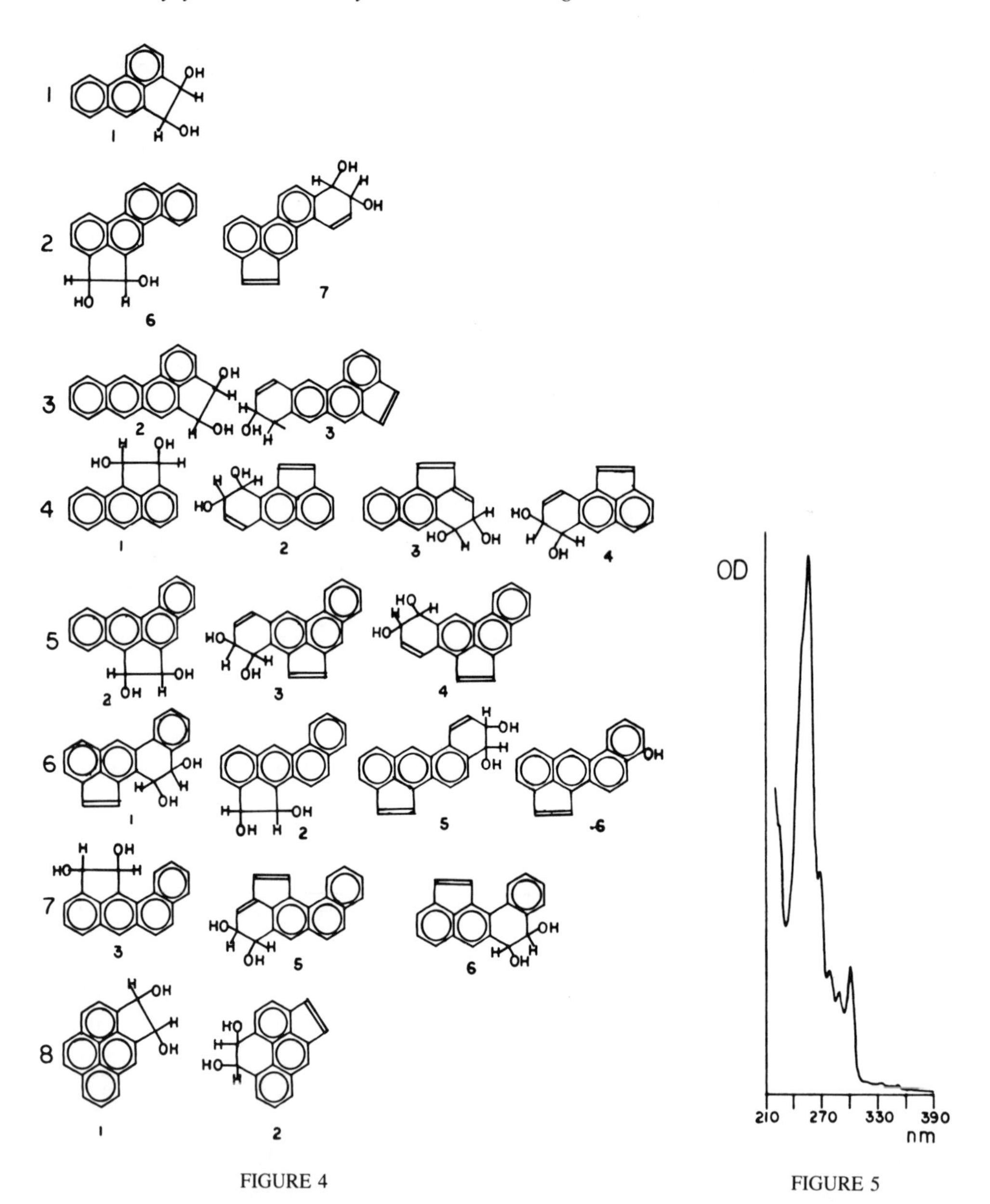

FIGURE 4

FIGURE 5

FIGURE 4. Structures of metabolites identified from Aroclor®-1254 induced metabolism of cyclopenta PAH: Row 1, acephenanthrylene; Row 2, benz[j]acephenanthrylene; Row 3, benz[k]acephenanthrylene; Row 4, aceanthrylene; Row 5, benz[e]aceanthrylene; Row 6, benz[j]aceanthrylene; Row 7, benz[l]aceanthrylene; Row 8, cyclopenta[cd]pyrene.

FIGURE 5. UV-vis spectrum (methanol) of acephenanthrylene-4,5-dihydrodiol.

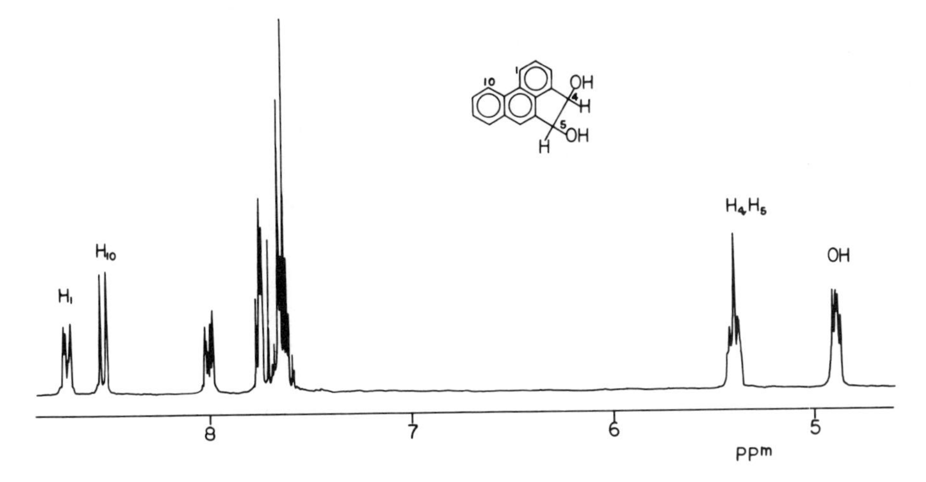

FIGURE 6. ^{1}H NMR (250 MHz, acetone-d_6) of acephenanthrylene-4,5-dihydrodiol. Proton assignments are indicated on spectrum.

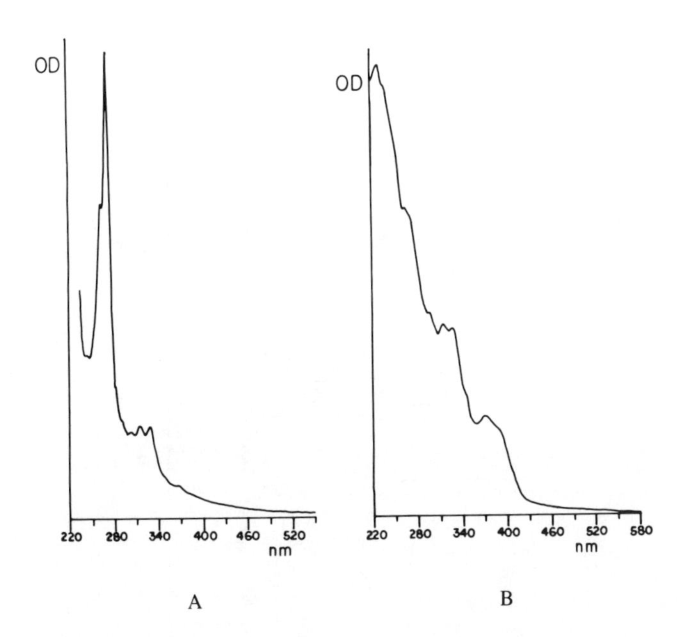

FIGURE 7. UV-vis spectra (methanol) of benz[j]acephenanthrylene metabolites. (A) Benz[j]acephenanthrylene-4,5-dihydrodiol (metabolite 6); (B) benz[j]acephenanthrylene-9,10-dihydrodiol (metabolite 7).

(Figure 8B), less deshielded than the bay-region resonances of metabolite 6, consistent with decrease of the ring current by partial saturation of the chrysene nucleus. The highly deshielded resonance at 7.42 ppm indicates the location of a vinylic double bond adjacent to a bay region. The only dihydrodiol structure consistent with this spectrum is the 9,10-dihydrodiol, distal to the bay region which is on the same side of the molecular periphery as the cyclopenta ring. Two dihydrodiols predominate the products of the Aroclor® 1254-induced rat-liver microsomal metabolism of benz[k]acephenanthrylene. The major dihydrodiol was identified as the 4,5-dihydrodiol (cyclopenta dihydrodiol) and the minor metabolite as the terminal-ring 8,9-dihydrodiol.[30]

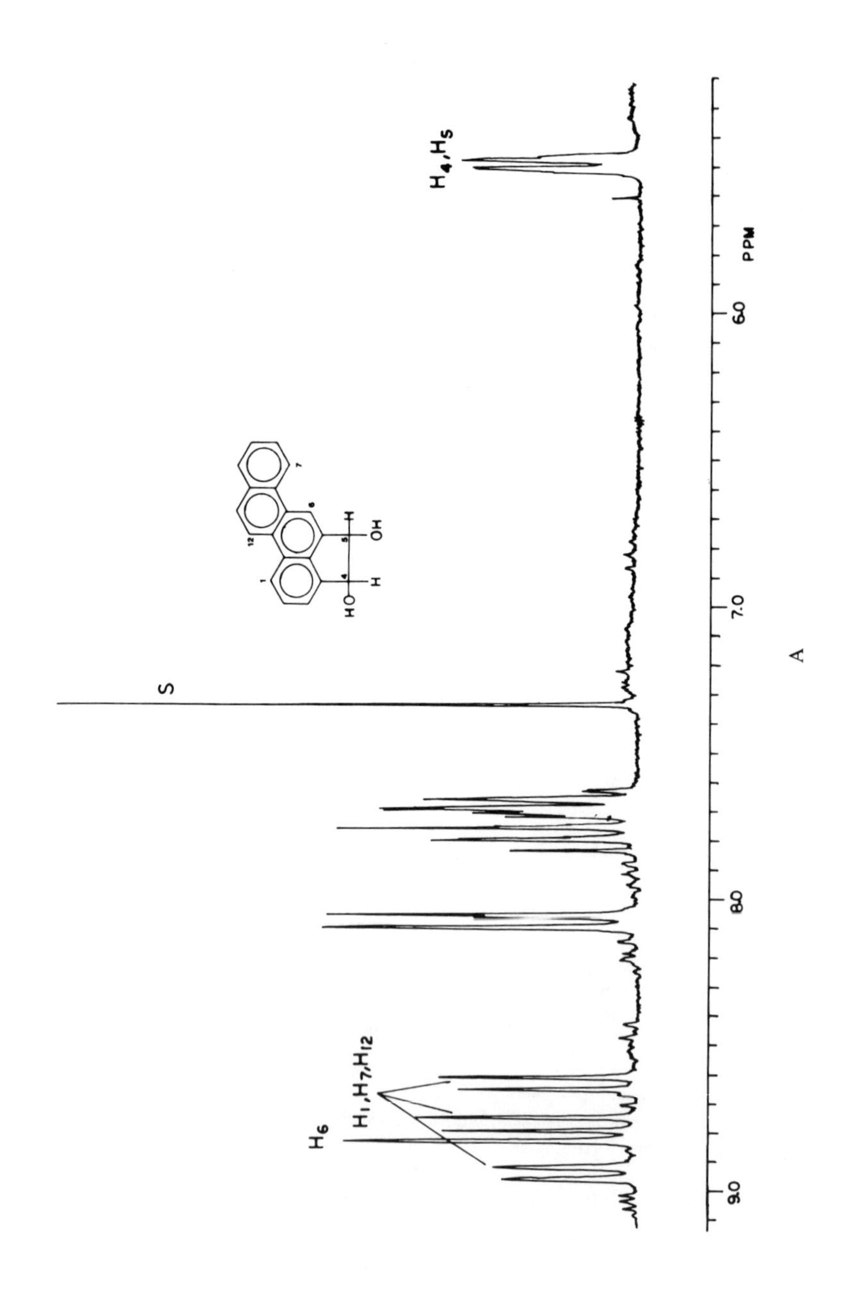
H4,H5
S
H6
H1,H7,H12
9.0
8.0
7.0
6.0
PPM
A

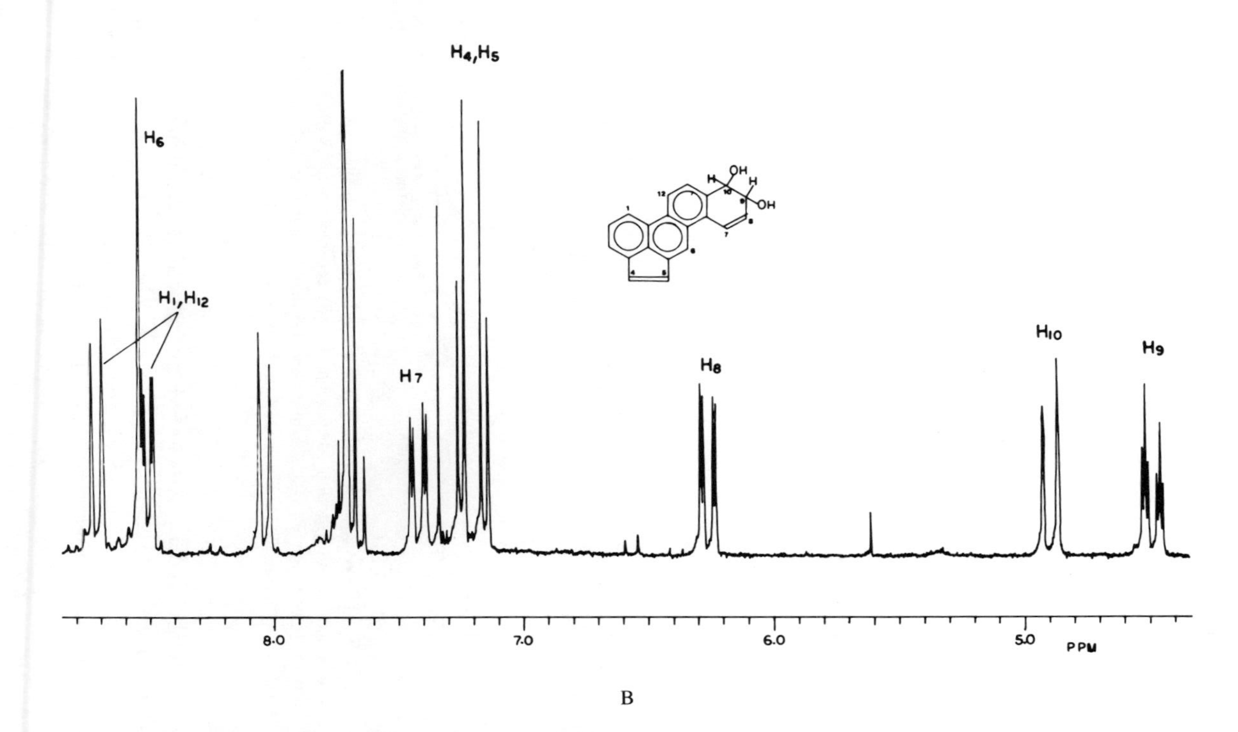

B

FIGURE 8. ^1H NMR (200 MHz, acetone-d_6) spectra of benz[j]acephenanthrylene metabolites. (A) Benz[j]acephenanthrylene-4,5-dihydrodiol (metabolite 6); (B) benz[j]acephenanthrylene-9,10-dihydrodiol (metabolite 7). Proton assignments are indicated on spectra.

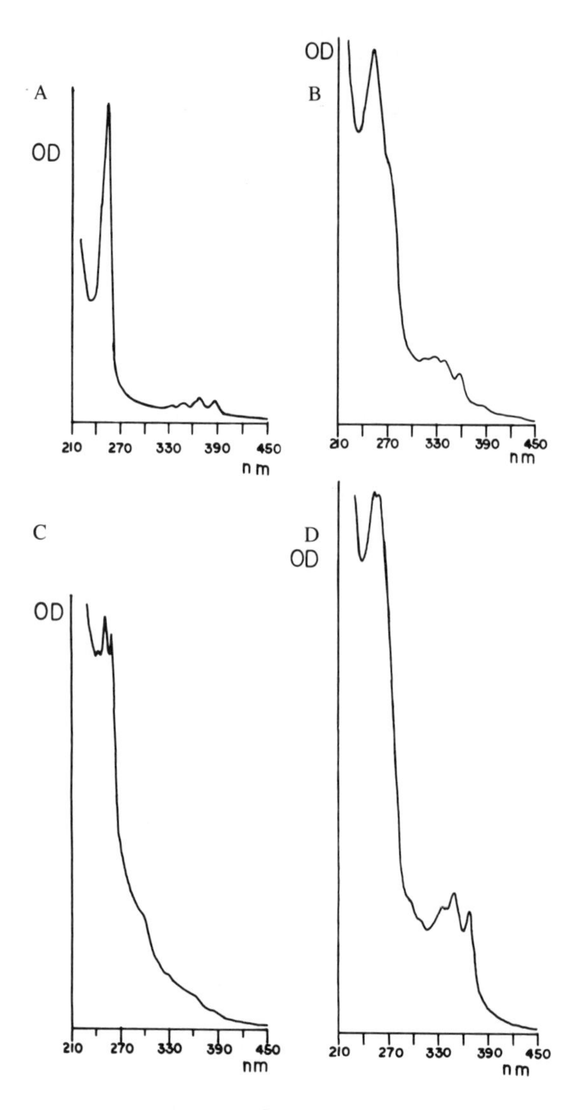

FIGURE 9. UV-vis spectra (methanol) of aceanthrylene metabolites. (A) Aceanthrylene-1,2-dihydrodiol (metabolite 1); (B) aceanthrylene-7,8-dihydrodiol (metabolite 2); (C) aceanthrylene-4,5-dihydrodiol (metabolite 3); (D) aceanthrylene-9,10-dihydrodiol (metabolite 4).

B. Aceanthrylenes

Aceanthrylene is metabolized by Aroclor® 1254-induced rat liver microsomes to four dihydrodiols with relative yields 1 >> 4 > 2 >> 3. The UV-vis spectrum of the major metabolite (Figure 9) was that of the anthracene chromophore, and the ^1H NMR (Figure 10A) showed no etheno AX quartet, consistent with the 1,2-dihydrodiol (cyclopenta dihydrodiol) structure. In the ^1H NMR of metabolite 4 (Figure 10B), the presence of two vinyl resonances, and the etheno AX quartet require the metabolized bond to be either C_7–C_8 or C_9–C_{10}. The large downfield shift of one vinyl resonance indicates the location of the vinyl bond adjacent to the ''pseudo bay region'', and on this basis the 7,8-dihydrodiol structure has been assigned. The presence of an AX quartet from the etheno bridge and two vinyl resonances at less deshielded positions (Figure 10C) leads to the assignment of metabolite 2 as the complimentary 9,10-dihydrodiol adjacent to the bay region. The structures of metabolites 2 and 4 are further supported by the less deshielded position of the resonance

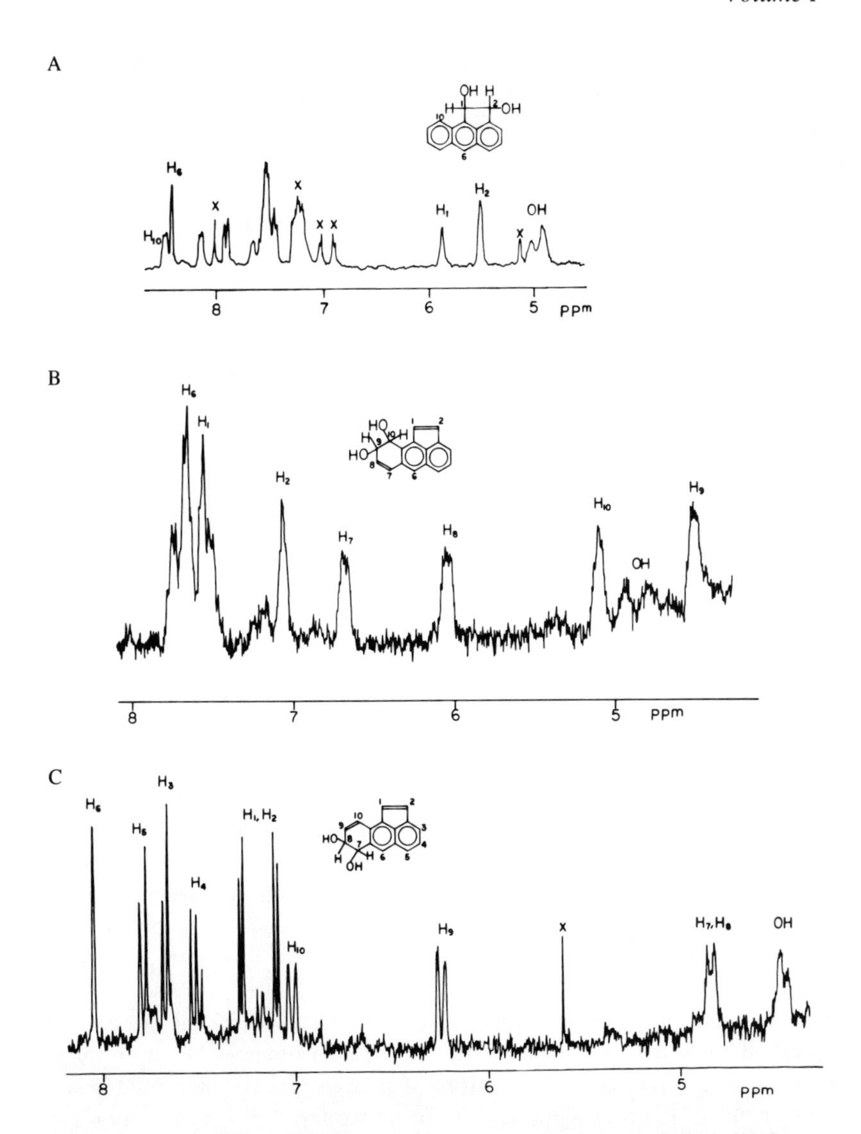

FIGURE 10. ^{1}H NMR (250 MHz, acetone-d_6) of aceanthrylene metabolites. (A) Aceanthrylene-1,2-dihydrodiol (metabolite 1); (B) aceanthrylene-9,10-dihydrodiol (metabolite 4); (C) aceanthrylene-7,8-dihydrodiol (metabolite 2). Proton assignments are indicated on spectra.

for the peri proton (H_6) of metabolite 2. A vinyl bond adjacent to a peri position, as in the 7,8-dihydrodiol structure, has been shown in structurally related dihydrodiols to cause an upfield shift of the peri resonance.[30,37] Peak 3 must represent the C_4–C_5 dihydrodiol — the only possibility remaining for metabolism of an arene bond not located at a ring-fusion site.

Three of the seven peaks in the metabolite profile generated from benz[e]aceanthrylene by Aroclor® 1254-induced rat-liver microsomes have been identified.[30] The structures and relative yields are 5,6-dihydrodiol (cyclopenta dihydrodiol) 3,4-dihydrodiol > 1,2-dihydrodiol.

Benz[j]aceanthrylene has been metabolized in Aroclor® 1254-induced rat-liver microsomes,[30] C3H10T1/2 cells, and Syrian hamster embryo (SHE) cells,[38] Four metabolites have been identified among the microsomal products, with the following relative yields: 1,2-dihydrodiol (cyclopenta dihydrodiol) >> 9,10-dihydrodiol > 10-phenol >> 11,12-dihydrodiol (K-region dihydrodiol). Quantitative yields were determined for a 24 hr incubation of the ^{3}H-labeled PAH in C3H10T1/2 cells, and the time-dependence of metabolic formation was followed in the SHE system over 48 hr. The C3H10T1/2 cells metabolized benz[j]aceanthrylene primarily to the 9,10-dihydrodiol (55%), with minor amounts of the 1,2-dihydrodiol (14%) and 11,12-dihydrodiol (5%), the remainder of the label being distributed among unidentified minor products. Through 24 hr, the SHE cells produced the 9,10-dihydrodiol at a rate 1.3 times that of the 1,2-dihydrodiol (including conjugates). Over the next 24 hr, the yield of 9,10-dihydrodiol increased, while the total yield of 1,2-dihydrodiol and conjugates remained unchanged, resulting in relative yields of 64% and 36%, respectively, after 48 hr.

Metabolite profiles of benz[l]aceanthrylene have been obtained in Aroclor® 1254-induced rat-liver microsomes[30] and C3H10T1/2 cells.[35] Three metabolites were identified in the microsomal products: 7,8-dihydrodiol > 1,2-dihydrodiol (cyclopenta dihydrodiol) >> 4,5-dihydrodiol. Quantitated yields for incubation of the ^{14}C-labeled PAH over 24 hr with C3H10T1/2 cells were 7,8-dihydrodiol, 45%; 4,5-dihydrodiol, 6%; 1,2-dihydrodiol, 5%.

C. Cyclopenta[cd]pyrene

^{3}H-Cyclopenta[cd]pyrene was metabolized by 3-methylcholanthrene-induced rat-liver microsomes[39] to a mixture containing primarily the 3,4-dihydrodiol (cyclopenta dihydrodiol) (73%) and 9,10-dihydrodiol (21%). Phenobarbital-induced rat-liver microsomes metabolized the cyclopenta ring even more selectively to the 3,4-dihydrodiol (81% to 19% for the 9,10-dihydrodiol).

V. STRUCTURE-ACTIVITY CONSIDERATIONS

A. Metabolism

The activation of promutagenic/procarcinogenic PAH proceeds in two steps: monooxygen transfer by P450 followed by formation of covalent adducts between reactive metabolites and cellular targets.[40] A rational approach to understanding pathways of bioactivation requires delineation of relationships between molecular structure, electronic structure, and reactivity of the PAH substrates for both steps of the activation process. A second key element in predicting regioselectivity of metabolism is a description of the molecular and electronic structure of the catalytically competent P450 intermediate.

Although the need for metabolism and the role of P450 in chemical carcinogenesis have been appreciated for less than two decades, considerable progress has been made towards a description of the complex actually involved in monooxygen transfer to P450 substrates.[41] The species involved is currently throught to be an oxoferryl porphyrin cation radical:

$$P^{\dot{+}}\ Fe^{IV} = 0 \qquad P = \text{porphyrin dianion}$$

Support for this conclusion rests on the assignment of the oxoferryl porphyrin cation radical structure to monooxygen complexes of the closely related peroxidase enzymes,[42,43] on the development of biomimetic model monooxygenase systems based on interactions of porphinatoiron complexes with monooxygen donors,[44,45] and also on molecular orbital calculations of magnetic properties and electronic structure of the putative oxo-P450 complex.[41] In certain of the model P450 systems, observable intermediates have been generated having physico-chemical properties consistent with predictions for the oxoferryl cation radical structure.[43,45,46] Both molecular orbital calculations and deuterium labeling studies suggest that the oxo ligand carries considerable unpaired spin density and lead to the expectation that oxygen transfer should have characteristics akin to nonconcerted electrophilic substitution.

For polycyclic aromatic hydrocarbons, susceptibility to electrophilic attack at peripheral sites has been found to correlate with the change in resonance energy resulting from localization of the carbon atom involved.[48,49] The change in energy is defined as localization energy (LE) and can be calculated by determining the difference between the π-resonance

Table 5
LOCALIZATION ENERGIES FOR ELECTROPHILIC ATTACK AT PERIPHERAL POSITIONS OF ACEPHENANTHRYLENE COMPARED TO PHENANTHRENE

Acephenanthrylene	|LE/β|		Phenanthrene	|LE/β|	
Position	Huckel π-localization energy	PMO approx.	Position	Huckel π-localization energy	PMO approx.
C_1	2.3627	2.14	C_4	2.3651	1.96
C_2	2.4879	2.04	C_3	2.4530	2.04
C_3	2.3912	2.49	C_2	2.4970	2.18
C_4	2.1198	1.34			
C_5	2.1298	1.36			
C_7	2.3630	1.86	$C_8(C_1)$	2.3173	1.86
C_8	2.4992	2.18	$C_7(C_2)$	2.4970	2.18
C_9	2.5196	2.03	$C_6(C_3)$	2.4530	2.04
C_{10}	2.3707	1.96	$C_5(C_4)$	2.3651	1.96

energy of the parent PAH and the π-resonance energy of the cationic system remaining after removal of the localized carbon form the π framework. Since there is a decrease in resonance energy for the localized system, LE is negative, and the smaller the absolute value of LE (|LE|), the less thermodynamically costly is the formation of the reaction intermediate. Hence, small |LE| indicate increased susceptibility to electrophilic attack. π-Energy calculations have been performed for aceanthrylene and acephenanthrylene by the simple Huckel method, and localization energies in units of |LE/β| for attack at peripheral carbon atoms leading to formation of dihydrodiols are given in Tables 4 and 5, respectively. In Tables 4 and 5, the effect of the fused five-membered ring on susceptibility to electrophilic attack is indicated by pairing |LE/β| values for the cyclopenta-PAH with corresponding positions of the related alternant PAH. Meso carbons (e.g., C_6 of acephenanthrylene and aceanthrylene; C_7, C_{12} of benz[k]acephenanthrylene) are not included in the correlation. Although meso positions are sites of high electron density, they do not appear to be readily oxidized by oxo-P450: for example, benzanthracene-7,12-quinone is not a metabolite of benzanthracene,[50,51] nor have any metabolites of cyclopenta-PAH oxidized at meso positions been identified.[30] Reversibility of attack of oxo-P450 at meso positions may result in kinetic control of the oxidation reaction and lead to other metabolites, analogous to the situation for Friedel-Crafts acylation of benzo[a]pyrene,[52,53] which gives only 1-acyl substitution with no acylation at the more reactive meso C_6.

A convenient perturbational molecular orbital (PMO) approximation can be used to estimate the |LE/β| values of cationic intermediates resulting from electrophilic attack on even-alternant PAH:

$$|LE/\beta| = 2(a_{or+1} + a_{or-1}) \quad (1)$$

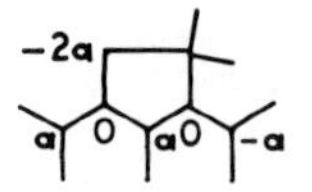

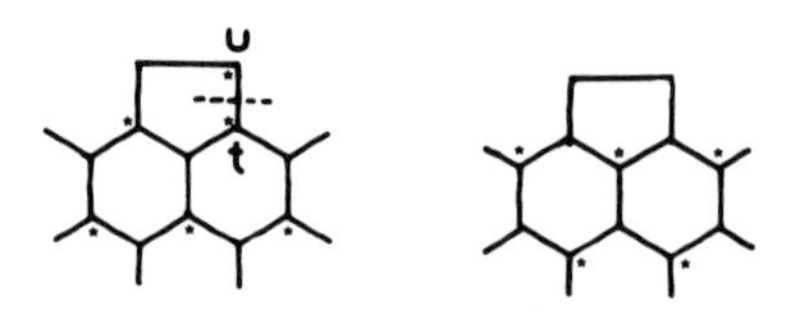

FIGURE 11. Alternant structure resulting from localization of an etheno carbon, showing the atoms having zero coefficients in the NBMO.

FIGURE 12. (A) Uncoupling C_u–C_t bond to approximate the PMO localization energy for a cyclopenta PAH in which the etheno bridge is fused to carbons having nonzero coefficients in the NBMO of the alternant nucleus after localization of a peripheral carbon atom; (B) Cyclopenta ring as an inactive branch when etheno bridge fusion occurs at carbons having zero coefficients in the NBMO of the alternant nucleus after localization of a peripheral carbon atom.

where a_{or+1}, a_{or-1} are the atomic orbital coefficients of the carbon atoms adjacent to the localized site in the nonbonding molecular orbital (NBMO) of the resultant odd-alternant cation and can be determined by inspection. Electrophilic attack at either carbon of the etheno bridge of a cyclopenta-PAH leaves an odd-alternant system that can be treated by the above approximation. Note that the topology of the cyclopenta systems requires the NBMO coefficient of the carbon in the aromatic nucleus adjacent to the localized etheno carbon to be zero (Figure 11).

Intuitively, this observation suggests that $|LE/\beta|$ for attack at the etheno bridge will tend to be small and that the cyclopenta ring will therefore be a highly reactive site.

Localization of any carbon which leaves the double bond of the etheno bridge intact yields a nonalternant ion which cannot be treated in as straightforward a manner. Nevertheless, PMO theory can still be used to estimate the ease of electrophilic attack. Any nonalternant system containing one odd-numbered ring (such as the cyclopenta PAH) can be treated[54] by uncoupling one bond of the odd ring to generate an alternant framework and then adjusting $|LE/\beta|$ by the quantity $-2a_{ot}a_{ou}$, where a_{ot},a_{ou} are the NBMO coefficients of the uncoupled carbon atoms and the product $-2\ a_{ot}a_{ou}$ represents a perturbational approximation of the *decrease* in resonance stabilization, $(\Delta E_{\pi}/\beta)$ resulting from reconstitution of the odd ring by rejoining carbon atoms C_u and C_t (Figure 12A). If localization of a peripheral carbon produces a nonalternant ion for which the above treatment results in the etheno bridge fused at nodes in the NBMO (Figure 12B), then the etheno bridge is equivalent to an inactive branch in a cross-conjugated system, and its atomic orbital coefficients vanish. $|LE/\beta|$ can then be calculated as if the etheno bridge were not present. Comparison of the ordering of $|LE/\beta|$ determined by Huckel π-energy calculations with those estimated by the PMO approximation (Tables 5 and 6) shows that both methods predict the same ordering of susceptibility to electrophilic attack. Note that for nonalternant ions in which PMO Theory treats the etheno bridge as an inactive branch, the Huckel π-localization energies are identical to the Huckel π-localization energies derived for the ions of the corresponding alternant PAH, justifying the PMO treatment. The PMO localization energies of the most reactive ring carbons of the cyclopenta-PAH discussed in the metabolism section are given in Table 7, in order of reactivity, along with localization energies for corresponding carbon atoms on the structurally related even-alternant parent ring systems for comparison. Attack at the sites indicated in Table 7 can yield either dihydrodiol or stable phenol metabolites depending

Table 6
LOCALIZATION ENERGIES FOR ELECTROPHILIC ATTACK AT PERIPHERAL POSITIONS OF ACEANTHRYLENE COMPARED TO ANTHRACENE

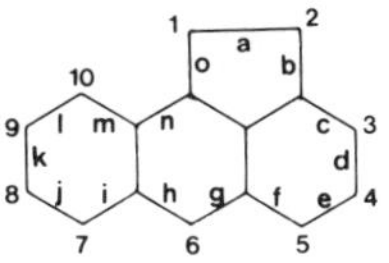

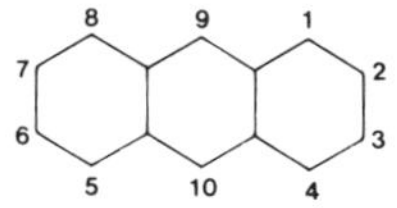

Aceanthrylene			**Anthracene**		
	$\lvert LE/\beta \rvert$			$\lvert LE/\beta \rvert$	
Position	**Huckel π-localization energy**	**PMO approx.**	**Position**	**Huckel π-localization energy**	**PMO approx.**
C_1	2.1388	1.23			
C_2	2.0357	1.07			
C_4	2.4754	1.89	C_3	2.4227	1.89
C_5	2.4342	1.86	C_4	2.2305	1.56
C_7	2.3110	1.59	$C_5(C_1)$	2.2305	1.56
C_8	2.4507	1.89	$C_6(C_3)$	2.4227	1.89
C_9	2.5141	1.89	$C_7(C_2)$	2.4227	1.89
C_{10}	2.2899	1.56	$C_8(C_1)$	2.2305	1.56

on the aromaticity of the ring involved. Tables 5, 6, and 7 show that for all compounds considered, the predicted effect of the cyclopenta ring on the reactivity of the molecular periphery is, in fact, to create a highly reactive center for electrophilic attack at the etheno bridge, leading to the further prediction that, to the extent regioselectivity is determined by electronic structure, the cyclopenta dihydrodiols should be the major metabolites isolated. Susceptibility to electrophilic attack at peripheral sites on the remaining aromatic nucleus should not, in general, be strongly perturbed with respect to the corresponding bonds of the even-alternant parent systems. An exception is the aceanthrylene series in which the terminal benzo-ring bound distal to the fused cyclopenta ring (e.g., C_4–C_5 bond of aceanthrylene) is strongly deactivated compared to the C_1–C_2 bond of anthracene.

However, steric considerations are obviously expected to play an important role in metabolism, and for certain PAH, will clearly be the major influence on regioselectivity. For example, severe steric crowding in the four-sided indentation between the cyclopenta and angular benzo rings (gulf region) of benz[l]aceanthrylene can be expected to force the compound into a helical configuration by analogy to the behavior of benzo[c]phenanthrene.[55] In addition to a possible decrease in access of the activated oxygen of the oxo-P450 complex to the etheno bond, the deviation from planarity might activate the K-region by decreasing π-overlap through torsion around C_7–C_8. Even for less-hindered arene bonds, the steric constraints imposed by the geometry at the active site of P450 is largely unknown. The importance of geometry at the active site is evident in the observation that the term P450 encompasses a family of isozymes differing in both substrate specificity and regioselectivity of monooxygen transfers depending upon animal species and tissue of origin and also upon induction regimen. Hence, a reasonable expectation for predicting metabolite distribution at the level of refinement possible by the treatment in this work is that reactive bonds and

Table 7
PMO LOCALIZATION ENERGIES FOR ELECTROPHILIC ATTACK AT PERIPHERAL POSITIONS OF CYCLOPENTA PAH ORDERED BY REACTIVITY

PAH	Carbon	|LE/β|	(Relative yield in S9)
Benz[j]acephenanthrylene	C_4	1.28	4,5-Dihydrodiol
	C_5	1.33	(major)
	C_{11}	1.69	11,12-Dihydrodiol (not identified)
	C_{10}	1.81	9,10-Dihydrodiol (major)
	C_7	1.90	7,8-Dihydrodiol (not identified)
Benz[k]acephenanthrylene	C_4	1.28	4,5-Dihydrodiol (major)
	C_5	1.35	
	C_8	1.64	8,9-Dihydrodiol (minor)
	C_{11}	1.66	10,11-Dihydrodiol (not identified)
Benz[e]aceanthrylene	C_6	1.12	5,6-Dihydrodiol
	C_5	1.29	(major)
	C_4	1.63	3,4-Dihydrodiol (major)
	C_1	1.64	1,2-Dihydrodiol (minor)
Benz[j]aceanthrylene	C_2	1.12	1,2-Dihydrodiol
	C_1	1.26	(major)
	C_{11}	1.64	11,12-Dihydrodiol (trace)
	C_{10}	1.84	9,10-Dihydrodiol (minor) 10-phenol (minor)
Benz[l]aceanthrylene	C_2	1.17	1,2-Dihydrodiol
	C_1	1.28	(minor)
	C_8	1.66	7,8-Dihydrodiol
	C_7	1.67	(major)
	C_5	1.90	4,5-Dihydrodiol (trace)
Cyclopenta[cd]pyrene	C_4	1.21	3,4-Dihydrodiol
	C_3	1.29	(major)
	C_9	1.65	9,10-Dihydrodiol (minor)
	C_1	1.51	Corresponding phenol
	C_6,C_8	1.66	(3 phenols as minor metabolites)

major metabolites will be correctly identified. As is evident from metabolite distributions determined in the in vitro experiments, this expectation is largely fulfilled.

Acephenanthrylene and aceanthrylene are the smallest members of the series of cyclopenta-PAH investigated and the compounds having the least-demanding steric requirements. The predicted and observed metabolite profiles generated by rat-liver S9 are in agreement. For acephenanthrylene (1), the only significant metabolite is the cyclopenta dihydrodiol, as expected from the large values of |LE/β| (>1.9) calculated for attack at other peripheral positions. The cyclopenta dihydrodiol is also correctly predicted to be major metabolite of aceanthrylene (2). Reactivity of C_7–C_8 and C_9–C_{10} bonds is equivalent and the observed ordering of dihydrodiols can be readily understood as a consequence of some steric hindrance of the C_9–C_{10} bond in the "gulf region", while oxidation to the 4,5-dihydrodiol occurs only in a trace amount in accord with the large $|LE/\beta|$ (= 1.86). Even though stereochemical considerations are expected to be more critical for the 5-ring PAH, agreement of observed metabolite distributions with predictions based on PMO treatment is generally good. The cyclopenta dihydrodiol of benz[j]acephenanthrylene (4) is predicted to be the major metabolite, with the C_{11}–C_{12} (K-region) > C_9–C_{10} bonds being the two other relatively reactive sites. The two major rat-liver S9 metabolites identified to date are the cyclopenta- and 9,10-dihydrodiols, which appear to be formed in nearly equivalent amounts. For benz[k]acephenanthrylene (5), the cyclopenta dihydrodiol and 8,9-dihydrodiol are the major and minor metabolites, respectively, with other positions unreactive as predicted. The distribution of benz[e]aceanthrylene (7) metabolites is predicted to be cyclopenta dihydrodiol > 1,2-dihydrodiols ~ 3,4-dihydrodiols. Metabolism by rat-liver S9 yielded the cyclopenta and 3,4-dihydrodiols in equivalent amounts while the 1,2-dihydrodiol, presumably hindered by the proximal bay region, was formed in smaller amount. Rat-liver S9 metabolized benz[j]aceanthrylene (8) primarily to the cyclopenta dihydrodiol as predicted; however, the 9,10-dihydrodiol was next in abundance rather than the K-region metabolite. Both C3H10T $^1/_2$ and Syrian hamster embryo cells transformed benz[j]aceanthrylene primarily to the 9,10-dihydrodiol. The changes in regioselectivity of benz[j]aceanthrylene metabolism may be attributed to metabolism by different isozymes. In order to more fully understand the observed differences, additional work on metabolism in these and other systems will be required, as well as development of knowledge about the active site of P450 and the effects of changes in protein on the physico-chemical properties of the enzyme.

Since the PMO approximation incorporates neither steric interactions nor geometric distortions, the failure to predict the metabolite profile of the highly strained, nonplanar benz[l]aceanthrylene (10) is not surprising. Both rat-liver S9 and C3H10T1/2 cells oxidize this PAH primarily at the K-region and produce only minor amounts of the cyclopenta dihydrodiol, which is predicted by the PMO $|LE/\beta|$ to be the major metabolite. The 4,5-dihydrodiol, as expected, is a minor metabolite. The cyclopenta dihydrodiol, as predicted, is the predominant metabolite of cyclopenta[cd]pyrene (3). According to calculated $|LE/\beta|$, attack at C_1 is slightly favored over attack at C_6, C_8, and the K-region carbons. Phenols would be expected from attack at C_1, C_6, and C_8 because of the highly aromatic character of the apical rings, while a dihydrodiol should result at the K-region. In fact, the K-region dihydrodiol has been reported to be the only metabolite other than the cyclopenta dihydrodiol that is formed to an appreciable extent, although phenolic products were present in small amounts.

B. Biological Activity of Metabolites

Considerable evidence supports epoxides or diol-epoxides as the reactive PAH metabolites responsible for chemical modification of DNA, which in turn has been shown to correlate with genotoxic activity.[5,6] Electrophilicity appears to be the critical feature determining the biological activity of metabolites, and this property can be predicted for epoxides (which are the primary metabolites corresponding to isolated dihydrodiols) by calculating the resonance stabilization energies of the benzylic carbonium ions derived from oxirane ring opening. Studies on alternant bay-region PAH have employed the PMO approximation of resonance energy, $\Delta E_{deloc}/\beta$, to correlate resonance stabilization with biological activity[10] and support the hypothesis that activity increases directly with increasing $\Delta E_{deloc}/\beta$. In alternant bay-region systems, bay-region diol-epoxides invariably yield the most highly stabilized oxirane ring-opened benzylic carbonium ions, and on the basis of biological data, a value of $\Delta E_{deloc}/\beta > 0.7$ appears to be a threshold for carcinogenic activity. The Bay-Region Theory of Carcinogenicity has been formulated, stating that bay-region diol-epoxides will be ultimate carcinogenic metabolites of carcinogenic PAH containing the bay-region feature.

Since it has become apparent that some PAH without bay-regions are carcinogenic, the PMO treatment has been extended[57] to other PAH in order to identify nonbay region metabolites capable of yielding highly stabilized benzylic carbonium ions and to test the relative importance of the bay-region geometric feature and high $\Delta E_{deloc}/\beta$ as determinants of carcinogenicity. The initial bioassay and metabolism studies on cyclopenta[cd]pyrene extended structure-activity considerations to nonalternant systems and demonstrated that the bay region was not an absolute requirement for carcinogenic potential. However, the weak activity of cyclopenta[cd]pyrene as a tumor initiator and a complete carcinogen[7,8] left unanswered the question of whether the bay region is necessary for high levels of carcinogenicity. The synthesis, metabolism, and bioassay studies on the series of cyclopenta compounds discussed in this chapter were designed to investigate this question in depth. The resonance stabilization energies of a number of carbonium ions from possible metabolites of cyclopenta[cd]pyrene[58] and the benzanthracene-derived cyclopenta-fused isomers[59] have recently been determined by an SCF calculation using optimized geometries. Except in the case of benz[l]aceanthrylene, the most stable carbonium ions were predicted to arise from cyclopenta epoxides. As a result of the distortion about the gulf region of benz[l]aceanthrylene, the resonance stabilization energies of the bay region and cyclopenta (C_1) carbonium ions were calculated to be equivalent. The SCF calculations support the generalized conclusion that qualitative ordering of delocalization energies can be based on certain proposed geometric relationships:[60] (1) carbonium ions fused to β, α, α' peripheral sites* increase in stability in the order $\beta < \alpha < \alpha'$ whether they are cyclopenta- or diol-epoxide-derived and, (2) peripheral fusion of an etheno bridge tends to destabilize carbonium ions at sites on the molecular periphery other than the etheno bridge carbons.

The benzylic carbonium ions derived from cyclopenta epoxides are odd-alternant ions and their delocalization energies can also be readily calculated by a PMO approximation:

$$\Delta E_{deloc}/\beta = 2(1 - a_0) \qquad (2)$$

Delocalization energies for nonalternant ions containing the cyclopenta ring can also be calculated using a PMO approximation by uncoupling one end of the etheno bridge, determining $\Delta E_{deloc}/\beta$ for the resulting odd-alternant ion, and subtracting the same perturbational correction $\Delta E_{\pi}/\beta$ to the stabilization energy that was used in calculating LE values. Table 8 compares the ordering of delocalization energies calculated by SCF and PMO methods and indicates that the destabilizing effect of the etheno bridge is considerably more important in the PMO approximation, resulting in changes in ordering for benz[e]aceanthrylene. Table 9 shows PMO delocalization energies of additional carbonium ions potentially derivable from diol-epoxide and cyclopenta epoxide precursors inferred from the metabolites identified among the products of the cyclopenta PAH discussed in the metabolism section. The cyclopenta epoxides are capable of yielding carbonium ions with the largest $\Delta E_{deloc}/\beta$ values for all the cyclopenta PAH metabolites; however, the value for the C_5 carbonium ion from acephenanthrylene-4,5-oxide is below the PMO threshold value of $\Delta E_{deloc}/\beta \sim 0.7$ that seems to be required for activity. Among other metabolites, only the bay-region diol-epoxide of benz[j]aceanthrylene is predicted to yield a biologically active carbonium ion. On the basis of calculated $\Delta E_{deloc}/\beta$, expectations are that acephenanthrylene will be inactive and cyclopenta epoxides will be the most important ultimate mutagens/carcinogens for all other cyclopenta-PAH, with the additional possibility that the bay-region diol-epoxide may also

* β, two carbons removed from a ring fusion; α, adjacent to a ring fusion; α', adjacent to two ring fusion sites.

Table 8
ORDERING OF SCF AND PMO STABILITIES OF DIOL-EPOXIDE AND CYCLOPENTA EPOXIDE-DERIVED CARBONIUM IONS OF CYCLOPENTA-FUSED BENZANTHRACENE ISOMERS

SCF (ΔE_π in Au)[59]	PMO ($\Delta E_{deloc}/\beta$)

Benz[e]aceanthrylene

Benz[j]aceanthrylene

Benz[l]aceanthrylene

Benz[k]acephenanthrylene

SCF	PMO
0.3802	0.7220
0.3680	0.5509
0.3637	0.5256

Table 9
PMO STABILITIES OF BENZYLIC CARBONIUM IONS DERIVED FROM DIOL-EPOXIDE AND CYCLOPENTA EPOXIDE METABOLITES IDENTIFIED IN PROFILES OF OTHER CYCLOPENTA-PAH STUDIED

Parent PAH	Carbonium ion	$\Delta E_{deloc}/\beta$
Acephenanthrylene	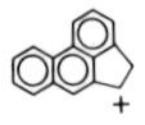	0.6637
Benz[j]acephenanthrylene		0.7190
	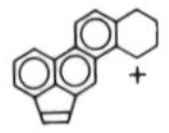	0.6179
Aceanthrylene	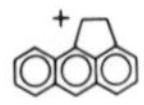	0.9310
	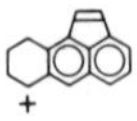	0.5448
	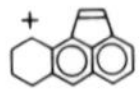	0.5002
Cyclopenta[cd]pyrene		0.794
	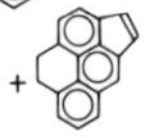	0.6090

contribute to the activity of benz[j]aceanthrylene. The bioassay data are in reasonable agreement with predictions based on $\Delta E_{deloc}/\beta$ values. Acephenanthrylene shows only marginal activity. The cyclopenta dihydrodiols are major metabolites of aceanthrylene (2) and benz[e]aceanthrylene (7), and both of these PAH are highly active bacterial mutagens; however, while benz[e]aceanthrylene is active in the C3H10T1/2 transformation assay, aceanthrylene is not. The lack of correlation between the activity of aceanthrylene in bacterial mutagenicity and mammalian cell transformation assays could result either from differences in metabolism or from intervention of other biochemical processes such as DNA repair; an explanation will require additional studies on aceanthrylene. The importance of the cyclopenta epoxides in activation pathways of other cyclopenta PAH is supported by the low concentration optima of the S9-dependence curves. Furthermore, all of the cyclopenta epoxides synthesized and tested are potent direct-acting mutagens in the Ames assay in TA98. Only for benz[j]acephenanthrylene does the cyclopenta epoxide appear to play a minor role in activity, despite the presence of the corresponding dihydrodiol as a major metabolite and $\Delta E_{deloc}/\beta > 0.7$ for the most stable cyclopenta epoxide-derived carbonium ion. The S9-dependence of benz[j]acephenanthrylene is virtually identical to the benzo[a]pyrene control, which is known to be activated in a multistep process involving the bay region diol-epoxide. Since a major metabolite, approximately equivalent in proportion to the cyclopenta dihydrodiol, is the bay-region diol-epoxide precursor, the S9 dependence curve of benz[j]acephenanthrylene implies that the bay-region diol-epoxide is involved in activation, although the calculated value of $\Delta E_{deloc}/\beta > 0.7$. Benz[k]acephenanthrylene shows intermediate behavior, suggesting that the cyclopenta epoxide is probably not completely responsible for the rat-liver S9-mediated mutagenicity. The prediction that the bay-region dio-epoxide of benz[j]aceanthrylene can yield a carbonium ion with large $\Delta E_{deloc}/\beta$ is interesting in light of the observation that, although the S9-mediated mutagenicity appears to involve the epoxidized cyclopenta ring (on the basis of the S9-dependence curve), benz[j]aceanthrylene also efficiently transforms C3H10T1/2 cells which produce the bay-region diol-epoxide precursor as the major metabolite. Hence, in accord with predictions, the bay-region diol-epoxide may also be an active transforming agent. The fact that Syrian hamster embryo cells, which also metabolize benz[j]aceanthrylene primarily to the bay-region diol-epoxide precursor, did not effectively activate the PAH to mutate V79 cells, points up the need for further work before any definitive conclusions can be reached. The cyclopenta dihydrodiol of benz[l]aceanthrylene is produced in sufficient amount in rat-liver S9-mediated metabolism to account for a significant proportion of activity in the Ames assay and the S9-dependence curve supports a one-step activation pathway via an arene oxide. However, benz[l]aceanthrylene is also active in transforming C3H10T1/2 cells which metabolize the cyclopenta ring only to a minor extent. Since oxidation at the K-region is the predominant metabolic pathway in both S9 and C3H10T1/2 and the K-region oxide might be highly reactive as a result of skeletal deformation, the possibility of activation via the K-region must be investigated. It is worthy of note that the K-region oxide, like the cyclopenta oxide, is an arene oxide and would show similar S9 dependence, characteristic of one-step activation.

Cyclopenta[cd]pyrene conforms to expectations based on the PMO treatment: the predominant metabolite in all metabolizing systems is the cyclopenta dihydrodiol, and the corresponding oxide is implicated as the ultimate active metabolite by its direct-acting mutagenicity both in bacterial and mammalian assays and also in the C3H10T1/2 cell transformation assay.

ACKNOWLEDGMENT

Portions of this work were supported by USPHS Grant No. ES 03433 and USEPA Grants No. CR 810849 and No. CR 811817.

REFERENCES

1. **Gold, A.,** Carbon black adsorbates: separation identification of a carcinogen and some oxygenated polyaromatics, *Anal. Chem.*, 47, 1469, 1975.
2. **Kaden, D. A., Hites, R. A., and Thilly, W. G.,** Mutagenicity of soot and associated polycyclic aromatic hydrocarbons to *S. typhimurium, Cancer Res.*, 39, 4152, 1979.
3. **Wallcave, L., Nagel, D., Smith, J. W., and Waniska, R. D.,** Two pyrene derivatives of widespread environmental distribution. Cyclopenta[cd]pyrene and acepyrene, *Environ. Sci. Technol.*, 9, 143, 1975.
4. **Grimmer, G.,** in *Air Pollution and Cancer in Man,* Mohr, Y., Schmel, D., and Tomatis, L., Eds., International Agency for Research on Cancer (IARC), Lyon, France, 1977, 193.
5. **Eisenstadt, E. and Gold, A.,** Cyclopenta[cd]pyrene: a highly mutagenic polycyclic aromatic hydrocarbon, *Proc. Natl. Acad. Sci. U.S.A.*, 75, 1667, 1978.
6. **Gold, A., Nesnow, S., Moore, M., Garland, H., Curtis, G., Howard, B., Graham, D., and Eisenstadt, E.,** Mutagenesis and morphological transformation of mammalian cells by a non-bay-region polycyclic cyclopenta[cd]pyrene pyrene and its 3,4-oxide, *Cancer Res.*, 40, 4482, 1980.
7. **Wood, A. W., Levin, W., Chang, R. L., Huang, M.-T., Ryan, D. F., Thomas, P. E., Lehr, R. E., Kumar, S., Koreeda, M., Akagi, H., Ittah, Y., Dansette, P., Yagi, H., Jerina, D. M., and Conney, A. H.,** Mutagenicity and tumor-initiating activity of cyclopenta[cd]pyrene and structurally related compounds, *Cancer Res.*, 40, 642, 1980.
8. **Cavalieri, E., Rogan, E., Toth, D., and Munhall, A.,** Carcinogenicity of the environmental pollutants cyclopenteno[cd]pyrene and cyclopenta[cd]pyrene in mouse skin, *Carcinogenesis,* 2, 277, 1981.
9. **Cavalieri, E., Munhall, A., Rogan, E., Salmasi, E., and Patil, K.,** Syncarcinogenic effects of the environmental pollutants cyclopenta[cd]pyrene and benzo[a]pyrene in mouse skin, *Carcinogenesis,* 4, 393, 1983.

10. **Jerina, D. M., Lehr, R. E., Yagi, H., Hernandez, O., Dansette, P. M., Wislocki, P. G., Wood, A. W., Chang, R. L., Levin, W., and Conney, A. H.,** Mutagenicity of benzo[a]pyrene derivatives and the description of a quantum mechanical model which predicts the use of carbonium ion formation from diol-epoxides, in *In Vitro Metabolic Activation in Mutagenesis Testing,* DeSerres, F. J., Fouts, J. R., Bend, J. R., and Philpot, R. M., Eds., Elsevier/North-Holland Biomedical, Amsterdam, 1976, 159.
11. **Krishnan, S. and Hites, R. A.,** Identification of acephenanthrylene in combustion effluents, *Anal. Chem.,* 53, 342, 1981.
12. **Grimmer, G.,** Personal communication.
13. **Laarhoven, W. H. and Cuppen, Th. J. H. M.,** Photodehydrocyclizations of stilbene-like compounds. XVI. Photo-reactions of 1-(9-phenanthryl)stilbene and 1-(9-phenanthryl)-1-phenylethylene, *Recl. Trav. Chim.,* 95, 165, 1976.
14. **Sangaiah, R., Gold, A., and Toney, G. E.,** Synthesis of a series of novel polycyclic aromatic systems: isomers of benz[a]anthracene containing a cyclopenta-fused ring, *J. Org. Chem.,* 48, 1632, 1983.
15. **Plummer, B. F., Al-Saigh, Z. Y., and Arfan, M.,** Synthesis of aceanthrylene, *J. Org. Chem.,* 49, 2069, 1984.
16. **Becker, H. D., Hansen, L., and Andersson, K.,** Aceanthrylene, *J. Org. Chem.,* 50, 277, 1985.
17. **Sangaiah, R. and Gold, A.,** A synthesis of aceanthrylene, *Org. Prep. Proced. Int.,* 17, 53, 1985.
18. **Stepan, V. and Vodehnal, J.,** Preparation and infrared spectra of anthraquinone and its methyl derivatives and of quinones related to polynuclear aromatic hydrocarbons, *Collect. Czech. Chem. Commun.,* 36, 3964, 1971.
19. **Fieser, L. F.,** Reduction products of naphthacene-quinone, *J. Am. Chem. Soc.,* 53, 2329, 1931.
20. **Schroeter, G.,** o-[Tetrahydronaphthoyl-2]-benzoic acid and its reduction and condensation products, *Ber.,* 54B, 2242, 1921; *Chem. Abstr.,* 16, 1091, 1922.
21. **Waldmann, H. and Mathiowetz, H.,** A new synthesis in the lin-2,3-benzanthraquinone series, *Ber.,* 64B, 1713, 1931; *Chem. Abstr.,* 25, 5162, 1931.
22. **Gold, A., Schultz, J., and Eisenstadt, E.,** Relative reactivities of pyrene ring positions: cyclopenta[cd]pyrene via an intramolecular Friedel-Crafts acylation, *Tetrahedron Lett.,* 4491, 1978.
23. **Ittah, Y. and Jerina, D. M.,** Cyclopenta[cd]pyrene, *Tetrahedron Lett.,* 4495, 1978.
24. **Ruehle, P. H., Fischer, D. L., and Wiley, J. C.,** Synthesis of cyclopenta[cd]pyrene, a ubiquitous environmental carcinogen, *J. Chem. Soc., Chem. Commun.,* 302, 1979.
25. **Konieczny, M. and Harvey, R. G.,** Synthesis of cyclopenta[cd]pyrene, *J. Org. Chem.,* 44, 2158, 1979.
26. **Tintel, K., Lugtenburg, J., and Cornelisse, J.,** Synthesis of cyclopenta[cd]pyrene via a high yield one-step preparation of pyren-4-ylacetic acid from pyrene, *J. Chem. Soc., Chem. Commun.,* 185, 1982.
27. **Tintel, K., Cornellise, J., and Lugtenburg, J.,** Convenient synthesis of cyclopenta[cd]pyrene and 3,4-dihydrocyclopenta[cd]pyrene. The reactivity of the pyrene dianion, *Recl. Trav. Chim.,* 102, 14, 1983.
28. **Sangaiah, R. and Gold, A.,** A short and convenient synthesis of cyclopenta[cd]pyrene and its oxygenated derivatives, in *Polynuclear Aromatic Hydrocarbons: Mechanisms, Methods and Metabolism,* Cooke, M. and Dennis, A. J., Eds., Battelle Press, Columbus, Ohio, 1985, 1145.
29. **Clar, E.,** *Polycyclic Hydrocarbons,* Vol. 1, Academic Press, New York, 1964, 386.
30. **Nesnow, S., Leavitt, S., Easterling, R., Watts, S. H., Toney, S. H., Claxton, L., Sangaiah, R., Toney, G. E., Wiley, J., Fraher, P., and Gold, A.,** Mutagenicity of cyclopenta-fused isomers of benz[a]anthracene in bacterial and rodent cells and identification of the major rat liver microsomal metabolites, *Cancer Res.,* 44, 4993, 1984.
31. **Sangaiah, R., Gold, A., Ball, L. M., Kohan, M., Bryant, B. J., Rudo, K., Claxton, L., and Nesnow, S.,** Biological activity and metabolism of aceanthrylene and acephenanthrylene, in *Polynuclear Aromatic Hydrocarbons: Chemistry and Carcinogenesis,* Cooke, M. and Dennis, A. J., Eds., Battelle Press, 1985, 795.
32. **Kohan, M. J., Sangaiah, R., Ball, L. M., and Gold, A.,** Bacterial mutagenicity of aceanthrylene: a novel cyclopenta-fused polycyclic aromatic hydrocarbon of low molecular weight, *Mutat. Res.,* 155, 95, 1985.
33. **Gold, A. and Eisenstadt, E.,** Metabolic activation of cyclopenta[cd]pyrene to 3,4-epoxycyclopenta[cd]pyrene by rat liver microsomes, *Cancer Res.,* 40, 3940, 1980.
34. **Mohapatra, N., MacNair, P., Bryant, B. J., Ellis, S., Rudo, K., Sangaiah, R., Gold, A., and Nesnow, S.,** Morphological transforming activity and metabolism of cyclopenta-fused isomers of benz(a)anthracene in mammalian cells, *Mutation Res.,* 188, 323, 1987.
35. **Nesnow, S., Gold, A., Mohapatra, N., Sangaiah, R., Bryant, B. J., Rudo, K., Macnair, P., and Ellis, S.,** Metabolic activation pathways of cyclopenta-fused PAH and their relationship to genetic and carcinogenic activity, in *Genetic Toxicology of Environmental Chemicals, Part A: Basic Principals and Mechanisms of Action,* Progress in Clinical and Biological Research Series, 209A, Ramel, C., Lambert, B., and Magnusson, J., Eds., Alan R. Liss, New York, 1986, 515.

36. **Nesnow, S., Gold, A., Sangaiah, R., Triplett, L. L., and Slaga, T. J.,** Mouse skin tumor-initiating activity of benz[e]aceanthrylene and benz[l]aceanthrylene in SENCAR mice, *Cancer Lett.*, 22, 263, 1984.
37. **Fu, P. P. and Harvey, R. G.,** Synthesis of diols and diolepoxides of carcinogenic hydrocarbons, *Tetrahedron Lett.*, 2059, 1977.
38. **Nesnow, S., Leavitt, S., Easterling, R., Watts, R., Macnair, P., Ellis, S., Bryant, B. J., Rudo, K., Toney, G. E., Sangaiah, R., and Gold, A.,** *Polynuclear Aromatic Hydrocarbons: Mechanisms, Methods and Metabolism,* Cooke, M. and Dennis, A. J., Eds., Battelle Press, Columbus, 1984, 949.
39. **Eisenstadt, E., Shpizner, B., and Gold, A.,** Metabolism of cyclopenta[cd]pyrene at the K-region by microsomes and a reconstituted Cytochrome P450 system from rat liver, *Biochem. Biophys. Res. Commun.*, 100, 965, 1981.
40. **Singer, B. and Grunberger, D.,** *Molecular Biology of Mutagens and Carcinogens,* Plenum Press, New York, 1983.
41. **Loew, G. H.,** Theoretical investigations of iron porphyrins, in *Iron Porphyrins,* Part I, Lever, A. B. P. and Gray, H. B., Eds., Addison-Wesley, Reading, Mass., 1983, Ch. 1.
42. **Hanson, L. K., Chang, C. K., Davis, M. S., and Fajer, J.,** Electron pathways in catalase and peroxidase enzymic catalysis. Metal and macrocycle oxidations of iron porphyrins and chlorins, *J. Am. Chem. Soc.*, 103, 663, 1981.
43. **Boso, B., Lang, G., McMurray, T. J., and Groves, J. T.,** Mossbauer effect of tight spin coupling in oxidized chloro-5,10,15,20-tetra(mesityl)porphinatoiron. III, *J. Chem. Phys.*, 79, 1122, 1983.
44. **Groves, J. T., Nemo, T. E., and Myers, R. S.,** Hydroxylation and epoxidation catalyzed by iron-porphine complexes. Oxygen transfer from iodosylbenzene, *J. Am. Chem. Soc.*, 101, 1032, 1979.
45. **Groves, J. T., Haushalter, R. C., Nakamura, M., Nemo, T. E., and Evans, B. J.,** High-valent iron-porphyrin complexes related to peroxidase and Cytochrome P450, *J. Am. Chem. Soc.*, 103, 2884, 1981.
46. **Balch, A. L., Latos-Grazynski, L., and Renner, M. W.,** Oxidation of red ferryl [$(Fe^{IV}O)^+$] porphyrin complexes to green ferryl [$(Fe^{IV}O)^{2+}$] porphyrin complexes, *J. Am. Chem. Soc.*, 107, 2983, 1985.
47. **Groves, J. T., McClusky, G. A., White, R. E., and Coon, M. J.,** Aliphatic hydroxylation by highly purified microsomal Cytochrome P450. Evidence for a carbon radical intermediate, *Biochem. Biophys. Res. Commun.*, 81, 154, 1978.
48. **Streitwieser, A.,** Aromatic substitution, in *Molecular Orbital Theory for Organic Chemists,* John Wiley & Sons, New York, 1967, 307.
49. **Dewar, M. J. S. and Dougherty, R. C.,** *The PMO Theory of Organic Chemistry,* Plenum Press, New York, 1975, 197-391.
50. **Tierney, B., Hewer, A., MacNicoll, A. D., Gervasi, P. G., Rattle, H., Walsh, C., Grover, P., and Sims, P.,** The formation of dihydrodiols by the chemical or enzymic oxidation of benz[a]anthracene and 7,12-dimethylbenz[a]anthracene, *Chem. Biol. Interact.*, 23, 243, 1978.
51. **Thakker, D. R., Levin, W., Yagi, H., Ryan, D., Thomas, P. E., Karle, J. M., Lehr, R. E., Jerina, D. M., and Conney, A. H.,** Metabolism of benzo[a]anthracene to its tumorigenic 3,4-dihydrodiol, *Mol. Pharmacol.*, 15, 138, 1979.
52. **Olah, G. A.,** *Friedel-Crafts and Related Reactions,* Vol. III, Part 1, Interscience, New York, 1965, 75.
53. **Buu-Hoi, N. P. and Lavit, D.,** Sur la synthese du dibenzo[a,i]pyrene, du tribenzo[a,e,i]pyrene et de nouveaux homologues du naphto[2,3-a]pyrene, *Tetrahedron,* 8, 1, 1960.
54. **Dewar, M. J. S. and Dougherty, R. C.,** *The PMO Theory of Organic Chemistry,* Plenum Press, New York, 1975, 131.
55. **Hirshfeld, F. L., Sandler, S., and Schmidt, G. M. J.,** The structure of overcrowded aromatic compounds. Part VI. The crystal structure of benzo[c]phenanthrene and 1,12-dimethylbenzo[c]phenanthrene, *J. Chem. Soc.*, 2108, 1963.
56. **Gelboin, H. V. and Ts'o, P. O. P., Eds.,** *Polycyclic Hydrocarbons and Cancer,* Vols. 2 and 3, Academic Press, New York, 1978, 1981.
57. **Fu, P. P., Harvey, R. G., and Beland, F. A.,** Molecular orbital theoretical prediction of the isomeric products formed from reactions of arene oxides and related metabolites of polycyclic aromatic hydrocarbons, *Tetrahedron,* 34, 857, 1978.
58. **Silverman, B. D.,** Calculated reactivity of the mutagenic cyclopenta-polycyclic aromatic hydrocarbon, 3,4-epoxycyclopenta[cd]pyrene, *Cancer Biochem. Biophys.*, 7, 83, 1984.
59. **Silverman, B. D. and Lowe, J. P.,** Calculated reactivities of potentially carcinogenic intermediates of the cyclopenta-fused isomers of benz[a]anthracene, *Cancer Biochem. Biophys.*, 7, 203, 1984.
60. **Lowe, J. P. and Silverman, B. D.,** MO theory of ease of formation of carbocations derived from nonalternant polycyclic aromatic hydrocarbons, *J. Am. Chem. Soc.*, 106, 5955, 1984.
61. **Kligerman, A. D., Moore, M. M., Erexson, G. E., Brock, K. H., Doerr, C. L., Allen, J. A., and Nesnow, S.,** Genotoxicity of benz(l)aceanthrylene, *Cancer Lett.*, 31, 123, 1986.

INDEX

A

B

C

D

E

F

O

P

Q

R

S

T

U

V

X